Hybrid Advanced Optimization Methods with Evolutionary Computation Techniques in Energy Forecasting

Hybrid Advanced Optimization Methods with Evolutionary Computation Techniques in Energy Forecasting

Special Issue Editor

Wei-Chiang Hong

MDPI • Basel • Beijing • Wuhan • Barcelona • Belgrade

MDPI

Special Issue Editor
Wei-Chiang Hong
School of Computer Science and Technology,
Jiangsu Normal University
China

Editorial Office
MDPI
St. Alban-Anlage 66
Basel, Switzerland

This is a reprint of articles from the Special Issue published online in the open access journal *Energies* (ISSN 1996-1073) from 2017 to 2018 (available at: https://www.mdpi.com/journal/energies/special_issues/energy_forecast_2017)

For citation purposes, cite each article independently as indicated on the article page online and as indicated below:

LastName, A.A.; LastName, B.B.; LastName, C.C. Article Title. *Journal Name* **Year**, *Article Number, Page Range.*

ISBN 978-3-03897-286-0 (Pbk)
ISBN 978-3-03897-287-7 (PDF)

Contents

About the Special Issue Editor . . . . **vii**

Preface to "Hybrid Advanced Optimization Methods with Evolutionary Computation
Techniques in Energy Forecasting" . . . . **ix**

Guo-Feng Fan, Li-Ling Peng, Xiangjun Zhao and Wei-Chiang Hong
Applications of Hybrid EMD with PSO and GA for an SVR-Based Load Forecasting Model
Reprinted from: *Energies* **2017**, *10*, 1713, doi: 10.3390/en10111713 . . . . **1**

Dongxiao Niu, Yi Liang, Haichao Wang, Meng Wang and Wei-Chiang Hong
Icing Forecasting of Transmission Lines with a Modified Back Propagation
Neural Network-Support Vector Machine-Extreme Learning Machine with Kernel
(BPNN-SVM-KELM) Based on the Variance-Covariance Weight Determination Method
Reprinted from: *Energies* **2017**, *10*, 1196, doi: 10.3390/en10081196 . . . . **23**

Xiaomin Xu, Dongxiao Niu, Lihui Zhang, Yongli Wang and Keke Wang
Ice Cover Prediction of a Power Grid Transmission Line Based on Two-Stage Data Processing
and Adaptive Support Vector Machine Optimized by Genetic Tabu Search
Reprinted from: *Energies* **2017**, *10*, 1862, doi: 10.3390/en10111862 . . . . **44**

Ming-Wei Li, Jing Geng, Shumei Wang and Wei-Chiang Hong
Hybrid Chaotic Quantum Bat Algorithm with SVR in Electric Load Forecasting
Reprinted from: *Energies* **2017**, *10*, 2180, doi: 10.3390/en10122180 . . . . **62**

Peng Lu, Lin Ye, Bohao Sun, Cihang Zhang, Yongning Zhao and Tengjing Zhu
A New Hybrid Prediction Method of Ultra-Short-Term Wind Power Forecasting Based on
EEMD-PE and LSSVM Optimized by the GSA
Reprinted from: *Energies* **2018**, *11*, 697, doi: 10.3390/en11040697 . . . . **80**

**Nadeem Javaid, Fahim Ahmed, Ibrar Ullah, Samia Abid, Wadood Abdul, Atif Alamri and
Ahmad S. Almogren**
Towards Cost and Comfort Based Hybrid Optimization for Residential Load Scheduling
in a Smart Grid
Reprinted from: *Energies* **2017**, *10*, 1546, doi: 10.3390/en10101546 . . . . **103**

Ping Jiang, Zeng Wang, Kequan Zhang and Wendong Yang
An Innovative Hybrid Model Based on Data Pre-Processing and Modified Optimization
Algorithm and Its Application in Wind Speed Forecasting
Reprinted from: *Energies* **2017**, *10*, 954, doi: 10.3390/en10070954 . . . . **130**

Dongxiao Niu, Yi Liang and Wei-Chiang Hong
Wind Speed Forecasting Based on EMD and GRNN Optimized by FOA
Reprinted from: *Energies* **2017**, *10*, 2001, doi: 10.3390/en10122001 . . . . **159**

Sen Guo, Haoran Zhao and Huiru Zhao
A New Hybrid Wind Power Forecaster Using the Beveridge-Nelson Decomposition Method
and a Relevance Vector Machine Optimized by the Ant Lion Optimizer
Reprinted from: *Energies* **2017**, *10*, 922, doi: 10.3390/en10070922 . . . . **177**

Dongxiao Niu, Haichao Wang, Hanyu Chen and Yi Liang
The General Regression Neural Network Based on the Fruit Fly Optimization Algorithm and
the Data Inconsistency Rate for Transmission Line Icing Prediction
Reprinted from: *Energies* **2017**, *10*, 2066, doi: 10.3390/en10122066 . . . . **197**

Cheng-Wen Lee and Bing-Yi Lin
Applications of the Chaotic Quantum Genetic Algorithm with Support Vector Regression in
Load Forecasting
Reprinted from: *Energies* **2017**, *10*, 1832, doi: 10.3390/en10111832 . . . . **217**

About the Special Issue Editor

Wei-Chiang Hong, Jiangsu Distinguished Professor, School of Computer Science and Technology, Jiangsu Normal University, China. His research interests mainly include computational intelligence (neural networks, evolutionary computation) and application of forecasting technology (ARIMA, support vector regression, and chaos theory). In May 2012, his paper was evaluated as a "Top Cited Article 2007–2011" by *Applied Mathematical Modelling* (Elsevier). In August 2014, he was elected to be awarded the "Outstanding Professor Award" by the Far Eastern Y. Z. Hsu Science and Technology Memorial Foundation (Taiwan). In November 2014, he was elected to be awarded the "Taiwan Inaugural Scopus Young Researcher Award—Computer Science", by Elsevier, in the Presidents' Forum of Southeast and South Asia and Taiwan Universities. He was awarded as one of the "Top 10 Best Reviewers" of Applied Energy in 2014, and as one of Applied Energy's "Best Reviewers" in 2016.

Preface to "Hybrid Advanced Optimization Methods with Evolutionary Computation Techniques in Energy Forecasting"

More accurate and precise energy demand forecasts are required when energy decisions are made in a competitive environment. Particularly in the Big Data era, forecasting models are always based on a complex function combination, and energy data are always complicated. Examples include seasonality, cyclicity, fluctuation, dynamic nonlinearity, and so on. These forecasting models have resulted in an over-reliance on the use of informal judgment and higher expenses when lacking the ability to determine data characteristics and patterns. The hybridization of optimization methods and superior evolutionary algorithms can provide important improvements via good parameter determinations in the optimization process, which is of great assistance to actions taken by energy decision-makers.

This book contains articles from the Special Issue titled "Hybrid Advanced Optimization Methods with Evolutionary Computation Techniques in Energy Forecasting", which aimed to attract researchers with an interest in the research areas described above. As Fan et al. [1] indicate, the research direction of energy forecasting in recent years has concentrated on proposing hybrid or combined models: (1) hybridizing or combining these artificial intelligence models with each other; (2) hybridizing or combining with traditional statistical tools; and (3) hybridizing or combining with those superior evolutionary algorithms. Therefore, this Special Issue sought contributions towards the development of any hybrid optimization methods (e.g., quadratic programming techniques, chaotic mapping, fuzzy inference theory, quantum computing, etc.) with superior evolutionary algorithms (e.g., genetic algorithms, ant colony optimization, particle swarm optimization algorithm, and so on) that have superior capabilities over the traditional optimization approaches to overcome some embedded drawbacks, and the application of these advanced hybrid approaches to significantly improve forecasting accuracy.

The 11 articles in this compendium all display a broad range of cutting-edge topics in hybrid optimization methods and superior hybrid evolutionary algorithms. The preface author believes that the hybridization of the advanced optimization methods and evolutionary computation techniques will play an important role in energy forecasting accuracy improvements, such as hybrid different evolutionary algorithms/models to overcome some critical shortcomings of single evolutionary algorithms/models or direct improvement of the shortcomings by innovative theoretical arrangements.

For Hybridizing Different Evolutionary Algorithms/Models

(1) Hybrid different evolutionary algorithms: It is known that the evolutionary algorithms have their theoretical drawbacks, such as a lack of knowledge, memory, or storage functions; they are time consuming in training; and become trapped in local optima. Therefore, the goal of hybridizing optimization methods to adjust their internal parameters (e.g., mutation rate, crossover rate, annealing temperature, etc.) is to overcome these shortcomings. For example, simulated annealing (SA) is a generic probabilistic search technique that simulates the material physical process of heating and controlled cooling. Each step of SA attempts to replace the current state by a random move. The new state can then be accepted with a probability that depends both on the difference between the corresponding function values and on a global parameter (temperature). Thus, SA can reach more ideal solutions. However, SA costs a great deal of computation time in the annealing process. To improve premature convergence and to receive more suitable objective function values, it is necessary to find an effective approach to

overcome these drawbacks of genetic algorithm (GA) and SA. The hybridization of a genetic algorithm with a simulated annealing (GA-SA) algorithm [2] is an innovative trial applying the superior capability of the SA algorithm to reach more ideal solutions, employing the mutation process of GA to enhance the search process.

(2) Different hybrid models: Each single model has its own drawbacks. For example, in Box–Jenkins' ARIMA model, the worst disadvantage is the inability to predict changes that are not clear in historical data, particularly for the nonlinearity of data patterns. The support vector regression (SVR) model cannot provide accurate forecasting performance when the data set reveals a cyclical (seasonal) tendency (e.g., caused by cyclic economic activities or seasonal nature hour to hour, day to day, week to week, month to month, and season to season), such as an hourly peak in a working day, a weekly peak in a business week, and a monthly peak in a demand-planned year. Therefore, the concepts of combined or hybrid models deserve consideration. Note that the term "hybrid" means that some process of the former model is integrated into the process of the later one. For example, hybrid A and B implies some processes of A are controlled by A, and some are controlled by B. On the other hand, for the so-called combined models, the output of the former model becomes the input of the latter one. Therefore, the classification results from combined models will be superior to a single model. Combined models are employed to further capture more data pattern information from the analyzed data series. For the mentioned shortcoming of the original SVR model, it is necessary to estimate this seasonal component (i.e., applying the seasonal mechanism to accomplish the goal of highly accurate forecasting performance). The preface author proposed a seasonal mechanism [3–5] with two steps for convenience in implementation: the first step is calculating the seasonal index (SI) for each cyclic point in a cycle length peak period; the second step is computing the forecasting value by multiplying the seasonal index (SI).

Another model hybridization example can be found in artificial neural network models. Inspired by the concept of recurrent neural networks (RNNs) where every unit is considered as an output of the network and the provision of adjusted information as input in a training process [6], the recurrent learning mechanism framework is also combined into the original analyzed model. For a feed-forward neural network, links can be established within layers of a neural network. These types of networks are called recurrent neural networks. RNNs include an additional information source from the output layer or the hidden layer. Therefore, they mainly use past information to capture detailed information, then improve their performances.

For Improvement by Innovative Theoretical Arrangements

Several disadvantages embedded in these evolutionary algorithms, such as their tendency to become trapped in local optima and evolutionary mechanism failure, can be improved by innovative theoretical arrangements to obtain more satisfactory performance.

(1) Chaotization of decision variables: Chaos is a ubiquitous phenomenon in nonlinear systems. Chaotic behaviors have characteristics such as high sensitivity to initial value, ergodicity, and randomness of motion trail, and can traverse each trail within a certain range according to its rule. Therefore, chaotic variables may be adopted by utilizing these characteristics of chaotic phenomena for global search and optimization to increase the particle diversity. Due to easy implementation process and a special mechanism to escape from local optima, chaos and chaos-based searching algorithms have received intense attention [7]. Any decision variable in an optimization problem can be chaotized by the chaotic sequence as a chaotic variable to carefully expand its search space (i.e., variables are allowed to travel ergodically over the search space). The critical factor influencing the performance improvement is the chaotic mapping function. There are several commonly adopted chaotic mapping functions for the chaotic sequence generator, such as the logistic mapping function, the tent mapping function, the An mapping function, and the cat mapping function.

(2) Adjustments by cloud theory: for example, based on the operation procedure of SA, subtle and skillful adjustment in the annealing schedule is required (e.g., the size of the temperature steps during annealing). Particularly, the temperature of each state is discrete and unchangeable, which does not meet the requirement of continuous decrease in temperature in actual physical annealing processes. In addition, SA easily accepts deteriorated solutions with high temperature, and it is difficult to escape from local minimum traps at low temperature [3]. To overcome these drawbacks of SA, cloud theory is considered. Cloud theory is a model of the uncertainty transformation between quantitative representation and qualitative concept using language value [3]. Based on the SA operation procedure, subtle and skillful adjustment in the annealing schedule is required (e.g., the size of the temperature steps during annealing, the temperature range, the number of re-starts and re-direction of the search). The annealing process is like a fuzzy system in which the molecules move from large-scale to small-scale randomly as the temperature decreases. In addition, due to its Monte Carlo scheme and lack of knowledge memory functions, its time-consuming nature is another problem. It is deserved to employ a chaotic simulated annealing (CSA) algorithm [3] to overcome these shortcomings.

In this, the transiently chaotic dynamics are temporarily generated for foraging and self-organizing. They are then gradually vanished with autonomous decrease of the temperature, and are accompanied by successive bifurcations and converged to a stable equilibrium. Therefore, CSA significantly improves the randomization of the Monte Carlo scheme, and controls the convergent process by bifurcation structures instead of stochastic "thermal" fluctuations, eventually performing efficient searching including a global optimum state. However, as mentioned above, the temperature of each state is discrete and unchangeable, which does not meet the requirement of continuous decrease in temperature in actual physical annealing processes. Even if some temperature annealing functions are exponential in general, the temperature gradually falls with a fixed value in every annealing step and the changing process of temperature between two neighbor steps is not continuous. This phenomenon also appears when other types of temperature update functions are implemented (e.g., arithmetical, geometrical, or logarithmic). In cloud theory, by introducing the Y condition normal cloud generator to the temperature generation process, it can randomly generate a group of new values that distribute around the given value like a "cloud". The fixed temperature point of each step becomes a changeable temperature zone in which the temperature of each state generation in every annealing step is chosen randomly, the course of temperature change in the whole annealing process is nearly continuous, and fits the physical annealing process better. Therefore, based on chaotic sequence and cloud theory, the chaotic cloud simulated annealing algorithm (CCSA) is employed to replace the stochastic "thermal" fluctuations control from traditional SA to enhance the continuous physical temperature annealing process from CSA. Cloud theory can realize the transformation between a qualitative concept in words and its numerical representation. It can be employed to avoid the problems mentioned above.

This discussion of the work by the author of this preface highlights work in an emerging area of hybrid optimization methods with superior evolutionary algorithms that has come to the forefront over the past decade. The articles collected in this text span many cutting-edge areas that are truly interdisciplinary in nature.

Wei-Chiang Hong

Guest Editor

Reference

1. Fan, G.F.; Peng, L.L.; Hong, W.C. Short term load forecasting based on phase space reconstruction algorithm and bi-square kernel regression model. *Appl. Energy* **2018**, *224*, 13–33.

2. Zhang, W.Y.; Hong, W.C.; Dong, Y.; Tsai, G.; Sung, J.T.; Fan, G. Application of SVR with chaotic GASA algorithm in cyclic electric load forecasting. *Energy* **2012**, *45*, 850–858.

3. Geng, J.; Huang, M.L.; Li, M.W.; Hong, W.C. Hybridization of seasonal chaotic cloud simulated annealing algorithm in a SVR-based load forecasting model. *Neurocomputing* **2015**, *151*, 1362–1373.

4. Hong, W.C.; Dong, Y.; Zhang, W.Y.; Chen, L.Y.; Panigrahi, B.K. Cyclic electric load forecasting by seasonal SVR with chaotic genetic algorithm. *Int. J. Electr. Power Energy Syst.* **2013**, *44*, 604–614.

5. Ju, F.Y.; Hong, W.C. Application of seasonal SVR with chaotic gravitational search algorithm in electricity forecasting. *Appl. Math. Modelling* **2013**, *37*, 9643–9651.

6. Hong, W.C. Electric load forecasting by seasonal recurrent support vector regression (SVR) with chaotic artificial bee colony algorithm. *Energy* **2011**, *36*, 5568–5578.

7. Li, M.; Hong, W.C.; Kang, H. Urban traffic flow forecasting using Gauss-SVR with cat mapping, cloud model and PSO hybrid algorithm. *Neurocomputing* **2013**, *99*, 230–240.

energies

MDPI

Article

Applications of Hybrid EMD with PSO and GA for an SVR-Based Load Forecasting Model

Guo-Feng Fan [1], Li-Ling Peng [1], Xiangjun Zhao [2] and Wei-Chiang Hong [2,*]

[1] College of Mathematics & Information Science, Ping Ding Shan University, Pingdingshan 467000, China; guofengtongzhi@163.com (G.-F.F.); plling1054@163.com (L.-L.P.)
[2] School of Education Intelligent Technology, Jiangsu Normal University, 101 Shanghai Rd., Tongshan District, Xuzhou 221116, China; xjzhao@jsnu.edu.cn
* Correspondence: samuelsonhong@gmail.com; Tel.: +86-516-8350-0307

Received: 30 September 2017; Accepted: 21 October 2017; Published: 26 October 2017

Abstract: Providing accurate load forecasting plays an important role for effective management operations of a power utility. When considering the superiority of support vector regression (SVR) in terms of non-linear optimization, this paper proposes a novel SVR-based load forecasting model, namely EMD-PSO-GA-SVR, by hybridizing the empirical mode decomposition (EMD) with two evolutionary algorithms, i.e., particle swarm optimization (PSO) and the genetic algorithm (GA). The EMD approach is applied to decompose the load data pattern into sequent elements, with higher and lower frequencies. The PSO, with global optimizing ability, is employed to determine the three parameters of a SVR model with higher frequencies. On the contrary, for lower frequencies, the GA, which is based on evolutionary rules of selection and crossover, is used to select suitable values of the three parameters. Finally, the load data collected from the New York Independent System Operator (NYISO) in the United States of America (USA) and the New South Wales (NSW) in the Australian electricity market are used to construct the proposed model and to compare the performances among different competitive forecasting models. The experimental results demonstrate the superiority of the proposed model that it can provide more accurate forecasting results and the interpretability than others.

Keywords: support vector regression; empirical mode decomposition (EMD); particle swarm optimization (PSO); genetic algorithm (GA); load forecasting

1. Introduction

Due to the difficult-reserved property of electricity, providing accurate load forecasting plays an important role for the effective management operations of a power utility, such as unit commitment, short-term maintenance, network power flow dispatched optimization, and security strategies. On the other hand, inaccurate load forecasting will increase operating costs: over forecasted loads lead to unnecessary reserved costs and an excess supply in the international energy networks; under forecasted loads result in high expenditures in the peaking unit. Therefore, it is essential that every utility can forecast its demands accurately.

There are lots of approaches, methodologies, and models proposed to improve forecasting accuracy in the literature recently. For example, Li et al. [1] propose a computationally efficient approach to forecast the quantiles of electricity load in the National Electricity Market of Australia. Arora and Taylor [2] present a case study on short-term load forecasting in France, by incorporating a rule-based methodology to generate forecasts for normal and special days, and by a seasonal autoregressive moving average (SARMA) model to deal with the intraday, intraweek, and intrayear seasonality in load. Takeda et al. [3] propose a novel framework for electricity load forecasting by combining the Kalman filter technique with multiple regression methods; Zhao and Guo [4] propose a hybrid optimized grey model (Rolling-ALO-GM (1,1)) to improve the accurate level of annual load forecasting. For those applications of neural networks in load forecasting, the authors of references [5–9] have proposed several useful short-term load forecasting models. For these applications of hybridizing popular methods with evolutionary algorithms, the authors of references [10–14] have demonstrated that the forecasting performance improvements can be made successfully. These proposed methods could receive obvious forecasting performance improvements in terms of accurate level in some cases, however, the issue of modeling with good interpretability should also be taken into account, as mentioned in [15]. Furthermore, these proposed models are almost embedded with strong intra-dependency association to experts' experiences, as well as, they often could not guarantee to receive satisfied accurate forecasting results. Therefore, it is essential to propose some kind of combined model, which hybridizes popular methods with advanced evolutionary algorithms, also combining expert systems and other techniques, to simultaneously receive high accuracy forecasting performances and interpretability.

Due to advanced higher dimensional mapping ability of kernel functions, support vector regression (SVR) is drastically applied to deal with the forecasting problem, which is with small but high dimensional data size. SVR has been an empirically popular model to provide satisfied performances in many forecasting problems [16–18]. As it is known that the biggest disadvantage of a SVR model is the premature problem, i.e., suffering from local minimum when evolutionary algorithms are used to determine its parameters. In addition, its robustness also could not receive a satisfied stable level. To look for effective novel algorithms or hybrid algorithms to avoid trapping into local minimum, and to simultaneously receive satisfied robustness is still the very hot point in the SVR modeling research fields [19]. In the meanwhile, to improve the forecasting accurate level, it is essential to extract the original data set with nonlinear or nonstationary components [20] and transfer them into single and conspicuous ones. The empirical mode decomposition (EMD) is dedicated to provide extracted components to demonstrate high accurate clustering performances, and it has also received lots of attention in relevant applications fields, such as communication, economics, engineering, and so on [21–23]. As aforementioned, the EMD could be applied to decompose the data set into some high frequency detailed parts and the low frequent approximate part. Therefore, it is easy to reduce the interactions among those singular points, thereby increasing the efficiency of the kernel function. It becomes a useful technique to help the kernel function to deal well with the tendencies of the data set, including the medium trends and the long term trends. For determining the values of the parameters in a SVR model well, it attracts lots of relevant researches during two past decades. The most effective approach is to employ evolutionary algorithms, such as GA [24,25], PSO [26,27], and so on [28,29]. Based on the authors' empirical experiences in applying the evolutionary algorithms, PSO is more suitable for solving real problems (with more details of data set), it is simple to be implemented, its shortcoming is trapped in the local optimum. For GA, it is more suitable for solving discrete problems (data set reveals stability), however one of its drawbacks is Hamming cliffs. In this paper, the data set will be divided into two parts by EMD (i.e., higher frequent detail parts and the lower frequent part), the higher frequency part is the so-called shock data which demonstrates the details of the data set, thus, SVR's parameters for this part are suitable to be determined by PSO due to its suitable for solving real problems. The lower frequency part is the so-called contour trend data which reveals its stability, thus, SVR's parameters for this part could be selected by GA due to its suitable for solving discrete problems.

Therefore, in this paper, the mentioned two parts divided by EMD are conducted by SVR-PSO and SVR-GA, respectively; eventually the improved forecasting performances of proposed model (namely EMD-PSO-GA-SVR model) would be demonstrated. The comprehensive framework could be shown in the following illustrations: (1) the data set is divided by EMD technique into high frequency part with more detailed information and low frequency part with more tendency information, respectively; (2) for the high frequency part, PSO is used to determine the SVR's parameters, i.e., SVR-PSO is implemented to forecast to receive higher accurate level; (3) for the low frequency part, due to stationary characteristics of tendency information, GA is employed to select suitable parameters in a SVR model, i.e., SVR-GA is implemented to forecast; and, (4) the final forecasting results are obtained from steps (2) and (3). There are also several advantages of the proposed EMD-PSO-GA-SVR model: (1) the proposed model is able to smooth and reduce the noise effects due to inheriting them from from EMD technique; (2) the proposed model is capable to filter data set with detail information and improve microscopic forecasting accurate level due to applying the PSO with the SVR model; and, (3) the proposed model is also capable of capturing the macroscopic outline and to provide accurate forecasting in future tendencies due to inherited from GA. The forecasting processes and superior results would be demonstrated in the next sections.

To demonstrate the advantages and suitability of the proposed model, 30-min electricity loads (i.e., 48 data collected daily) from New South Wales are selected to construct model and to compare the forecasted accurate level with other competitive models, namely, original SVR model, SVR-PSO model (SVR parameters determined by PSO), SVR-GA (SVR parameters selected by GA), and the AFCM model (an adaptive fuzzy model based on a self-organizing map). The second example is from the New York Independent System Operator (NYISO, New York, NY, USA), similarly, 1-h electricity loads (i.e., only 24 data collected daily) are collected to model and to compare the forecasting accurate level. The results demonstrate that the proposed EMD-PSO-GA-SVR model could receive a higher forecasting accuracy level and more comprehensive interpretability. In addition, due to employing the EMD technique, the proposed model could consider more information during the modeling process; thus, it is able to provide more generalization in modeling.

The remainder of this paper is organized as follows. Section 2 provide the modeling details of the proposed EMD-PSO-GA-SVR model. Section 3 provides the description of the data set and relevant modeling design. Section 4 investigates the forecasting results and compares with other competitive models, some insightful discussions are also provided. Section 5 concludes the study.

2. The EMD-PSO-GA-SVR Model

2.1. The Empirical Mode Decomposition (EMD) Technique

The principal assumption of the EMD technique is that any data set contains several simple intrinsic modes of fluctuations. For every linear or non-linear mode, it would have only one extreme value among continuous zero-crossings. Thereby, each data set is theoretically able to be decomposed into several intrinsic mode functions (IMFs) [30]. The decomposition steps of a data set ($x(t)$) are briefed as follows.

Step 1   Identify. Determine all of the local extremes (including all maxima and minima) of the data set.

Step 2   Produce Envelope. Connect all of the local maxima and minima by two cubic spline lines as the upper envelope and lower envelope, respectively. The mean envelope, m_1, is set by the mean of upper envelope and lower envelope.

Step 3   IMF Decomposition. Define the first component, h_1, as the difference between the data set $x(t)$ and m_1, and is displayed in Equation (1),

$$h_1 = x(t) - m_1 \qquad (1)$$

Notice that h_1 does not have to be a standard IMF, thus, it is unnecessary with the conditions of the IMF. m1 would be approximated to zero after k times evolutions, then, the kth component, h_{1k}, could be shown as Equation (2),

$$h_{1k} = h_{1(k-1)} - m_{1k} \qquad (2)$$

where h_{1k} is the kth component after k times evolutions; $h_{1(k-1)}$ is the $(k-1)$th component after $k-1$ times evolutions. The standard deviation (*SD*) for the kth component is given in Equation (3),

$$SD = \sum_{k=1}^{L} \frac{\left| h_{1(k-1)}(t) - h_{1k}(t) \right|^2}{h_{1(k-1)}^2(t)} \in (0.2, 0.3) \tag{3}$$

where L is the total number of the data set.

Step 4   New IMF Component. If h_{1k} reaches the conditions of *SD*, then, the first IMF component, c_1, could be obtained, i.e., $c_1 = h_{1k}$. A new series, r_1, could be decomposed after deleting the first IMF, as shown in Equation (4),

$$r_1 = x(t) - c_1 \tag{4}$$

Step 5   IMF Composition. Repeat steps 1–4 until no any new IMF component could be decomposed. The process is demonstrated in Equation (5). The series, r_n, is the remainder of the original data set $x(t)$, as Equation (6).

$$r_1 = x(t) - c_1$$
$$r_2 = r_1 - c_2$$
$$\cdots\cdots \tag{5}$$
$$r_n = r_{n-1} - c_n$$

$$x(t) = \sum_{i=1}^{n} c_i + r_n \tag{6}$$

2.2. The Support Vector Regression Model

By introducing the concept of ε-insensitive loss function, support vector machines have successfully been applied to deal with nonlinear regression problems [31], the so-called support vector regression (SVR). The principal idea of SVR is mapping the non-linear data set into a higher dimensional feature space to receive more satisfied forecasting performances. Thus, given a data set, $G = \{(x_i, a_i)\}_{i=1}^{N}$, with total N input data, x_i, and actual values, a_i, the SVR function could be shown as Equation (7),

$$f(\mathbf{x}) = \mathbf{w}^T \varphi(\mathbf{x}) + b \tag{7}$$

where $\varphi(\mathbf{x})$ is the so-called feature mapping function, which is capable to conduct nonlinear mapping to feature space from the input space $\mathbf{x}$. The $\mathbf{w}$ and b are coefficients that are estimated by optimizing the regularized risk function, as shown in Equation (8),

$$R(C) = \frac{C}{N} \sum_{i=1}^{N} L_\varepsilon(d_i, y_i) + \frac{\|w^2\|}{2} \tag{8}$$

where

$$L_\varepsilon(d, y) = \begin{cases} 0 & if \ |d - y| \le \varepsilon \\ |d - y| - \varepsilon & otherwise \end{cases} \tag{9}$$

In Equation (8), $L_\varepsilon(d, y)$ is the so-called ε-insensitive loss function, the loss could be viewed as zero only if the forecasted value is within the defined ε-tube (refer Equation (9)); the second term, $\frac{\|w^2\|}{2}$, could be used to measure the flatness of the function. Obviously, parameter C is used to determine the trade-off between the loss function (empirical risk) and the flatness of the function. Two positive slack variables, ζ and ζ^*, which measure the distance from actual values to the associate boundary of ε-tube, are employed to help in solving the problem. Then, Equation (8) could be transformed to the optimization problem with inequality constraints as Equation (10),

$$\text{Minimize } R(w, \xi, \xi^*) = \frac{\|w^2\|}{2} + C \sum_{i=1}^{N} (\xi_i + \xi_i^*) \tag{10}$$

with the constraints,

$$w\varphi(x_i) + b_i - d_i \leq \varepsilon + \xi_i^*$$
$$d_i - w\varphi(x_i) - b_i \leq \varepsilon + \xi_i$$
$$\xi_i, \xi_i^* \geq 0$$
$$i = 1, 2 \ldots, N$$

After solving Equation (10), the solution of the weight, w, in Equation (7) is calculated as Equation (11),

$$w^* = \sum_{i=1}^{N} (\beta_i - \beta_i^*) \varphi(x) \tag{11}$$

Hence, the SVR function has been constructed as Equation (12),

$$f(x, \beta, \beta^*) = \sum_{i=1}^{l} (\beta_i - \beta_i^*) K(x, x_i) + b \tag{12}$$

$K(x, x_i)$ is the so-called kernel function. The value of the kernel is equal to the inner product of two vectors x and x_i in the feature space $\varphi(x)$ and $\varphi(x_i)$, i.e., $K(x, x_i) = \varphi(x) \times \varphi(x_i)$. It is difficult to decide which type of kernel functions is suitable for which specific data pattern. However, only if the function satisfies Mercer's condition [31], it can be employed as the kernel function. Due to its easiness to implement and the non-linear mapping capability, the Gaussian function, $K(x, x_i) = e^{-\frac{\|x_i - x^2\|}{2\sigma^2}}$, is used in this paper. The selection of three parameters, ε, σ, and C, in a SVR model would influence the forecasting accurate level. Authors have conducted a series researches to test the suitability of different evolutionary algorithms hybridizing with a SVR model. Based on authors' research results, in this paper, two evolutionary algorithms, PSO and GA, are applied to determine the parameters of SVR models from high frequency and low frequency, respectively.

2.3. Particle Swarm Optimization (PSO) Algorithm

Due to its simple framework and easy fulfillment, particle swarm optimization [32] has become a famous optimization algorithm [33]. While PSO modeling, each particle adjusts its searching direction not only based on its search experiences (local search), but also on its social learning experiences from neighboring particles (global search) to eventually find out the global best position of the system, i.e., successful the exploration-exploitation trade off would guarantee to succeed in position searching.

In the meanwhile, during the PSO modeling process, the degree of inertia also plays the critical role in terms of the speed of convergence. With a larger inertia, it would facilitate implementation of the global exploration, the convergent speed would be hence slowed down; on the contrary, with a smaller inertia, it would lead well to asearch for the current range, the convergence is speedy but might be suffering from a local optimum. There are several novel approaches to well tune the inertia weight in literature [34,35].

The procedure of SVR-PSO is briefly demonstrated as following steps. Interested readers could refer to [25] for more detail.

Step 1   Initialization. Initialize the population of three particles (σ, ε, C), the random positions and velocities.

Step 2   Compute Initial Objective Values. Compute the objective values by using the three particles. Set the local initial objective values, f_{besti}, based on their own best position, and set the global initial objective values, $f_{globalbesti}$, based on their global best position.

Step 3   Update Inertia, Velocity, and Position. According to Equations (13)–(15) to update the inertia, velocity, and position for these three particles. Evaluate the objective values by using these particles. By the way, the inertia weight is employed the most popular linear decreasing function [34], as shown in Equation (13).

$$l_i = \alpha * l_{i-1} \tag{13}$$

where α is a constant, its value is less than but closed to 1; $i = 1, 2, \ldots, N$.

$$V_i = l_i \times V_{i-1} + q_1 \times rand(\cdot) \times (p_{i-1} - X_{i-1}) + q_2 \times Rand(\cdot) \times (P_{i-1} - X_{i-1}) \tag{14}$$

where q_1 and q_2 are positive acceleration constants; $rand(\cdot)$ and $Rand(\cdot)$ are independent random distributed within range [0, 1]; p_i is the best position for each particle itself; P_i is the global best position; X_i is the position for each particle itself; $i = 1, 2, \ldots, N$.

$$X_i = X_{i-1} + V_i \tag{15}$$

Step 4   Update Objective Values. For each iteration, by using its current position of the three particles to compare the current objective value with f_{besti}. If the current objective value is better (i.e., with smaller forecasting errors), then, update its objective value. In this paper, mean absolute percentage error (*MAPE*) and the root mean square error (*RMSE*), as shown in Equations (16) and (17), respectively, are used to measure the forecasting errors. In the meanwhile, a switching between *MAPE* and *RSME* is employed once found the error of *MAPE* is less than those of *RMSE* and vice versa, to ensure the smallest objective values that could be selected.

$$MAPE = \frac{1}{N} \sum_{i=1}^{N} \left| \frac{a_i - f_i}{a_i} \right| \times 100\% \tag{16}$$

$$RMSE = \sqrt{\frac{\sum_{i=1}^{N} (a_i - f_i)^2}{N}} \tag{17}$$

where N is the total number of data; a_i is the actual load value at point i; f_i is the forecasted load value at point i.

Step 5   Determine Best Particles. If the current objective value is also smaller than $f_{globalbesti}$, then the best particles could be determined in this iteration.

Step 6   Stopping Criteria. If the stopping criteria (forecasting error) are reached, the final $f_{globalbesti}$ would be the solution; otherwise, go back to Step 3.

The detail procedure of the SVR-PSO is illustrated in Figure 1.

2.4. Genetic Algorithm (GA)

Genetic algorithm is the most famous evolutionary algorithm. Its primary concept is its elitist selection principle to keep best gene to be well survival from generation to generation during the evolutionary process in the natural systems. GA has several important operations, including selection, crossover, and mutation, to generate new individuals. It is able to effectively avoid trapping into local optima if more satisfied objective values could be successfully found. GA has been applied in many optimization problems.

The procedure of SVR-GA is briefly demonstrated as following steps and illustrated in Figure 2. Interested readers could refer to [27] for more detail.

Step 1   Initialization. Randomly generate the initial population of chromosomes.

Step 2   Fitness Evaluation. The fitness values of each chromosome would be evaluated by the fitness function. Similarly, in SVR-GA modeling process, *MAPE* and *RMSE* are used to measure the

Step 3   Selection. Elitism policy is employed while selecting good chromosomes, which receive satisfied fitness values for yielding new offspring in the next generation. Practically, the useful roulette wheel selection technique is applied to choose chromosomes for yielding new offspring.

fitness (i.e., forecasting errors), and a switching between *MAPE* and *RSME* is also employed to ensure the smallest objective values could be selected.

Step 4   Mutation and Crossover. For crossover operation, this paper uses the single point technique, i.e., any paired chromosomes are exchanged at the same breaking points. For mutation operation, it is implemented randomly. The both techniques in GA, i.e., (a) crossover followed by mutation, and (b) mutation followed by crossover are tested in both of the series. In this paper, the rates of crossover and mutation operations are set as 0.8 and 0.05, as Hong et al. [26] suggested, respectively.

Step 5   New Generation. Generate new population for next new generation.

Step 6   Stopping Criterion. If the number of generation meets the stopping criterion, the current chromosome should be the best solution; otherwise, get back to Step 2 and repeat the procedure.

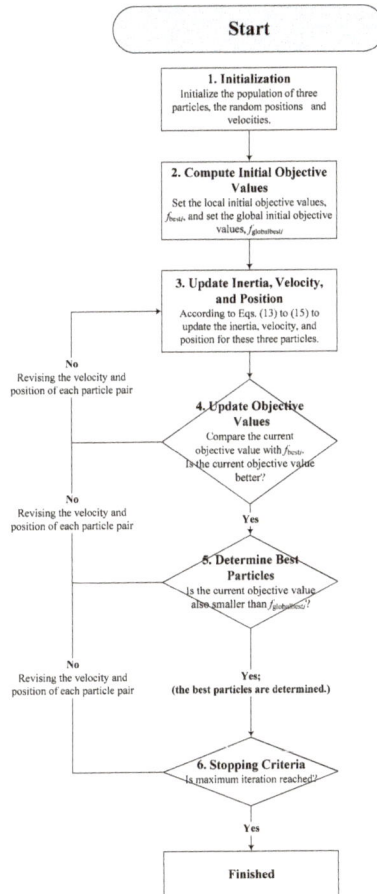

Figure 1. Support vector regression (SVR)- particle swarm optimization (PSO) algorithm flowchart.

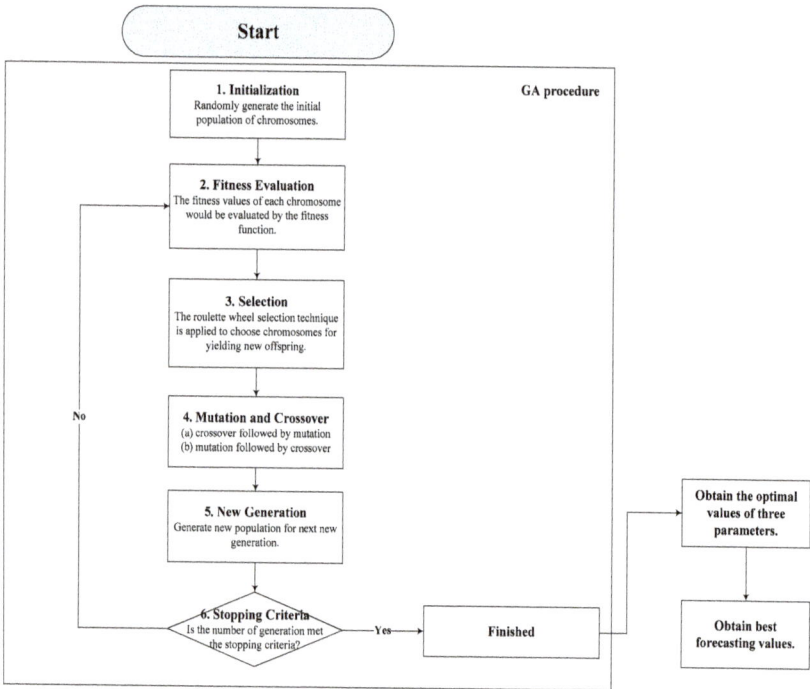

Figure 2. SVR- genetic algorithm (GA) algorithm flowchart.

2.5. The Complete Processes of the Proposed Model

The complete processes of the proposed EMD-PSO-GA-SVR model is indicated as follows, the associate total flowchart is demonstrated in Figure 3.

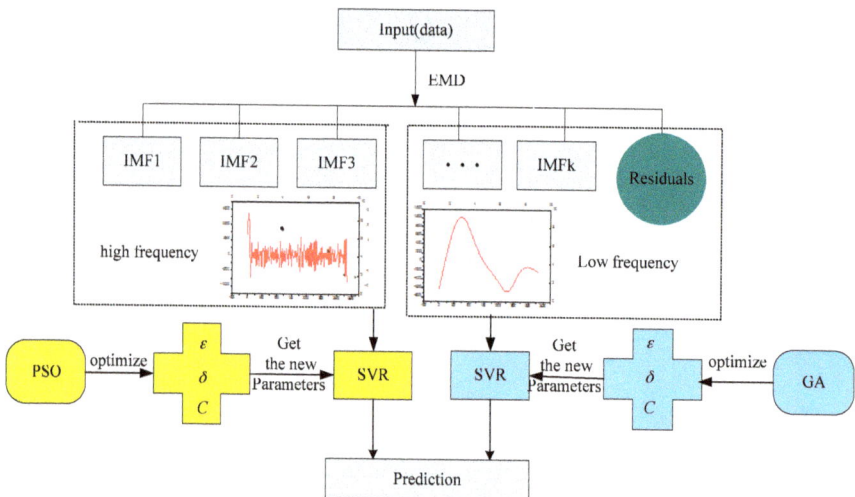

Figure 3. The total flowchart of the proposed empirical mode decomposition (EMD)-PSO-GA-SVR model.

Step 1 Data Decomposed. Apply the EMD technique to decompose the load data (training data) into intrinsic mode functions (IMFs) with higher and lower frequency parts, respectively. The details could be found in Section 2.1.

Step 2 Higher Frequency IMFs Modeling (by SVR-PSO). Higher frequency IMFs are modeled by an SVR model, and, as abovementioned that its parameters are determined by the PSO algorithm. Most suitable parameter combinations will be finalized with only the smallest forecasting errors (i.e., smallest *MAPE* value). The modeling details could be found in the relevant Sections 2.2 and 2.3, and Figure 1.

Step 3 Lower Frequency IMFs Modeling (by SVR-GA). On the contrary, lower frequency IMFs are modeled by an SVR model, and, its parameters are determined by the GA algorithm. Similarly, only with smallest *MAPE* value, the associate parameters in the SVR model will be determined to be the most suitable combination. The relevant details could be referred Section 2.4 and Figure 2.

Step 4 EMD-PSO-GA-SVR Modeling. While the forecasting values for higher and lower frequencies IMFs have been calculated, respectively, then, the final forecasting results will also be performed.

3. Experimental Examples

Two experimental examples are used to demonstrate the advantages of the proposed model in terms of applicability, superiority, and generality. The data of the first example (Example 1, in Section 3.1) is obtained from Australian New South Wales (NSW) electricity market; the data of the second example (Example 2, in Section 3.2) is from American New York Independent System Operator (NYISO). Furthermore, to demonstrate the overtraining effect for different data sizes, in this paper, two kinds of data sizes, i.e., small data size and large data size, are employed to modeling and analysis, respectively.

To ensure the feasibility of employing EMD to decompose the target data sets (both Examples 1 and 2) into higher and lower frequent parts, it is necessary to verify whether they are nonlinear and non-stationary. This paper applies the recurrence plot (RP) theory [36,37] to analyze these two characteristics. As it is known that RP reveals all of the times when the phase space trajectory of the dynamical system visits roughly the same area in the phase space, therefore, it is suitable to analyze the nonlinear and non-stationary characteristics of a data set. The RP analysis for Examples 1 and 2, as shown in Figure 4a,b, indicate: (1) it is clearly to see the parallel diagonal lines in both figures, i.e., both data sets reveal periodicity and deterministic, and this is the reason that authors could use these data sets to conduct forecasting; (2) it is also clearly to see the checkerboard structures in both figures, i.e., both data sets reveal that, after a transient, they go into slow oscillations that are superimposed on the chaotic motion; and, (3) it obviously demonstrates that vertical and horizontal lines cross at the cross, both data sets reveal laminar states, i.e., non-stationary characteristics. The relevant recurrence rate (the lower rate implies the nonlinear characteristic) and laminarity (the smaller value represents the non-stationary characteristic) for these two data sets also support the abovementioned RP analysis results. Based on the RP analysis, both the electricity load data set demonstrate macroscopic periodicity tendency (i.e., lower frequent part) and the microscopic chaotic oscillations tendency (i.e., higher frequent part), therefore, it is useful to employ EMD to decompose these two data sets into higher and lower frequent part, respectively.

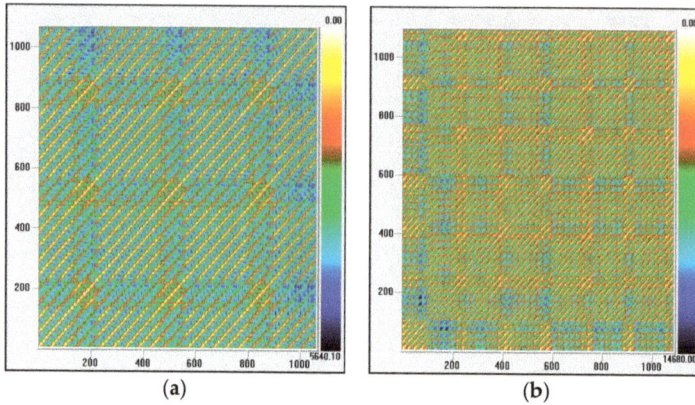

Figure 4. The recurrence plot for Examples 1 and 2. (**a**) Example 1 embedding dimensions = 4; time delays = 13; recurrence rate = 33.3%; laminarity = 51.0% (Small data size from Australian New South Wales, NSW); (**b**) Example 2 embedding dimensions = 4; time delays = 4; recurrence rate = 38.1%; laminarity = 54.7% (Small data size from American New York Independent System Operator, NYISO).

3.1. The Forecasting Results of Example 1

3.1.1. Data Sets for Small and Large Sizes

In Example 1, for small data size, there are totally 336 electricity load data per 30 min for seven days, i.e., from 2 to 8 May 2007. In which, the former 288 load data are used as the training data set, the latter 48 load data are as testing data set. The original data set is shown in Figure 5a.

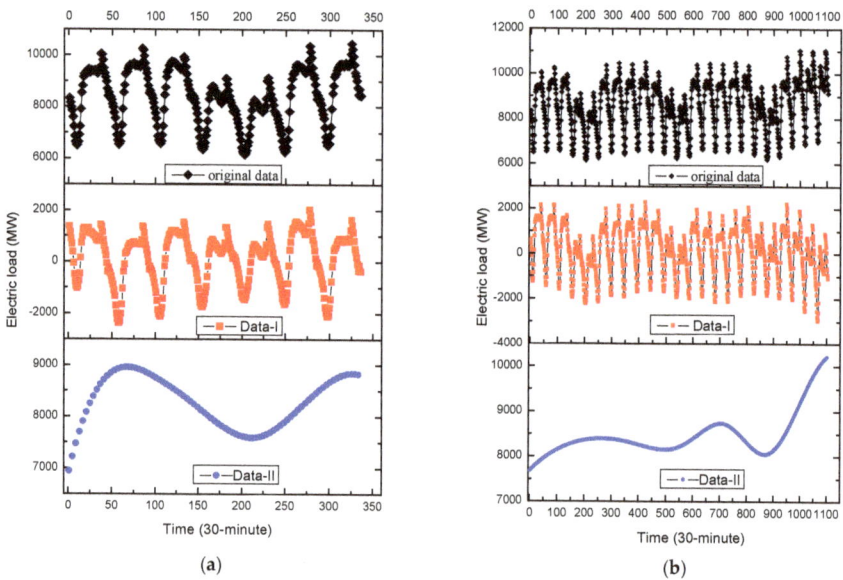

Figure 5. The 30-min based original data, Data-I, Data-II (Example 1). (**a**) Small data size (2–8 May 2007); (**b**) Large data size (2–24 May 2007).

On the other hand, for large data size, there are totally 1104 electricity load data from 2 to 24 May 2007, based on 30-min scale. In which, the former 768 load data (i.e., from 2 to 17 May 2007) are used as the training data set, the remainder 336 load data are used as the testing data set. The original data set is shown in Figure 5b.

3.1.2. Decomposition Results by EMD

The EMD is employed to decompose the original data set into higher and lower frequency items. For small data size, it could be divided into eight groups, as demonstrated in Figure 6a–h. In the meanwhile, the time delay of the data set in RP analysis (which value is 4) is simultaneously considered to select the higher and lower frequent parts. Consequently, the former four groups with much more frequency are classified as higher frequency items; the continued four groups with less frequency are classified as lower frequency items. The latest figure represents the original data; by the way, Figure 6h represents the trend term (i.e., residuals). As also shown in Figure 5a,b, the fluctuation characteristics of higher frequency item (namely Data-I) are almost the same with the original data, on the contrary, the macrostructure of lower frequency item (namely Data-II) is more stable. Data-I and Data-II will be further analyzed by SVR-PSO and SVR-GA models, respectively, to receive satisfied regression results. The details are illustrated in the next sub-section.

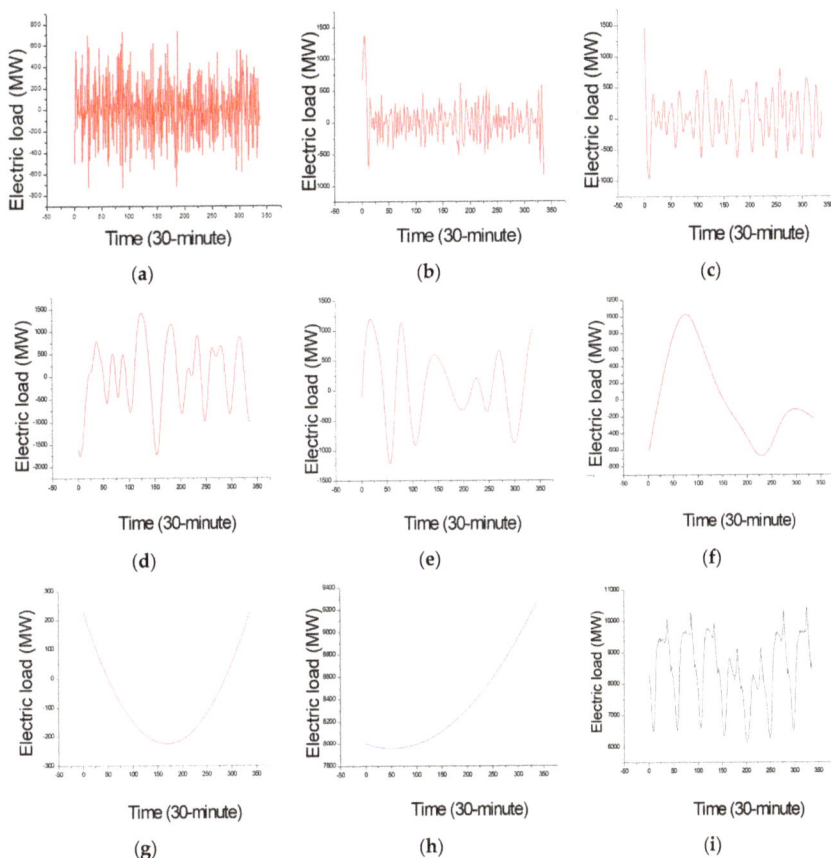

Figure 6. The decomposed different items for small data size (Example 1). (**a**) IMF (intrinsic modulo function) 1; (**b**) IMF 2; (**c**) IMF 3; (**d**) IMF 4; (**e**) IMF 5; (**f**) IMF 6; (**g**) IMF 7; (**h**) residuals; (**i**) the raw data.

3.1.3. SVR-PSO for Data-I

As shown in Figure 5, Data-I almost presents as periodical stability every 24 h, which is consistent with people's production and life. It also reflects the details of continuous changes. In this sub-section, SVR model is employed to conduct forecasting of Data-I, and as it is known that SVR requires evolutionary algorithms to appropriately determine its parameters to improve its forecasting accurate level. When considering that PSO is capable of solving the parameter determination problem from the data set with mentioned continuous change details, therefore, PSO is applied to be hybridized with the SVR model to forecast Data-I.

Firstly, the higher frequency items (i.e., Data-I) from small and large data sizes are both used for SVR-PSO modeling. Then, the best modeled results in training and testing stages are received, as shown in Figure 7a,b, respectively. As demonstrated in Figure 7, it is obvious to learn about the forecasting accuracy improvements from the hybridization of PSO.

The parameters settings of the SVR model for small and large data sizes are shown in Table 1. The determined parameters of the SVR-PSO model are shown in Table 2.

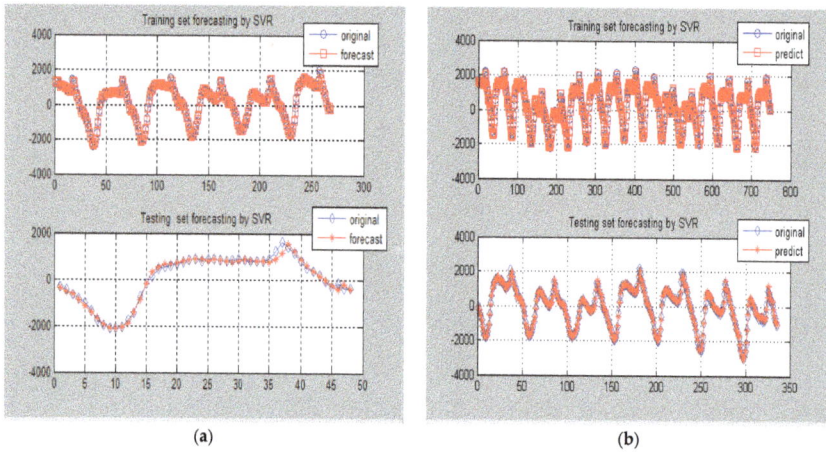

(a)　　　　　　　　　　　　(b)

Figure 7. Comparison of forecasting results by SVR-PSO model for Data-I (Example 1). (**a**) One-day ahead forecasting on 8 May 2007; (**b**) One-week ahead forecasting from 18 to 24 May 2007.

Table 1. The parameters settings of SVR-PSO model for original data and Data-I (Examples 1 and 2).

Data Types	Number of Particle	Length of Particle	Constant q_1	Constant q_2	Maximum of Iteration	C_{min}	C_{max}	σ_{min}	σ_{max}
Original data	30	3	2	2	300	0	200	0	200
Small data size	20	3	2	2	50	0	200	0	200
Large data size	5	3	2	2	20	0	200	0	200

Table 2. The SVR-PSO's parameters for Data-I (Example 1).

Data Size	σ	C	ε	Testing *MAPE*	Testing *RMSE*
Samall data size	0.15	91	0.0025	9.14	101.76
Large data size	0.20	99	0.0012	4.15	102.57

3.1.4. SVR-GA for Data-II

As shown in Figure 6e–h, the lower frequency item has not only less frequency, but also demonstrates more stability, particularly for the residuals, Figure 6h. In addition, in the long term, there would suffer from nonlinear mechanical changes, which are relatively discrete. In this sub-section, the SVR model is also used to conduct forecasting of Data-II, and also requires appropriate algorithm to well determine its three parameters. Therefore, it is more suitable to apply GA, while the SVR model is modeling. As abovementioned, GA is familiar to solve discrete problems, thus, the parameters settings of the SVR model for small and large data sizes are the same as shown in Table 3.

Table 3. The parameters settings of SVR-GA model for original data and Data-II (Examples 1 and 2).

Data Types	Population Size	Mutation Rate	Crossover Rate	Maximum of Generation	C_{min}	C_{max}	σ_{min}	σ_{max}
Original data	100	0.05	0.8	200	0	100	0	1000
Small data size	100	0.05	0.8	200	0	100	0	1000
Large data size	100	0.05	0.8	200	0	100	0	1000

Similarly, the lower frequency items (i.e., Data-II) from small and large data sizes are used for SVR-GA modeling. Then, the best modeled results in training and testing stages are received, as shown in Figure 8a,b, respectively. In which, it has demonstrated the superiority from the hybridization of GA. The determined parameters of the SVR-GA model are shown in Table 4.

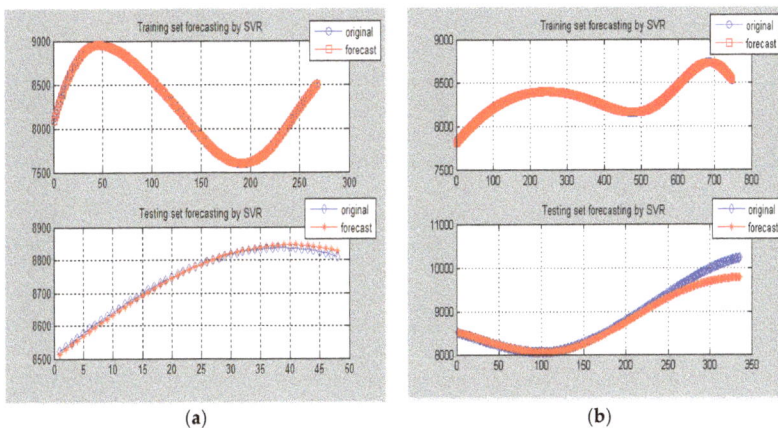

(a)   (b)

Figure 8. Comparison of forecasting results by SVR-GA model for Data-II (Example 1). (**a**) One-day ahead forecasting on 8 May 2007; (**b**) One-week ahead forecasting from 18 to 24 May 2007.

Table 4. The SVR-GA's parameters for Data-II (Example 1).

Data Size	σ	C	ε	Testing *MAPE*	Testing *RMSE*
Samall data size	0.15	90	0.0022	8.70	69.79
Large data size	0.20	95	0.0015	3.83	98.79

3.2. The Forecasting Results of Example 2

3.2.1. Data Sets for Small and Large Sizes

In Example 2, small data size, also totaling 336 hourly load data for 14 days (from 1 to 14 January 2015) are collected. In which, the former 288 load data are used as the training data set, the latter 48 load data are as testing data set. The original data set is shown in Figure 9a.

For large data size, totally 1104 hourly load data for 46 days (from 1 January to 15 February 2015) are collected to model. The former 768 load data are employed as the training data set, and the remainder 336 load data are used as the testing data set. The original data set is illustrated in Figure 9b.

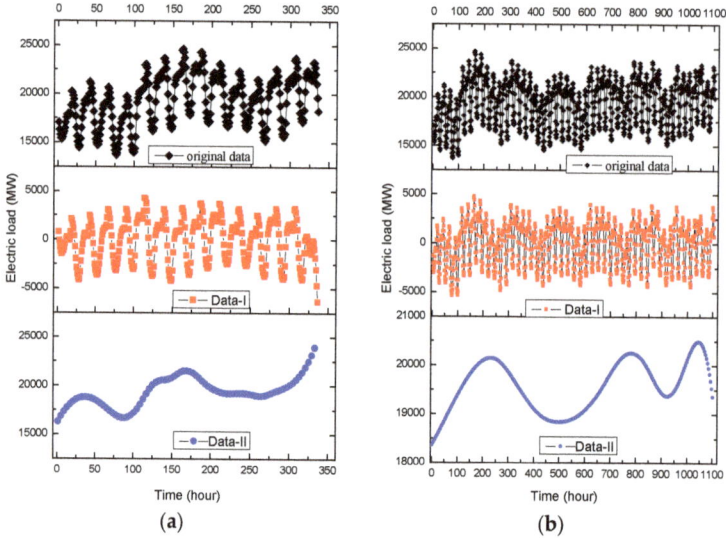

Figure 9. The hour based original data, Data-I, Data-II (Example 2). (**a**) Small data size (1–14 January 2015); (**b**) Large data size (1 January–15 February 2015).

3.2.2. Decomposition Results by EMD

Similar as in Example 1, EMD is used to decompose the original data set into higher and lower frequency items. For small data size, it could be divided into nine groups, as shown in Figure 10a–i. In the meanwhile, the time delay of the data set in RP analysis (which value is 4) is simultaneously considered to select the higher and lower frequent parts. Consequently, the former five groups with much more frequency are classified as higher frequency items; the continued four groups with less frequency are classified as lower frequency items. Figure 10i also represents the trend term (i.e., residuals). The fluctuation characteristics of Data-I are also obviously the same as the original data, as demonstrated in Figure 9a,b. On the contrary, the macrostructure of Data-II is more stable. Data-I and Data-II will also be further analyzed by SVR-PSO and SVR-GA models, respectively, to receive satisfied regression results.

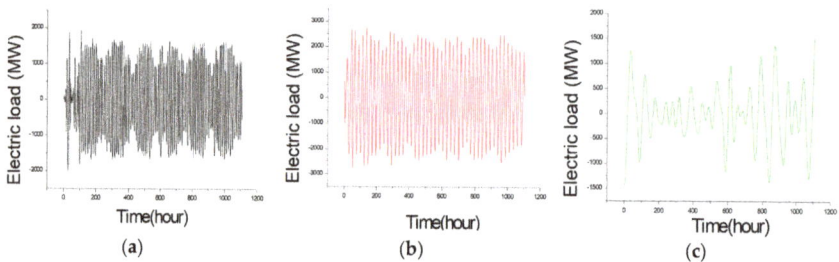

Figure 10. *Cont.*

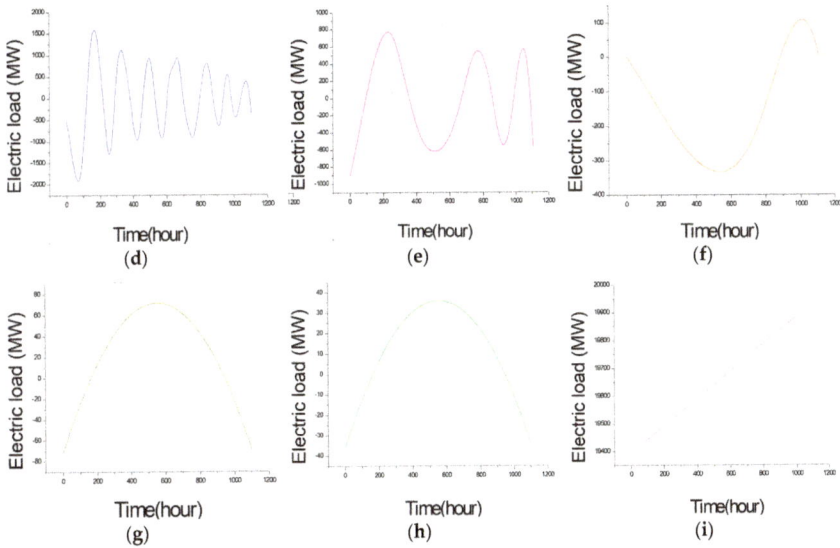

Figure 10. The decomposed different items for the small data size (Example 2). (**a**) IMF (intrinsic modulo function) 1; (**b**) IMF 2; (**c**) IMF 3; (**d**) IMF 4; (**e**) IMF 5; (**f**) IMF 6; (**g**) IMF 7; (**h**) IMF 8; (**i**) residuals.

3.2.3. SVR-PSO for Data-I

As shown in Figure 9, Data-I also presents as periodical stability every 24 h, which is the same with the original data. Therefore, it is as similar as in Example 1, PSO is applied to be hybridized with the SVR model to forecast the Data-I.

Firstly, the higher frequency items (i.e., Data-I) from small and large data sizes are both used for SVR-PSO modeling. The best modeled results in the training and testing stages are received, as shown in Figure 11a,b, respectively. In which, it is obvious to observe that the forecasting accuracy improvements from the hybridization of PSO. The parameters settings of the SVR model for small and large data sizes are as the same as in Example 1, i.e., as shown in Table 1; the determined parameters of the SVR-PSO model are illustrated Table 5.

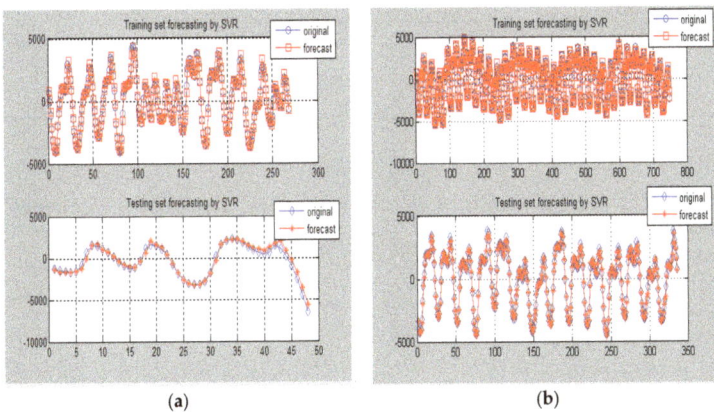

Figure 11. Comparison of forecasting results by SVR-PSO model for Data-I (Example 2). (**a**) One-day ahead forecasting from 13 to 14 January 2015; (**b**) One-week ahead forecasting from 2 to 15 February 2015.

15

Table 5. The SVR-PSO's parameters for Data-I (Example 2).

Data Size	σ	C	ε	Testing *MAPE*	Testing *RMSE*
Small data size	0.12	80	0.0021	7.21	110.04
Large data size	0.24	89	0.0012	4.70	115.63

3.2.4. SVR-GA for Data-II

As shown in Figure 10f–i, the lower frequency item (i.e., Data-II) has less frequency and stability. Therefore, it is as the same as in Example 1, GA is also hybridized with the SVR model to forecast the Data-II. The parameters settings of the SVR model for small and large data sizes are also as the same as shown in Table 3. The best modeled results in training and testing stages are received, as shown in Figure 12a,b, respectively. In which, it has demonstrated the superiority from the hybridization of GA. The determined parameters of the SVR-GA model are shown in Table 6.

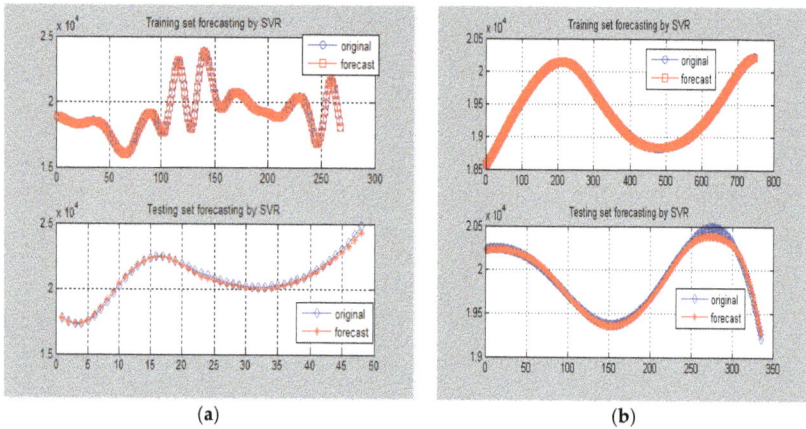

Figure 12. Comparison of forecasting results by SVR-GA model for Data-II (Example 2). (**a**) One-day ahead forecasting from 13 to 14 January 2015; (**b**) One-week ahead forecasting from 2 to 15 February 2015.

Table 6. The SVR-GA's parameters for Data-II (Example 2).

Data Size	σ	C	ε	Testing *MAPE*	Testing *RMSE*
Samall data size	0.15	90	0.0023	7.02	82.11
Large data size	0.22	98	0.0012	4.41	96.83

4. Forecasting Results and Analyses

This section will completely address the forecasting results of these two employed examples with two data sizes and associate higher and lower frequency items (Data-I and Data-II), to further demonstrate the efficiency of the proposed EMD-PSO-GA-SVR model in terms of forecasting accuracy and interpretability.

4.1. Forecasting Accuracy Evaluation Indexes

To completely reflect the forecasting performances of the proposed model, three representative forecasting accuracy evaluation indexes are employed to conduct the evaluation. Except the mean absolute percentage error (*MAPE*) and the root mean square error (*RMSE*), as introduced in Equations (16) and (17), the other one is the mean absolute error (*MAE*), which is calculated by Equation (18),

$$MAE = \frac{\sum_{i=1}^{N}|a_i - f_i|}{N} \tag{18}$$

where a_i is the actual load value at point i; f_i is the forecasted load value at point i; and, N is the total number of data.

In addition, two famous criteria for model selection among a finite set of models, i.e., Akaike information criterion (*AIC*) [38,39] and Bayesian information criterion (*BIC*) [39,40], are employed to further compare which model performs (fits) better. The *AIC* estimates the quality of each of the collected models for fitting the data, relative to each of the other models. The *AIC* value of the selected model is calculated as Equation (19). During the model fitting processes, it is possible to increase the accuracy by adding parameters, and it eventually results in overfitting problems. The *BIC* value of the selected model is calculated as Equation (20).

The model with the lowest *AIC* and *BIC* is preferred. They both attempt to resolve the overfitting problem by introducing a penalty term for the number of parameters in the model.

$$AIC = N\ln(RMSE) + 2N(k+1)\ln(N) \tag{19}$$

$$BIC = N\ln(RMSE) + (k+1)\ln(N) - N\ln(N) \tag{20}$$

where N is the total number of data; k is the number of parameters in a model.

4.2. Forecasting Performances

In Example 1, the forecasting performances of the proposed EMD-PSO-GA-SVR model for small data size and large data size are illustrated in Figure 13, in which other competitive models are also shown, such as the original SVR model, SVR-PSO model, and SVR-GA model. Figure 13 demonstrates clearly that the proposed model provides better fitness than other competitive models for both small and large data sizes.

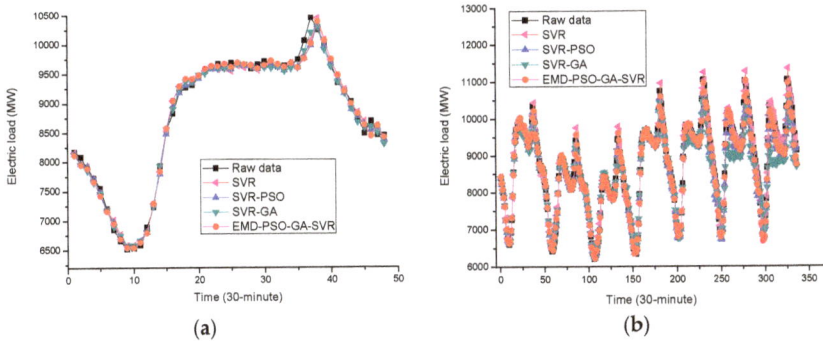

Figure 13. Forecasting performances comparisons (Example 1). (**a**) One-day ahead forecasting on 8 May 2007; (**b**) One-week ahead forecasting from 18 to 24 May 2007.

In Example 2, for both small and large data sizes, the forecasting performances of the proposed model and other competitive models are shown in Figure 14. Once again, the proposed model also receives better fitness, particularly within the period peak loads occurs, i.e., the proposed EMD-PSO-GA-SVR model demonstrates a better generalization capability than other competitive models. To verify this result clearly, the local enlargement around the period of peak load in Figure 14a,b are enlarged in Figure 15a,b, respectively. It is shown that the extraction of microscopic detail can express its periodic characteristics very well, and the macroscopic structure is also in the lower frequency range. The discretization characteristic is expressed by GA, especially at the sharp point; in addition, 24-h periodicity has also been reflected. It is clearly to learn that the forecasting curve of the proposed model can fit closer to the actual load curve than other competitive models. It is powerful to capture the changing tendency of the data, including the nonlinear fluctuation tendency.

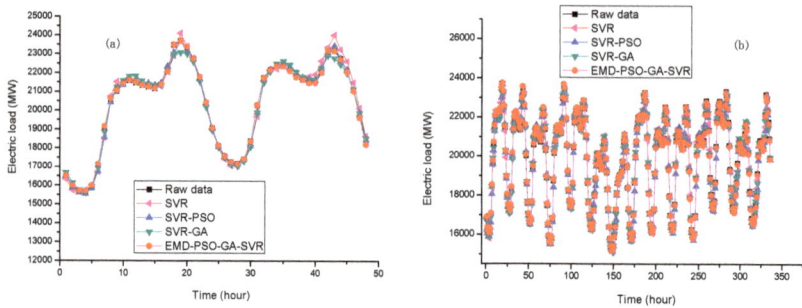

Figure 14. Forecasting performance comparisons (Example 2). (**a**) One-day ahead forecasting from 13 to 14 January 2015; (**b**) One-week ahead forecasting from 2 to 15 February 2015.

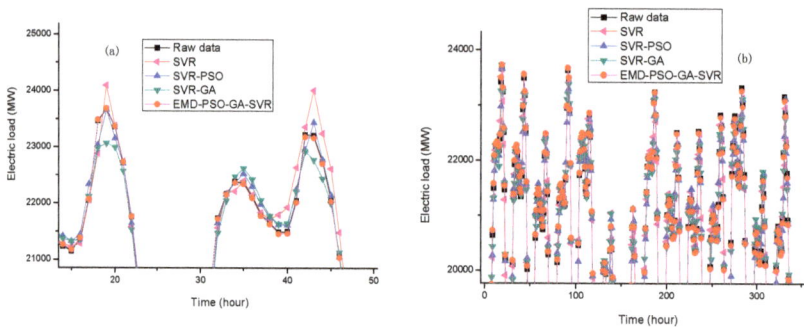

Figure 15. The local enlargement (peak) comparisons (Example 2). (**a**) One-day ahead forecasting from 13 to 14 January 2015; (**b**) One-week ahead forecasting from 2 to15 February 2015.

The forecasting detail results in both Examples 1 and 2 are listed in Tables 7 and 8, respectively. When comparing the proposed EMD-PSO-GA-SVR model with other competitive models, it could be found that the proposed model is superior in terms of all of the forecasting performance evaluation indexes. Therefore, it could be concluded that the proposed model performs with effectiveness and efficiency more than other competitive models, and eventually, the proposed hybrid model could receive better forecasting accuracy levels and statistical interpretation. Particularly, as illustrated in Figure 15, the proposed model demonstrates a higher fitness capability and excellent flexibility during the period of peak point or inflection point, due to capturing redundant information by GA in the modeling process, and by significantly increasing the forecasting accurate level.

Some advantages of the proposed model could be concluded based on the forecasting performances, as abovementioned. The first one should be that the proposed model is superior to other competitive models based on the comparisons (Figures 14 and 15, Tables 7 and 8). Secondly, based on Figures 14 and 15 in Example 2, the proposed EMD-PSO-GA-SVR model has been trained to receive a better generalization ability while dealing with different input patterns. In addition, based on the forecasting performance comparisons from small and large data sizes in Examples 1 and 2, the proposed model is capable learning about more redundant information and to successfully model the model with large data size. Eventually, as illustrated in Tables 7 and 8, the proposed model performance satisfied forecasting accuracy and well interpretability parameters, i.e., the robustness and effectiveness are received. Thus, the proposed model is a remarkable approach to easily forecast electricity load. Particularly, in terms of *AIC* and *BIC* criteria, the proposed model also receives smallest values of *AIC* and *BIC* than the other competitive models, it indicates that the proposed model fits better and avoid overfitting problems.

Table 7. Forecasting performances of competitive models for both small and large data sizes (Example 1).

Models	MAPE	RMSE	MAE	AIC	BIC
Small Data Size					
Original SVR	11.70	145.87	10.92	−79.76	−68.71
SVR-PSO	11.41	145.69	10.67	−79.94	−68.89
SVR-GA	13.52	150.38	11.88	−75.31	−64.26
AFCM [35]	9.95	125.32	9.26	−101.92	−90.87
EMD-PSO-GA-SVR	**9.09**	**123.38**	9.19	−104.19	−93.14
Large Data Size					
Original SVR	12.88	181.62	12.05	−823.32	−812.27
SVR-PSO	13.50	271.43	13.07	−630.68	−619.63
SVR-GA	14.31	183.57	15.31	−818.20	−807.15
AFCM [35]	11.10	158.75	10.44	−887.85	−876.80
EMD-PSO-GA-SVR	**3.92**	**142.41**	**9.04**	**−939.93**	**−928.88**

Table 8. Forecasting performances of competitive models for both small and large data sizes (Example 2).

Models	MAPE	RMSE	MAE	AIC	BIC
Small Data Size					
Original SVR	19.43	220.92	22.33	−19.19	−8.14
SVR-PSO	18.76	200.53	21.19	−33.32	−22.27
SVR-GA	17.98	201.74	22.58	−32.44	−21.39
AFCM [35]	14.31	158.11	17.44	−68.00	−56.95
EMD-PSO-GA-SVR	**7.15**	**137.30**	**14.44**	**−88.59**	**−77.54**
Large Data Size					
Original SVR	33. 72	321.44	32.05	−549.60	−538.55
SVR-PSO	37.51	300.32	31.39	−582.18	−571.13
SVR-GA	34.20	298.11	26.31	−585.73	−574.68
AFCM [35]	11.29	289.21	20.76	−600.26	−589.21
EMD-PSO-GA-SVR	**4.63**	**150.82**	**15.20**	**−912.42**	**−901.37**

Finally, to ensure the significance of forecasting accuracy improvements among the proposed model and other competitive models, a statistical test, namely Wilcoxon signed-rank test, is implemented at the 0.05 significant level under one-tail-test. The test results are illustrated in Table 9. Obviously, the proposed model outperforms the other competitive models significantly.

Table 9. Wilcoxon signed-rank test.

Examples	Compared Models	Wilcoxon Signed-Rank Test $\alpha = 0.05$; $W = 6$
Example 1	EMD-PSO-GA-SVR vs. Original SVR	3 [a]
	EMD-PSO-GA-SVR vs. SVR-PSO	2 [a]
	EMD-PSO-GA-SVR vs. SVR-GA	2 [a]
	EMD-PSO-GA-SVR vs. AFCM	2 [a]
Example 2	EMD-PSO-GA-SVR vs. Original SVR	2 [a]
	EMD-PSO-GA-SVR vs. SVR-PSO	2 [a]
	EMD-PSO-GA-SVR vs. SVR-GA	2 [a]
	EMD-PSO-GA-SVR vs. AFCM	2 [a]

[a] Denotes that the proposed EMD-PSO-GA-SVR model is significantly superior to other competitive models.

5. Conclusions

This paper proposes a novel SVR-based electricity load forecasting model, by hybridizing EMD to decompose the time series data set into higher and lower frequency parts, and, by hybridizing PSO and GA algorithms to determine the three parameters of the SVR models for these two parts, respectively. Via two experimental examples from Australian and American electricity market open data sets, the proposed EMD-PSO-GA-SVR model receives significant forecasting performances rather than other competitive forecasting models in published papers, such as original SVR, SVR-PSO, SVR-GA, and AFCM models.

The most significant contribution of this paper is to overcome the practical drawbacks of an SVR model: the SVR model could only provide poor forecasting for other data patterns, if it is over trained to some data pattern with overwhelming size. Authors firstly apply EMD to decompose the data set into two sub-sets with different data patterns, the higher frequency part and the lower frequency part, to take into account both the accuracy and interpretability of the forecast results. Secondly, authors employ two suitable evolutionary algorithms to reduce the performance volatility of an SVR model with different parameters. PSO and GA are implemented to determine the parameter combination during the SVR modeling process. The results indicate that the proposed EMD-PSO-GA-SVR model demonstrates a better generalization capability than the other competitive models in terms of forecasting capability. It is shown that the extraction of microscopic detail can express its periodic characteristics well, and the macroscopic structure is also in the lower frequency range. The discretization characteristic is expressed by GA, especially at the sharp point, i.e., GA can effectively capture the exact sharp characteristics while the embedded effects of noise and the other factors intertwined. Eventually receiving more satisfying forecasting performances than the other competitive models.

Acknowledgments: This paper is sponsored by the Start-up Foundation for Doctors (No. PXY-BSQD-2014001), Educational Commission of Henan Province of China (No. 15A530010), The Youth Foundation of Ping Ding Shan University (No. PXY-QNJJ-2014008), Jiangsu Distinguished Professor Project by Jiangsu Provincial Department of Education, and Ministry of Science and Technology, Taiwan (MOST 106-2221-E-161-005-MY2).

Author Contributions: Guo-Feng Fan and Wei-Chiang Hong conceived and designed the experiments; Li-Ling Peng and Xiangjun Zhao performed the experiments; Guo-Feng Fan and Wei-Chiang Hong analyzed the data and wrote the paper.

Conflicts of Interest: The authors declare no conflict of interest.

References

1. Li, Z.; Hurn, A.S.; Clements, A.E. Forecasting quantiles of day-ahead electricity load. *Energy Econ.* **2017**, *67*, 60–71. [CrossRef]
2. Arora, S.; Taylor, J.W. Rule-based autoregressive moving average models for forecasting load on special days: A case study for France. *Eur. J. Oper. Res.* **2017**. [CrossRef]
3. Takeda, H.; Tamura, Y.; Sato, S. Using the ensemble Kalman filter for electricity load forecasting and analysis. *Energy* **2016**, *104*, 184–198. [CrossRef]
4. Zhao, H.; Guo, S. An optimized grey model for annual power load forecasting. *Energy* **2016**, *107*, 272–286. [CrossRef]
5. Dedinec, A.; Filiposka, S.; Dedinec, A.; Kocarev, L. Deep belief network based electricity load forecasting: An analysis of Macedonian case. *Energy* **2016**, *115*, 1688–1700. [CrossRef]
6. Mordjaoui, M.; Haddad, S.; Medoued, A.; Laouafi, A. Electric load forecasting by using dynamic neural network. *Int. J. Hydrogen Energy* **2017**, *42*, 17655–17663. [CrossRef]
7. Khwaja, A.S.; Zhang, X.; Anpalagan, A.; Venkatesh, B. Boosted neural networks for improved short-term electric load forecasting. *Electr. Power Syst. Res.* **2017**, *143*, 431–437. [CrossRef]
8. Hua, R.; Wen, S.; Zeng, Z.; Huang, T. A short-term power load forecasting model based on the generalized regression neural network with decreasing step fruit fly optimization algorithm. *Neurocomputing* **2017**, *221*, 24–31. [CrossRef]

9. Mahmoud, T.S.; Habibi, D.; Hassan, M.Y.; Bass, O. Modelling self-optimised short term load forecasting for medium voltage loads using tunning fuzzy systems and artificial neural networks. *Energy Convers. Manag.* **2015**, *106*, 1396–1408. [CrossRef]

10. Bessec, M.; Fouquau, J. Short-run electricity load forecasting with combinations of stationary wavelet transforms. *Eur. J. Oper. Res.* **2018**, *264*, 149–164. [CrossRef]

11. Zeng, N.; Zhang, H.; Liu, W.; Liang, J.; Alsaadi, F.E. A switching delayed PSO optimized extreme learning machine for short-term load forecasting. *Neurocomputing* **2017**, *240*, 175–182. [CrossRef]

12. Liu, N.; Tang, Q.; Zhang, J.; Fan, W.; Liu, J. A hybrid forecasting model with parameter optimization for short-term load forecasting of micro-grids. *Appl. Energy* **2014**, *129*, 336–345. [CrossRef]

13. Kouhi, S.; Keynia, F.; Ravadanegh, S.N. A new short-term load forecast method based on neuro-evolutionary algorithm and chaotic feature selection. *Int. J. Electr. Power Energy Syst.* **2014**, *62*, 862–867. [CrossRef]

14. Bahrami, S.; Hooshmand, R.-A.; Parastegari, M. Short term electric load forecasting by wavelet transform and grey model improved by PSO (particle swarm optimization) algorithm. *Energy* **2014**, *72*, 434–442. [CrossRef]

15. An, X.; Jiang, D.; Liu, C.; Zhao, M. Wind farm power prediction based on wavelet decomposition and chaotic time series. *Expert Syst. Appl.* **2011**, *38*, 11280–11285. [CrossRef]

16. Hung, W.M.; Hong, W.C. Application of SVR with improved ant colony optimization algorithms in exchange rate forecasting. *Control Cybern.* **2009**, *38*, 863–891.

17. Hong, W.C. Traffic flow forecasting by seasonal SVR with chaotic simulated annealing algorithm. *Neurocomputing* **2011**, *74*, 2096–2107. [CrossRef]

18. Pai, P.F.; Hong, W.C. A recurrent support vector regression model in rainfall forecasting. *Hydrol. Process* **2007**, *21*, 819–827. [CrossRef]

19. Hong, W.C. *Intelligent Energy Demand Forecasting*; Springer: London, UK, 2013.

20. An, X.; Jiang, D.; Zhao, M.; Liu, C. Short-term prediction of wind power using EMD and chaotic theory. *Commun. Nonlinear Sci. Numer. Simul.* **2012**, *17*, 1036–1042. [CrossRef]

21. Huang, B.; Kunoth, A. An optimization based empirical mode decomposition scheme. *J. Comput. Appl. Math.* **2013**, *240*, 174–183. [CrossRef]

22. Fan, G.; Qing, S.; Wang, S.Z.; Hong, W.C.; Dai, L. Study on apparent kinetic prediction model of the smelting reduction based on the time series. *Math. Probl. Eng.* **2012**, 720849. [CrossRef]

23. Premanode, B.; Toumazou, C. Improving prediction of exchange rates using differential EMD. *Expert Syst. Appl.* **2013**, *40*, 377–384. [CrossRef]

24. Li, M.; Hong, W.-C.; Kang, H. Urban traffic flow forecasting using Gauss-SVR with cat mapping, cloud model and PSO hybrid algorithm. *Neurocomputing* **2013**, *99*, 230–240. [CrossRef]

25. Hong, W.-C. Chaotic particle swarm optimization algorithm in a support vector regression electric load forecasting model. *Energy Convers. Manag.* **2009**, *50*, 105–117. [CrossRef]

26. Hong, W.-C.; Dong, Y.; Zhang, W.; Chen, L.-Y.; Panigrahi, B.K. Cyclic electric load forecasting by seasonal SVR with chaotic genetic algorithm. *Int. J. Electr. Power Energy Syst.* **2013**, *44*, 604–614. [CrossRef]

27. Chen, R.; Liang, C.; Hong, W.-C.; Gu, D. Forecasting holiday daily tourist flow based on seasonal support vector regression with adaptive genetic algorithm. *Appl. Soft Comput.* **2015**, *26*, 435–443. [CrossRef]

28. Geng, J.; Huang, M.-L.; Li, M.-W.; Hong, W.-C. Hybridization of seasonal chaotic cloud simulated annealing algorithm in a SVR-based load forecasting model. *Neurocomputing* **2015**, *151*, 1362–1373. [CrossRef]

29. Ju, F.-Y.; Hong, W.-C. Application of seasonal SVR with chaotic gravitational search algorithm in electricity forecasting. *Appl. Math. Model.* **2013**, *37*, 9643–9651. [CrossRef]

30. Huang, Y.; Schmitt, F.G. Time dependent intrinsic correlation analysis of temperature and dissolved oxygen time series using empirical mode decomposition. *J. Mar. Syst.* **2014**, *130*, 90–100. [CrossRef]

31. Vapnik, V.; Golowich, S.; Smola, A. Support vector machine for function approximation, regression estimation, and signal processing. *Adv. Neural Inf. Process. Syst.* **1996**, *9*, 281–287.

32. Kennedy, J.; Eberhart, R.C. Particle swarm optimization. In Proceedings of the IEEE International Conference Neural Networks, Perth, Australia, 27 November–1 December 1995; pp. 1942–1948.

33. Eberhart, R.C.; Shi, Y. Particle swarm optimization: Developments, applications and resources. In Proceedings of the 2001 Congress on Evolutionary Computation, Seoul, Korea, 27–30 May 2001; pp. 81–86.

34. Sun, T.H. Applying particle swarm optimization algorithm to roundness measurement. *Expert Syst. Appl.* **2009**, *36*, 3428–3438. [CrossRef]

35. Coelho, L.S. An efficient particle swarm approach for mixed-integer programming in reliability–redundancy optimization applications. *Reliab. Eng. Syst. Saf.* **2009**, *94*, 830–837. [CrossRef]

36. Eckmann, J.P.; Kamphorst, S.O.; Ruelle, D. Recurrence plots of dynamical systems. *Europhys. Lett.* **1987**, *5*, 973–977. [CrossRef]

37. Marwan, N.; Romano, M.C.; Thiel, M.; Kurths, J. Recurrence plots for the analysis of complex systems. *Phys. Rep.* **2007**, *438*, 237–329. [CrossRef]

38. Akaike, H. A new look at the statistical model identification. *IEEE Trans. Autom. Control* **1974**, *19*, 716–723. [CrossRef]

39. Aho, K.; Derryberry, D.; Peterson, T. Model selection for ecologists: The worldviews of AIC and BIC. *Ecology* **2014**, *95*, 631–636. [CrossRef] [PubMed]

40. Schwarz, G.E. Estimating the dimension of a model. *Ann. Stat.* **1978**, *6*, 461–464. [CrossRef]

energies

MDPI

Article

Icing Forecasting of Transmission Lines with a Modified Back Propagation Neural Network-Support Vector Machine-Extreme Learning Machine with Kernel (BPNN-SVM-KELM) Based on the Variance-Covariance Weight Determination Method

Dongxiao Niu [1], Yi Liang [1,*], Haichao Wang [1], Meng Wang [1] and Wei-Chiang Hong [2,3]

[1] School of Economics and Management, North China Electric Power University, Beijing 102206, China; niudx@126.com (D.N.); ncepuwhc@ncepu.edu.cn (H.W.); wmeng3007@163.com (M.W.)

[2] School of Education Intelligent Technology, Jiangsu Normal University, Xuzhou 221116, China; samuelsonhong@gmail.com

[3] Department of Information Management, Oriental Institute of Technology, New Taipei 220, Taiwan

* Correspondence: louisliang@ncepu.edu.cn; Tel.: +86-10-61773079

Received: 26 May 2017; Accepted: 11 August 2017; Published: 13 August 2017

Abstract: Stable and accurate forecasting of icing thickness is of great significance for the safe operation of the power grid. In order to improve the robustness and accuracy of such forecasting, this paper proposes an innovative combination forecasting model using a modified Back Propagation Neural Network-Support Vector Machine-Extreme Learning Machine with Kernel (BPNN-SVM-KELM) based on the variance-covariance (VC) weight determination method. Firstly, the initial weights and thresholds of BPNN are optimized by mind evolutionary computation (MEC) to prevent the BPNN from falling into local optima and speed up its convergence. Secondly, a bat algorithm (BA) is utilized to optimize the key parameters of SVM. Thirdly, the kernel function is introduced into an extreme learning machine (ELM) to improve the regression prediction accuracy of the model. Lastly, after adopting the above three modified models to predict, the variance-covariance weight determination method is applied to combine the forecasting results. Through performance verification of the model by real-world examples, the results show that the forecasting accuracy of the three individual modified models proposed in this paper has been improved, but the stability is poor, whereas the combination forecasting method proposed in this paper is not only accurate, but also stable. As a result, it can provide technical reference for the safety management of power grid.

Keywords: icing forecasting; back propagation neural network; mind evolutionary computation; bat algorithm; support vector machine; extreme learning machine with kernel; variance-covariance

1. Introduction

Transmission line icing has a significant impact on the safe operation of power systems. In severe cases, it can even cause trips, disconnections, tower collapses, insulator ice flashovers, communication interruptions and other problems, which bring about great economic losses [1]. For example, a large cold wave area occurred in southeastern Canada and the northeastern United States in 1998, resulting in the collapse of more than 1000 power transmission towers, 4.7 million people couldn't use electricity properly and the direct economic losses reached $5.4 billion [2]. In 2008, severe line icing accidents happened in South China, and caused forced-outages of 7541 10 KV lines and the power shortfall reached 14.82 GW [3]. The construction of a reasonable and scientific transmission line icing prediction model would be helpful for the power sector to deal with icing accidents in advance so as to effectively reduce the potential accident losses. Therefore, the study of icing prediction is of great practical significance and value.

In recent years, many scholars have been carrying out research on transmission line icing prediction. Some experts have developed sensor systems for direct measurement of icing events on transmission lines and they obtained real-time and intuitive icing thickness monitoring information [4–6]. However, the prediction accuracy of this method is poor, and this method is more suitable for collection equipment as raw data for it often needs icing model algorithms to predict the future icing trends. As a result, it is necessary to study model algorithms to predict icing thickness. Generally, the icing forecasting models can be divided into two categories, which include traditional models and modern intelligent models. Traditional models are further divided into two methods: physical models and statistical models. The physical prediction models are based on heat transmission science and fluid mechanics and other physics theories to analyze the icing thickness, such as Imai model [7], Goodwin model [8], Lenhard model [9], and hydrodynamic model [10]. However, icing is caused by many factors with too many uncertainties, which leads to the fact that the final forecasts provided by physical prediction models cannot live up to expectations. The statistical prediction models use the notion of mathematical statistics to predict the icing thickness based on icing records and extreme value theory. They include the time series model [11], extreme value model [12] and so on. However the application of statistical forecasting models needs to meet a variety of statistical assumptions, and they cannot consider the factors that influence icing thickness, which greatly limits the scope of application of the statistical models and improvement of the forecasting accuracy.

Therefore, it is more important to adopt the modern intelligent prediction models to predict transmission line icing with the development of big data, further research on artificial intelligence and constantly emerging optimization algorithms. Modern intelligent prediction models can handle nonlinear and uncertain problems scientifically and efficiently with computer technology and mathematical tools to improve prediction accuracy and speed. The back propagation neural network model and the support vector machine model are commonly-used intelligent models in the field of transmission line icing prediction. Li et al. [13] proposed a model based on BP neural networks for forecasting the ice thickness and the forecasting results showed that this model had good accuracy of prediction whether in the same icing process or in a different one. Wang et al. [14] put forward a prediction model of icing thickness and weight based on a BP neural network. The orthogonal least squares (OLS) method was used for the number of network hidden layer units and center vector so that the forecasting error could be controlled in a smaller range. However, the BP algorithm has a very slow convergence speed and it falls into local minima easily, so some scholars use the genetic algorithm and the particle swarm optimization to optimize the BP neural network. Zheng and Liu [15] proposed a forecast model based on genetic algorithm (GA) and BP, and the predication results proved that the GA-BP model was more effective than BP to forecast transmission line icing. Wang [16] structured a prediction model which used improved particle swarm optimization algorithm to optimize a normalized radial basis function (NRBF) neural network, and the training speed of the network was improved. In addition, some scholars used SVM to avoid the selection of neural network structure and local optimization problems, [17] and [18] built the icing prediction model based on a SVM algorithm with better accuracy, but the SVM algorithm is hard to implement for large-scale training samples, and there are difficulties in solving multiple classification problems, so some scholars have addressed these defects of SVM using the ant colony (ACO) [19], particle swarm (PSO) [19], fireworks algorithm (FA) [20] and quantum fireworks algorithm (QFA) [21]. Xu et al. [19] introduced a weighted support vector machine regression model that was optimized by the particle swarm and ant colony algorithms, and the proposed method obtained a higher forecasting accuracy. Ma and Niu [20] combined a weighted least squares support vector machine (W-LSSVM) with a fireworks algorithm to forecast icing thickness, which improved the prediction accuracy and robustness. Ma et al. [21] proposed a combination model based on the wavelet support vector machine (w-SVM) and the quantum fireworks algorithm (QFA) for icing forecasting.

GA, PSO, ACO and other algorithms are applied to advance the performance of BP neural networks and SVM, but these algorithms require a large initial population to solve large-scale optimization problems, and the solving efficiency and the ability to solve local optimization problems are still relatively general. Both the mind evolutionary computation (MEC) [22] and the bat algorithm (BA) [23] have high solving efficiency and strong competence in global optimization. Two new operators are added to MEC on the basis of genetic algorithm: convergence and dissimilation. They are responsible for local and global optimization, respectively, which greatly enhances the overall search efficiency and global optimization algorithm ability. The BA algorithm is a meta-heuristic algorithm proposed by Yang in 2010. Many scholars at home and abroad have studied the proposed algorithm, indicating that this algorithm takes into account both local and global aspects of solving a problem compared with other algorithms. In the search process, both of them can be interconverted into each other so that they can avoid falling into local optimal solutions and achieve better convergence.

As a new feed forward neural network, ELM can overcome the shortcomings of the traditional BP neural network and SVM. The algorithm not only reduces the risk of falling into a local optimum but also greatly improves the learning speed and generalization ability of the network. It has been applied in several prediction fields and obtained relatively accurate prediction results [24–26]. However, its prediction robustness is relatively poor due to the random initialization of the input weights and hidden layer bias characteristics, so Huang [27] proposed the kernel extreme learning machine algorithm (KELM) and thus overcame the weakness of poor stability and improved the algorithm precision.

The different forecasting methods reflect the change tendency of the object and its influencing factors from different aspects, respectively, and provide different information because of the respective principles, so any single forecasting method confronts the obstacle that the information is not comprehensive and the fluctuation of prediction accuracy is larger. Based on this, Bates and Granger [28] put forward the combination forecasting method for the first time in 1969 and it has achieved good results in many fields. For example, Liang et al. [29] proposed the optimal combination forecasting model combined the extreme learning machine and the multiple regression forecasting model to predict the power demand. The result indicated that this method effectively combined the advantages of the single forecasting models, thus its global instability was reduced and the prediction precision was satisfactory. Reference [30] introduced a combination model that included five single prediction models for probabilistic short-term wind speed forecasting and the proposed combination model generated a more reliable and accurate forecast. Few scholars have applied combination prediction methods in the field of the transmission line icing forecasting, so in this paper, we decided to adopt the combination forecasting method to predict line icing thickness. How to determine the weighted average coefficients of individual methods is the key problem. Compared to the arithmetic mean method [31] and induced ordered weighted averaging (IOWA) [32], the biggest advantage of the variance-covariance combined method [33] is that it can improve the robustness of prediction, which is more suitable for forecasting icing thickness.

In summary, this paper adopts three models, including the BPNN optimized by the mind evolutionary algorithm (MEC-BPNN), the SVM optimized by the bat algorithm (BA-SVM) and the extreme learning machine with kernel based on single-hidden layer feed-forward neural network, to predict icing thickness using the historical icing thickness data and related meteorological data. The weighted average coefficients of individual forecasting methods are determined by a variance-covariance combined method to solve the problem of dynamic weight distribution. Then a modified BP-SVM-KELM combination forecasting model based on the VC combined method solving the problem of dynamic weight distribution method is constructed. The reason why we combine the three modified models is that their individual robustness is still poor, especially the BA-SVM and KELM. Furthermore, MEC-BPNN and KELM have the defects of underfitting and overfitting, respectively, and BA-SVM has difficulties dealing with large-scale training samples. Therefore, the combination model can give full play to the advantages of various prediction models, complement each other, and offer better robustness, stronger adaptability and higher prediction accuracy.

The rest of this paper is organized as follows: in Section 2, the MEC-BP, BA-SVM, KELM and VC combined method are presented in detail. Also, in this section, the integrated prediction framework is built. In Section 3, several real-world cases are selected to verify the robustness and accuracy of this model. In Section 4, another case is used to test the prediction performance of the proposed model. Section 5 concludes this paper.

2. Methodology

2.1. Mind Evolutionary Computation (MEC) to Optimize BPNN

MEC is a new evolutionary algorithm aiming at solving the defects of genetic algorithm and imitating the evolutionary process of human thinking. It inherits some ideas from the genetic algorithm and introduces two new operation operators, namely convergence and dissimilation, which are responsible for local and global optimization. The two operators are independent and coordinated, so improvement of any one can increase the whole search efficiency of the algorithm. Besides, there is strong ability of global optimization with a directed learning and memory mechanism. Using MEC to optimize the initial weights and thresholds of BPNN can make up for the defects that BPNN often falls into local optima and converges slowly. At present, MEC-BPNN is still rare in the field of the transmission line icing prediction though it has been widely used in other fields. The steps of using MEC to optimize the initial weights and thresholds of BPNN for forecasting are as follows:

(1) Select the training set and test set. The training set is not only used for BPNN but also serves for the initialization of MEC. The test dataset is used for examining the model prediction accuracy. In order to make MEC-BPNN have good generalization performance, the training samples should be enough and representative.

(2) MEC initialization. Set the population size of MEC, the number of superior races, the number of temporary population, the size of the sub-population, the number of iterations and the parameters of the BPNN interface.

(3) Population generation. The initial population, superior sub-population and temporary sub-population are generated here serving for the convergent operation and dissimilation operation.

(4) Convergence operation. The process of individuals' competition for winners within a sub-population is called convergence. The end of the convergence process is the absence of winners within the population. That's a process of iteration.

(5) Dissimilation operation. In the course of global competition among the sub-populations, if the score of a temporary sub-population is higher than that of a mature dominant population, the latter will be replaced and dissolved, or the former will be eliminated and disbanded. The new sub-population will be supplied with constant iteration.

(6) Get the best individual. The MEC stops optimizing when the terminate condition of the iteration is reached. Then the optimal individual is parsed according to the encoding rules so that the weights and thresholds of the corresponding BPNN are obtained.

(7) BPNN training. Set the initial input layer, hidden layer and output layer neuron number in initial settings of BPNN, and use the training set samples to train BP neural network with the optimized initial weights and thresholds.

(8) Simulation prediction. Carry out the transmission lines icing forecasting if the simulation testing of the training result meets the expected goal, and to analyze the results.

2.2. Bat Algorithm (BA) to Optimize SVM

SVM is a machine learning algorithm based on statistical learning theory that can avoid the lack of learning ability of BPNN. SVM maps linear non-separable low dimensional space data into a linearly separable high-dimensional feature space by introducing a nonlinear inner product kernel function and the classification or regression fitting is carried out in this space. The regression fitting of SVM is called support vector regression (SVR). In this paper, ε-SVR is used to study the nonlinear icing thickness prediction. The nonlinear SVR needs to map the raw data into high-dimensional feature space by kernel function, and then apply linear regression in high-dimensional feature space. The specific algorithm flow of ε-SVR refers to the reference [34].

The selections of the penalty parameter c, kernel function parameter g and ε loss function parameter p are crucial because the prediction performance of SVM is influenced by these key parameters. Compared with other algorithms, the global optimization ability of BA is stronger, and it can avoid falling into local optimization. Therefore, this paper adopts BA algorithm to optimize the three key parameters.

The bat algorithm is a new intelligent optimization algorithm inspired by the echo localization of micro bats in nature. In nature, most bats use echolocation method to hunt their prey, and they can emit dozens of sounds at up to 110 dB ultrasonic pulses per second. When the bats come near the prey, the pulse intensity decreases and frequency increases. Bats usually produce higher frequency sound waves and wider bands for hunting prey in complex environments. If the bat is simulated as agent in the search space, the good or bad of the agent's position is measured by the quality of objective functions, and the process of bats finding prey is just like the process of searching for the optimal solution in solution space. Then the behavior of bats using ultrasonic positioning can be described using the following equations. Suppose the bat population is n, the speed and position of the bat i are updated according to Equations (2) and (3):

$$f_i = f_{\min} + (f_{\max} - f_{\min})\alpha \tag{1}$$

$$v_i^t = v_i^{t-1} + (l_i^{t-1} - l_*)f_i \tag{2}$$

$$l_i^t = l_i^{t-1} + v_i^t \tag{3}$$

where f_i is the frequency of sound waves generated by the i bat; $f_{\min}$ and $f_{\max}$ are the minimum and maximum frequency of sound waves respectively; α is a random number within [0, 1]; v_i^{t-1} and v_i^t are the velocity at time $t-1$ and the time t of the i bat; l_i^{t-1} and l_i^t are the position at time $t-1$ and the time t of the i bat; l_* is the position of the bat when the target function is optimal in the current global search. In the initialization process, each bat should be assigned a random frequency, but the frequency should be within the set range. In the local search, the position of the bat is updated according to the new formula if a solution is selected from the optimal set:

$$l_{new} = l_{old} + \varepsilon A^t \tag{4}$$

where ε is a random number within [0, 1]; A^t is the average loudness of all bats at time t; l_{old} is a solution that is randomly selected from the set of optimal solutions.

The pulse loudness A_i and frequency R_i emitted by i bat will change continuously. During searching, for example, $A_{min} = 0$ indicates that the bat has discovered the prey at this time and pauses the ultrasonic wave; $A_{max} = 10$ indicates that bats increase the pulse loudness as much as possible to obtain more information in order to search for prey. Pulse loudness and pulse frequency can be updated by Equations (5) and (6):

$$A_i^{t+1} = \tau A_i^t \tag{5}$$

$$R_i^{t+1} = R_i^0[1 - e^{-\gamma t}] \tag{6}$$

where the value of the pulse loudness increasing coefficient τ and pulse frequency attenuation coefficient are selected according to the subjects. The range of τ is in [0, 1]; $\gamma > 0$. The optimal solution is similar to the prey of the bat in BA algorithm, and the variation of pulse loudness and frequency represents, to some extent, the closeness to the optimal solution.

The fitness function used by the BA algorithm is the root-mean-square error (RMSE) under k-fold cross validation (K-CV). The RMSE can be obtained by Equation (22). K-CV randomly divides the training samples into k disjoint subsets, each of which is roughly equal in size. Using k-1 training subsets, a regression model is established for a given set of parameters, and the RMSE of the remaining last subset is used to evaluate the performance of the parameters. Repeat the procedure K times, and each subset has the opportunity to be tested. The accuracy of cross validation is the average value of the percentage of data correctly predicted for K times. The expected generalization error is estimated according to the average value of RMSE obtained after the K iteration, and finally a set of optimal parameters is selected [35].

2.3. Extreme Learning Machine with Kernel (KELM)

ELM was put forward by Huang et al. in 2006. Based on this theory, the basic extreme learning machines, online sequential extreme learning machines and KELM algorithms have been derived [36]. KELM is a single layer feedforward neural network algorithm. Compared with ELM, its ability to solve regression prediction is stronger, and compared with BPNN and SVM, its calculation speed is faster when the prediction accuracy is better or similar, which greatly improve the generalization ability of network [37]. The KELM algorithm has been proved to have excellent forecasting performance in many fields.

First, the neural network construction mechanism of the basic ELM algorithm is briefly described, and its neural network function is shown as follows:

$$g(x) = h_i(x) \cdot \beta_i \tag{7}$$

where $g(x)$ is the network output value, $h_i(x)$ is the output of the i hidden layer neurons which corresponds to the input x; β_i is the connection weights between the i hidden layer neurons and the output neurons.

ELM's precision of regression forecasting is guaranteed by minimizing the output error as follows:

$$\lim_{L \to \infty} \|g(x) - g_O(x)\| = \lim_{L \to \infty} \|\sum_{i=1}^{L} \beta_i h_i(x) - g_O(x)\| = 0 \tag{8}$$

where L is the number of neurons in the hidden layer; $g_O(x)$ is the predictive function of the target value.

At the same time, the ELM algorithm guarantees the generalization ability of neural networks by minimizing the output weight β. The β usually takes its least square solution, and the calculation method is shown as follows:

$$\begin{aligned} \beta &= \mathbf{H}^+ \mathbf{O} = \mathbf{H}^T (\mathbf{H}\mathbf{H}^T)^{-1} \mathbf{O} \\ &= \mathbf{H}^T (\tfrac{1}{C} + \mathbf{H}\mathbf{H}^T)^{-1} \mathbf{O} \end{aligned} \tag{9}$$

where H is the hidden layer matrix of neural network; H$^+$ is the generalized inverse matrix of H matrix; O is predictive target vector. According to ridge regression theory, the results will be more stable and provide better generalization ability by increasing the normal number $1/C$.

The KELM algorithm introduces the kernel function for obtaining better regression prediction accuracy. The kernel matrix is defined by applying Mercer's condition as follows:

$$\begin{cases} \Omega_{ELM} = \mathbf{H}\mathbf{H}^T \\ \Omega_{i,j} = h(x_i) \cdot h(x_j) = K(x_i, x_j) \end{cases} \tag{10}$$

The random matrix HH^T of ELM is replaced by the kernel matrix Ω, then all the input samples are mapped from the n-dimensional input space to a high dimensional implicit feature space by kernel function. The mapping value of the kernel matrix Ω is fixed after setting the nuclear parameter. The kernel functions include Radical Basis Function (RBF) kernel functions, linear kernel functions and polynomial kernel functions. It is usually set as RBF kernel, and the formula is as follows:

$$K(\mu, v) = \exp[-(\mu - v^2/\sigma)] \tag{11}$$

The parameter $1/C$ is added to the main diagonal of the unit diagonal HH^T so that the eigenvalues is not 0, and then the weight vector β^* is obtained. It makes ELM more stable and has better generalization. The output weight of the ELM network here is as follows:

$$\beta^* = \mathbf{H}^T(\mathbf{I}/C + \mathbf{HH}^T)^{-1}\mathbf{O} \tag{12}$$

where I is diagonal matrix; C is penalty coefficient for weighing the proportion between structural risk and empirical risk; HH^T is generated by mapping input samples from kernel functions.

From the above formulas, the output of the KELM model is described as follows:

$$
\begin{aligned}
f(x) &= h(x)\mathbf{H}^T(\mathbf{I}/C + \mathbf{HH}^T)^{-1}\mathbf{O} \\
&= \begin{bmatrix} K(x, x_1) \\ \vdots \\ K(x, x_N) \end{bmatrix}^T (\mathbf{I}/C + \Omega_{ELM})^{-1}\mathbf{O}
\end{aligned} \tag{13}
$$

In the KELM algorithm based on kernel, the specific form of feature mapping function $h(x)$ of hidden nodes is not given specially, and the output function value can be obtained only by the concrete form of kernel function. In addition, since the kernel function uses the inner product directly, it is unnecessary to set the number of hidden layer nodes when solving the output function value, so the initial weight and bias of hidden layer needn't be set.

2.4. VC Combined Method Solved the Problem of Dynamic Weight Distribution

The combination forecasting model can integrate the advantages of each single model and improve the prediction precision. The merit of VC combined method solved the problem of dynamic weight distribution is that the optimum combination weight coefficient can be found, so the robustness and accuracy can be improved.

The variance of each prediction model is calculated by the following formula:

$$\delta_i = \frac{1}{n} \cdot \left[(e_1 - \bar{e})^2 + (e_2 - \bar{e})^2 + \cdots + (e_n - \bar{e})^2\right] \quad i = 1, 2, 3 \tag{14}$$

where n is the number of training samples; $e_1, e_2, \ldots, e_n$ are the absolute percentage error for each training sample; $\bar{e}$ is the average absolute percentage error of the n training sample.

The weights are derived from the variance according to the following formula:

$$\omega_1 = 1/[\delta_1(1/\delta_1 + 1/\delta_2 + 1/\delta_3)] \tag{15}$$

$$\omega_2 = 1/[\delta_2(1/\delta_1 + 1/\delta_2 + 1/\delta_3)] \tag{16}$$

$$\omega_3 = 1/[\delta_3(1/\delta_1 + 1/\delta_2 + 1/\delta_3)] \tag{17}$$

The weights are multiplied by the corresponding prediction results, and the combined prediction results are shown as follows:

$$g = \omega_1 g_1 + \omega_2 g_2 + \omega_3 g_3 \tag{18}$$

where g is the combined forecasting result; g_1, g_2 and g_3 are the individual prediction results of each model. The result of the combination is that the corresponding weights are adjusted dynamically with the different training and test results for better adaptability.

2.5. Combination Forecasting Model

Firstly, the original relevant data is selected and preprocessed, after which the data is divided into test samples and training samples. Then, three single modified models are utilized to forecast respectively, including MEC-BP, BA-SVM and KELM. Finally, the forecasting results are combined by VC combined method solving the problem of dynamic weight distribution. The proposed combination forecasting model is shown in Figure 1.

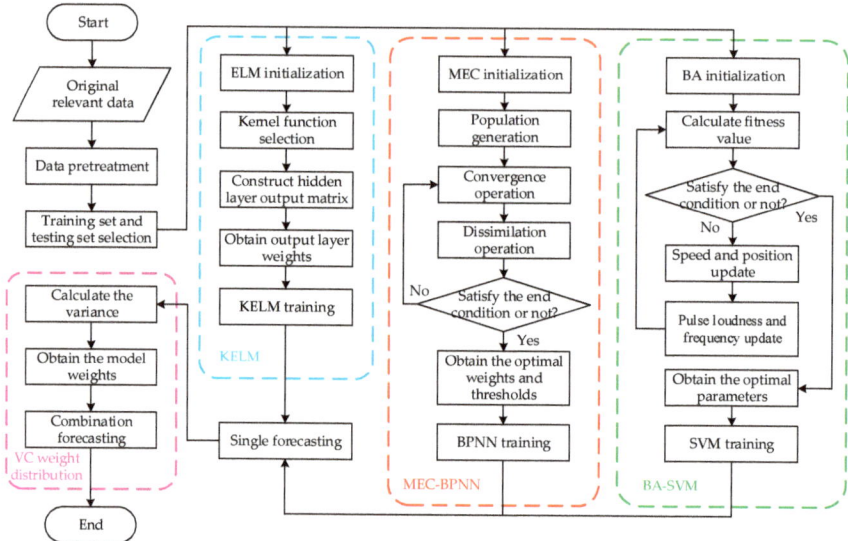

Figure 1. Combination forecasting model.

3. Case Study and Results Analysis

3.1. Data Selection

There are many factors affecting the transmission line icing thickness. According to [38], the temperature, relative air humidity, wind speed and wind direction are the major factors. The temperature must be less than or equal to 0 °C. If the relative air humidity is above 85%, it is easier for icing to occur on transmission lines. When the icing temperature and vapor conditions are present the wind plays an important role in the icing of the wires. It can deposit large amounts of supercooled water droplets continuously onto the lines, and then collide with the wire and gradually increase these deposits to cause icing phenomens. It is observed that icing first grows on the windward side of the line, and the wire is twisted due to gravity when the windward side reaches a certain icing thickness, so a new windward surface appears. In this way, the icing gradually increases by constantly twisting, and eventually circular or elliptical icing is formed, so the wind speed should exceed 1 m/s in the process. In addition, the wind direction also affects the lines' icing. The angle of wind direction is measured by taking the direction of the wire as the benchmark, i.e., the direction of the wire is set to be horizontal 0°. When the wind rotates counterclockwise around the wire, if the angle between the wind and the wire is in the range of [0°, 180°), the closer the angle is to 0° or 180°, the lighter the icing degree is, and the closer the angle is to 90°, the more serious the degree of icing is, or the closer the angle is to 180° or 360°, the lighter the icing degree is, and the closer the angle is to 270°, the more serious the degree of icing.

Affected by the abnormal atmospheric circulation and La Niña weather patterns, cold air entered Hunan in 11 January 2008 making the area cool rapidly. The strength of the frontal inversion formed by the confluence of cold and warm air was great. What's more, the terrain in Hunan is lower in the north and higher in the south, which made the strength of the frontal inversion greater. The stronger the strength of the frontal inversion, the stronger the strength of the rain and snow, so a continuous glaze was formed with the continuous supplement of warm wet air, and Hunan power grid suffered a record disaster accompanied by a large area of rain and snow, freezing rain incidents and large scale icing of transmission lines and substations. During the freezing disaster, the number of collapsed transmission towers reached 2242 and the number of deformed transmission towers reached 692, causing serious damage to the power grid. Therefore, we selected the "Dong-Chao line" of Hunan, which was the hardest-hit area during the Chinese icing incident in 2008 as a case study to verify the effectiveness of the proposed model. The example chooses some data including the transmission lines icing thickness, regional temperature, relative air humidity, wind speed and wind direction from 0:00 12 January 2008 to 24:00 6 February 2008. Here we take 2 h as the data collection frequency, and each indicator collects 312 sets of data where the first 192 are used as the training samples and the latter 120 are test samples in Case 1.

The original data is shown in Figure 2. All of the data were provided by Key Laboratory of Disaster Prevention and Mitigation of Power Transmission and Transformation Equipment (Changsha, Hunan Province, China), where all the data are collected by professional instruments and can reflect the state changes during the icing process. As we can see from Figure 2, the temperature and wind speed data present a cyclical downward trend, while the relative air humidity data present a cyclical upward trend. In addition, there is no exceptional data or missing data. Hence these data can be used directly as data sources.

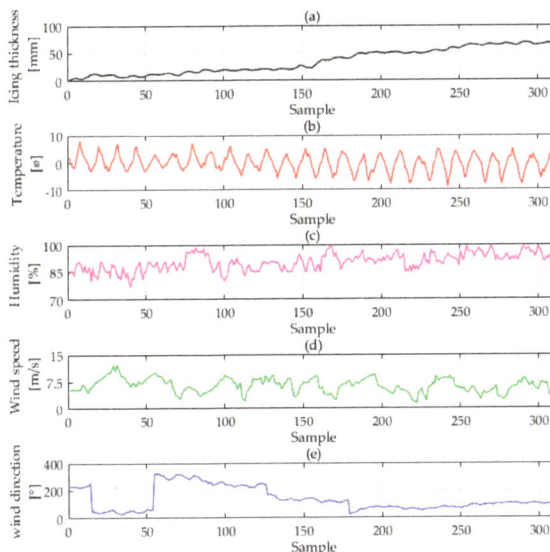

Figure 2. Original data chart of icing thickness, temperature, humidity, wind speed, and wind direction. Note: (**a**) represents the original data of icing thickness; (**b**) represents the original data of temperature; (**c**) represents the original data of humidity; (**d**) represents the original data of wind speed; (**e**) represents the original data of wind direction.

In addition to the icing thickness, the data of temperature, relative air humidity, wind speed and wind direction at the forecast point T was selected as input data. However, ice accretion phenomenon is a continuous process and the prior T-z's icing thickness can influence the transmission lines icing thickness at the forecast point T. When selecting the different prior T-z's icing thickness as input data, the forecasting effectiveness is different. Hence this paper selects different prior T-z's icing thicknesses as input data. For example, when z equals 3, the input icing thickness data includes the icing thickness at T-1, T-2, and T-3. After selecting the input data, the proposed model is applied to check the exact input icing thickness by using the training samples, whose experimental results are shown in Table 1.

From Table 1, it can be found that the proposed model obtains different error values when the icing thickness values are selected at different time points. However when z equals 4, the RMSE of proposed model reaches the minimum value, thus the input icing thickness data includes the icing thickness at T-1, T-2, T-3 and T-4, while the other input data includes the temperature, relative air humidity, wind speed and wind direction at the forecast point T.

Table 1. RMSE of proposed model when selecting the input icing thickness data at different time point.

z	1	2	3	4	5
RMSE of Proposed Model	0.0321	0.0201	0.0156	0.0115	0.0115
z	6	7	8	9	10
RMSE of Proposed Model	0.0115	0.0115	0.0115	0.0115	0.0115

3.2. Data Pretreatment

The data preprocessing steps are as follows:

(1) Wind direction data clustering processing

The large fluctuation range of wind direction data will reduce the accuracy of prediction results, and the clustering of wind direction data can make the fluctuation smaller which can improve the prediction accuracy. Thus, the paper uses clustering to process wind direction data according to the degree of influence of wind direction on icing thickness; which formula is as follows:

$$J = \begin{cases} ceil(0.1\theta), & 0 \leq \theta < 90 \\ ceil(18 - 0.1\theta), & 90 \leq \theta < 180 \\ ceil(0.1\theta - 18), & 180 \leq \theta < 270 \\ ceil(36 - 0.1\theta), & 270 \leq \theta < 360 \end{cases} \tag{19}$$

where J is the result value of clustering process for wind direction; θ is the angle between the wind and the wire when the wind rotates counterclockwise around the wire, i.e., the direction of the transmission line is set to be horizontal $0°$; Ceil is the bracket function. The wind direction processing data for "Dong-Chao line" is shown in Figure 3.

(2) Standardized processing of all data

Due to the different nature of each evaluation index, and they usually have different dimensions and orders of magnitude. In order to ensure the accuracy of the prediction results, it is necessary to standardize the original index data. The data are processed by the following equation:

$$Z = \{z_i\} = \frac{x_i - x_{min}}{x_{max} - x_{min}} \quad i = 1, 2, 3, \ldots, n \tag{20}$$

where x_i is actual value and the actual value of wind direction data is the result of clustering; x_{min} and x_{max} are the minimum and maximum values of the sample data.

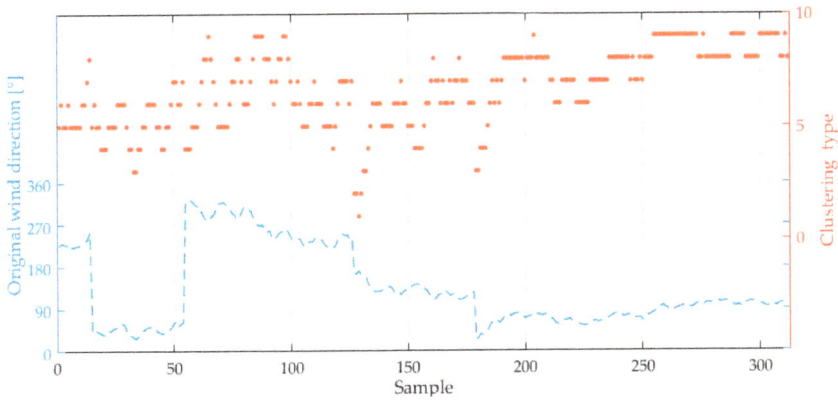

Figure 3. The wind direction processing data of "Dong-Chao line".

3.3. Performance Evaluation Index

The evaluation index of the icing prediction result used in this paper are:

(1)   Relative error (RE):

$$RE = \frac{x_i - \hat{x}_i}{x_i} \times 100\% \tag{21}$$

(2)   Root-mean-square error (RMSE):

$$RMSE = \sqrt{\frac{1}{n}\sum_{i=1}^{n}\left(\frac{x_i - \hat{x}_i}{x_i}\right)^2} \tag{22}$$

(3)   Mean Absolute Percentage Error (MAPE):

$$MAPE = \frac{1}{n}\sum_{i=1}^{n}|(x_i - \hat{x}_i)/x_i| \cdot 100\% \tag{23}$$

(4)   Average absolute error (AAE):

$$AAE = \frac{1}{n}\left(\sum_{i=1}^{n}|x_i - \hat{x}_i|\right)/\left(\frac{1}{n}\sum_{i=1}^{n}x_i\right) \tag{24}$$

where x is the actual value of the icing thickness; $\hat{x}$ is the forecasting value; n is groups of data. The smaller the above index value is, the higher the prediction accuracy is.

3.4. Modified BPNN-SVM-KELM for Icing Forecasting

The paper's experiment and modeling platform is Matlab R2014a, and the operating environment is an Intel Core i5-6300U CPU with 4G memory and a 500 G hard disk. The topology structure of BPNN in MEC-BPNN is 9-7-1. The transfer function of the hidden layer uses the expression $f(x) = 2/(1 + e^{-2x}) - 1$ which is a tansig function. The output layer transfer function takes the form $f(x) = x$ which is a purelin function. The maximum training time is 100, the training target minimum error is 0.0001, and the training speed is 0.1. In addition, the population size of MEC is 200, the sub-population is 20, the dominant sub-population's number is 5, the quantity of the temporary population is 5, the number of iterations is 10. The parameters of the BA algorithm in the BA-SVM prediction are set as follows: the dimension of search space is 7; the size of the bat population is

30; the pulse frequency R_i of bats is 0.5; loudness A_i is 0.25; the acoustic frequency range is [0, 2]; the termination condition of the algorithm is when the calculation reaches the maximum number of iterations (300). In SVM, the penalty parameter that needs to be optimized is C whose range of variation is [0.01, 100]; the range of kernel parameter g is [0.01, 100]; the range of the ε loss function parameter p is set as [0.01, 100]. The SVM optimal penalty parameter c is 1.971, the kernel parameter g is 0.010 and the ε loss function parameter p is 0.01 by BA optimization. The kernel function of KELM algorithm uses RBF kernel function whose input and output data processing interval is [−1, 1].

In order to show whether the forecasting results of the three modified models were a local optimal or global optimal location and whether these models can be generalized to other unseen data, a K-CV test is conducted here. According to the K-CV method described in Section 2.2, the data set is substituted into the model for testing and analysis. The 312 sets of data are randomly divided into 12 datasets, each of which has 26 groups of data and do not intersect each other. After 12 operations, each sub-data set is tested and the RMSE of the sample is obtained, which can be seen in the Table 2.

Table 2. Results of the k-fold cross validation.

Fold Number	RMSE of MEC-BPNN	RMSE of BA-SVM	RMSE of KELM
1	0.0126	0.0132	0.0136
2	0.0131	0.0205	0.0142
3	0.0126	0.0141	0.0135
4	0.0112	0.0126	0.0128
5	0.0118	0.0139	0.0142
6	0.0139	0.0149	0.0151
7	0.0143	0.0152	0.0137
8	0.0125	0.0133	0.0132
9	0.0101	0.0124	0.0152
10	0.0117	0.0128	0.0139
11	0.0109	0.0151	0.0149
12	0.0102	0.0118	0.0122
Average Value	0.0121	0.0142	0.0139
Standard Deviation	0.00129	0.00219	0.00087

From Table 2, it can be known that the average RMSE values of MEC-BPNN, BA-SVM and KELM are 0.0121, 0.0142 and 0.0139, respectively. The RMSE standard deviations of MEC-BPNN, BA-SVM and KELM are 0.00129, 0.00219 and 0.00087, respectively. It is indicated that the validation error of the each modified model proposed in this paper can obtain its global minimum.

After the prediction of the three individual improved models, the VC combined method to solve the problem of dynamic weight distribution is adopted to combine these models. The result of the combination is that the corresponding weights are adjusted dynamically according to the different training results whose adaptability is better. The combination weights of three individual models of MEC-BPNN, BA-SVM and KELM are 0.42, 0.34 and 0.24, respectively.

The paper uses the mature BP neural network model and SVM model to do a comparative experiment based on the sample data mentioned in Section 3.1 in order to verify the performance of the proposed combination forecasting model. The initial weights and thresholds of a single BPNN model are obtained by their own training, and other parameter settings are consistent with the MEC-BPNN. Besides, in the single SVM model, the penalty parameter c is 9.063, the kernel function parameter g is 0.256, and the ε loss function parameter p is 3.185.

The forecasting values and original values of BPNN, SVM, MEC-BPNN, BA-SVM, KELM, and improved BP-SVM-KELM based on VC, part of which are given in Table 3, are shown in Figure 4. The relative forecasting error of each model is revealed in Figure 5. This paper divides the test set samples into four groups to show the forecasting effect of each model owing to the large model quantities and sample points. Figures 4 and 5 therefore consist of three sub-graphs, respectively. The RMSE, MAPE and AAE of models are demonstrated in Figure 6.

The deviation between the icing thickness forecasting value and the original value of BPNN and SVM is large by contrasting the results of the six forecasting methods in Figure 4. In addition, the curves of forecasting value and original value are suddenly far or suddenly near, indicating that the forecasting accuracy and robustness of these two methods are poor. Besides, the deviations of MEC-BPNN, BA-SVM and KELM are smaller than the above two models so that the precision is improved, but the curves are still like those with poor stability. However, the deviation which is obtained from improved BP-SVM-KELM based on VC is smaller and it is between the most accurate and the most inaccurate single improvement model. The value is the closest to the most precise model which indicates the accuracy of the combined forecasting model is guaranteed. Then the curves distance between the forecasting value and the actual value of the composite model is basically distributed near the actual value curve, indicating that the combination forecasting model has the strongest robustness.

Table 3. Part of the forecasting value and relative errors of each model.

Data Point Number	Actual Value (mm)	BPNN		SVM		MEC-BPNN		BA-SVM		KELM		Proposed Model	
		Forecast Value	Error %	Forecast Value	Error %	Forecast Value	Error %	Forecast Value	Error %	Forecast Value	Error %	Forecast Value	Error %
1	50.01	50.5	0.98	50.38	−0.74	50.17	−0.32	50.18	−0.34	50.11	−0.20	50.16	−0.30
2	50.08	50.45	−0.74	50.38	−0.60	50.2	−0.24	50.2	−0.24	50.25	−0.34	50.21	−0.26
3	50.11	49.61	1.00	50.47	−0.72	50	0.22	50	0.22	50.27	−0.32	50.07	0.09
4	50.35	49.85	0.99	50	0.70%	50.5	−0.30	50.16	0.38	50.11	0.48	50.29	0.12
5	50.89	51.34	−0.88	50.53	0.71	51	−0.22	50.7	0.37	50.69	0.39	50.82	0.13
6	51.01	50.51	0.98	51.46	−0.88	50.87	0.27	51.18	−0.33	51.21	−0.39	51.06	−0.09
7	50.35	49.85	0.99	50	0.70%	50.25	0.20	50.5	−0.30	50.15	0.40	50.31	0.08
8	49.67	50.12	−0.91	50.05	−0.77	49.8	−0.26	49.8	−0.26	49.53	0.28	49.73	−0.13
9	49.53	50	−0.95	49.91	−0.77	49.43	0.20	49.43	0.20	49.64	−0.22	49.48	0.10
10	48.87	48.37	1.02%	49.25	−0.78	49	−0.27	49	−0.27	48.77	0.20	48.94	−0.15

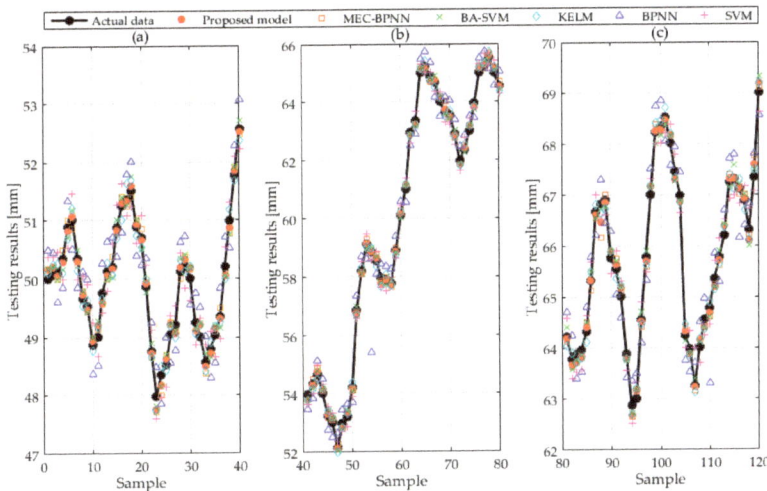

Figure 4. The forecasting values of different methods: (**a**) the forecasting value from 1 to 40 sample point; (**b**) the forecasting value from 41 to 80 sample point; (**c**) the forecasting value from 81 to 120 sample point.

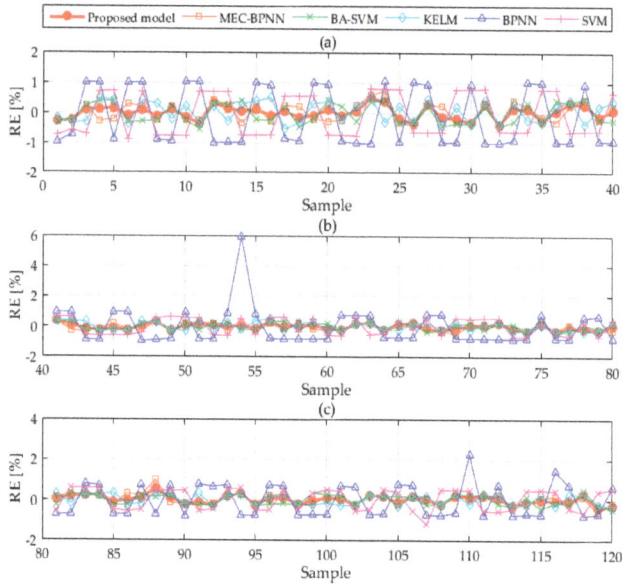

Figure 5. The relative errors curve of each method: (**a**) the relative error from 1 to 40 sample point; (**b**) the relative error from 41 to 80 sample point; (**c**) the relative error from 81 to 120 sample point.

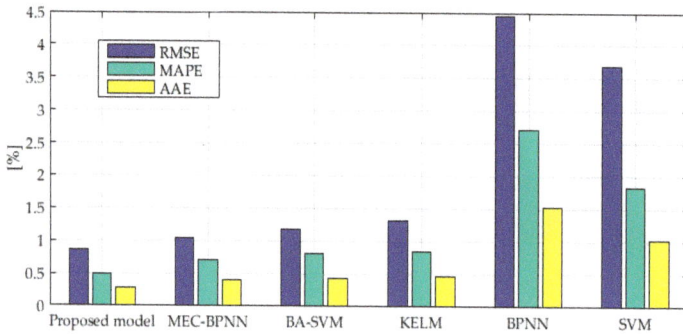

Figure 6. Values of root-mean-square error (RMSE), mean absolute percentage error (MAPE) and average absolute error (AAE).

Figure 5 compares the relative errors of the six forecasting methods. By counting the maximum and minimum relative errors, it can be found that the maximum relative errors of BPNN, SVM, MEC-BPNN, BA-SVM, KELM and the combined forecasting model are 5.910%, 1.186%, 1.003%, 0.551%, 0.545% and 0.543%, the minimum values are 0.611%, 0.170%, 0.135%, 0.119% and 0.002%. The maximum and minimum values in the three improved models are less than the two basic models, which shows that the prediction accuracy is better than the BPNN model and the SVM model. The maximum and minimum values in combined forecasting model are less than the three individual improved models, indicating that its prediction value is the nearest to optimal single improved model. The fluctuation ranges of the RE curves of the two basic models are the largest showing their stabilities are the poorest, and the stabilities of the three improved models have improved with a relatively small range of fluctuation. However, compared with the combination forecasting model, the fluctuation range is still large, which shows that the combination forecasting model plays a role in avoiding the weaknesses and improves the stability of prediction under the premise of ensuring the accuracy.

As we can see from Figure 6, the RMSE value of the combination forecasting model is 0.86%, and the RMSE values of the BPNN, SVM, MEC-BPNN, BA-SVM and KELM models are 4.44%, 3.68%, 1.04% and 1.32%, respectively. The proposed combination forecasting model has a lower error and a higher accuracy than the other models which makes the accuracy of single points be higher than the worst though the VC combined method solved the problem of dynamic weight distribution. The value is infinitely close to the most accurate single model at this point, so its stability and accuracy can be fully guaranteed. The prediction results of the three improved models are better than SVM and BPNN models indicating that their prediction performance has been improved by intelligent algorithms. The MAPE values of the BPNN, SVM, MEC-BPNN, BA-SVM, KELM and the combination forecasting models are 2.71%, 1.83%, 0.70%, 0.81%,0.84% and 0.50%. The evaluation index also shows that the combination forecasting method has the best overall prediction effect, the three improved models are second, and the two basic models have the worst prediction performance. The AAE value of the combination forecasting method is the smallest which enough shows the overall prediction performance of the proposed model is the best.

In conclusion, the prediction accuracy of the three improved models is advanced through the improvement of the basic model, but the robustness is still poor. The combination prediction model is not the most accurate at each point, but it is the closest to the most accurate predictions because the weights tend to the model with the highest accuracy according to the weight distribution. In the unknown prediction, the combination method can make best use of the advantages and bypass the disadvantages, so the flexibility, adaptability and accuracy are guaranteed.

4. Further Simulation

This paper now selects another line in Hunan, the "Tianshang line", as a case to further verify the performance of the proposed model. The data of "Tianshang line" are from 17 January 2008 to 15 February 2008, and have a total of 360 data groups. The first 240 are training samples and the latter 120 are test samples. All data of icing thickness, temperature, humidity, wind speed and wind direction clustering are shown in Figure 7. Like Case 1, all of the data were provided by the Key Laboratory of Disaster Prevention and Mitigation of Power Transmission and Transformation Equipment (Changsha, China), where all the data are collected by professional instruments and can reflect the state changes in the icing process. As we can see from Figure 7, the temperature data first decreases periodically, then rises periodically. The data of relative air humidity and wind speed present a cyclical upward trend. What's more, there is no exception data or missing data. Hence these data can be used directly as data sources.

BPNN, SVM, MEC-BPNN, BA-SVM, KELM and thw improved BP-SVM-KELM combination model based on VC are utilized to compare and analyze in this section. The parameter setting of BPNN, MEC-BPNN, and KELM are consistent with the previous case. In the single SVM model, the penalty parameter c is 10.307, the kernel function parameter g is 0.328, and the ε loss function parameter p is 2.261. For the same case, the SVM optimal penalty parameter c is 2.083, the kernel parameter g is 0.012 and the ε loss function parameter p is 0.011 by BA optimization. The combination weights of the three individual models of MEC-BPNN, BA-SVM and KELM are 0.40, 0.25 and 0.35, respectively, through the VC combined method that solved the problem of dynamic weight distribution.

The results of the k-fold cross validation for the three modified models proposed in this paper are described in Table 4. A part of the forecasting values and relative errors of each model is listed in Table 5. The forecasting results of each model are described in Figure 8, and the relative errors are shown in Figure 9. The RMSE, MAPE and AAE are shown in Figure 10.

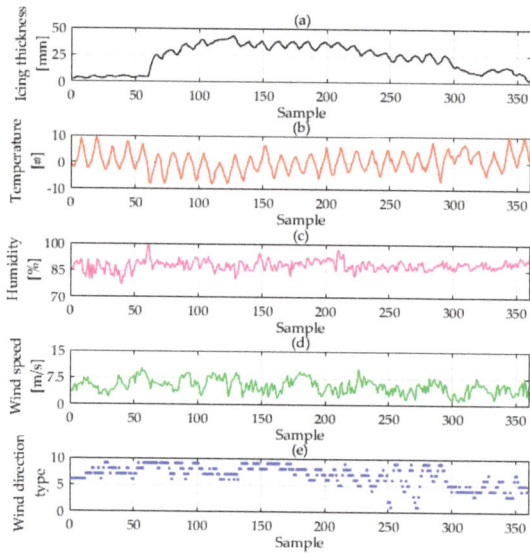

Figure 7. Original data of icing thickness, temperature, humidity, wind speed, and the processing data of wind direction. Note: (**a**) represents the original data of icing thickness; (**b**) represents the original data of temperature; (**c**) represents the original data of humidity; (**d**) represents the original data of wind speed; (**e**) represents the processing data of wind direction.

Table 4. Results of the k-fold cross validation.

Fold Number	RMSE of MEC-BPNN	RMSE of BA-SVM	RMSE of KELM
1	0.0178	0.0217	0.0182
2	0.0156	0.0233	0.0181
3	0.0182	0.0202	0.0192
4	0.0166	0.0193	0.0169
5	0.0167	0.0205	0.0175
6	0.0198	0.0202	0.0196
7	0.0202	0.0192	0.0185
8	0.0186	0.0178	0.0179
9	0.0191	0.0191	0.0192
10	0.0182	0.0203	0.0197
11	0.0175	0.0195	0.0186
12	0.0190	0.0212	0.0192
Average Value	0.0181	0.0202	0.0186
Standard Deviation	0.00130	0.00136	0.00083

Table 5. Part of the forecasting value and relative errors of each model.

Data Point Number	Actual Value (mm)	BPNN		SVM		MEC-BPNN		BA-SVM		KELM		Proposed Model	
		Forecast Value	Error %	Forecast Value	Error %	Forecast Value	Error %	Forecast Value	Error %	Forecast Value	Error %	Forecast Value	Error %
1	26.38	26.729	−1.32	26.02	1.36	26.54	−0.61	26.37	0.04	26.51	−0.49	26.49	−0.40
2	26.97	27.426	−1.69	26.62	1.30	26.86	0.41	26.86	0.41	27.16	−0.70	26.96	0.02
3	27.32	27.801	−1.76	26.969	1.28	27.21	0.40	27.432	−0.41	27.43	−0.40	27.34	−0.09
4	27.68	28.136	−1.65	27.37	1.12	27.47	0.76	27.8	−0.43	27.87	−0.69	27.69	−0.05
5	28.01	27.55	1.64	28.37	−1.29	27.9	0.39	28.14	−0.46	28.09	−0.29	28.03	−0.06
6	28.32	27.86	1.62	28.68	−1.27	28.05	0.95	28.14	0.64	28.18	0.49	28.12	0.71
7	27.87	27.33	1.94	28.176	−1.10	27.8	0.25	27.69	0.65	27.71	0.57	27.74	0.46
8	25.51	25.12	1.53	25.84	−1.29	25.68	−0.67	25.38	0.51	25.36	0.59	25.49	0.07
9	23.65	23.22	1.82	24.01	−1.52	23.82	−0.72	23.53	0.51	23.54	0.47	23.65	0.01
10	22.35	21.92	1.92	22.74	−1.74	22.57	−0.98	22.17	0.81	22.51	−0.72	22.45	−0.43

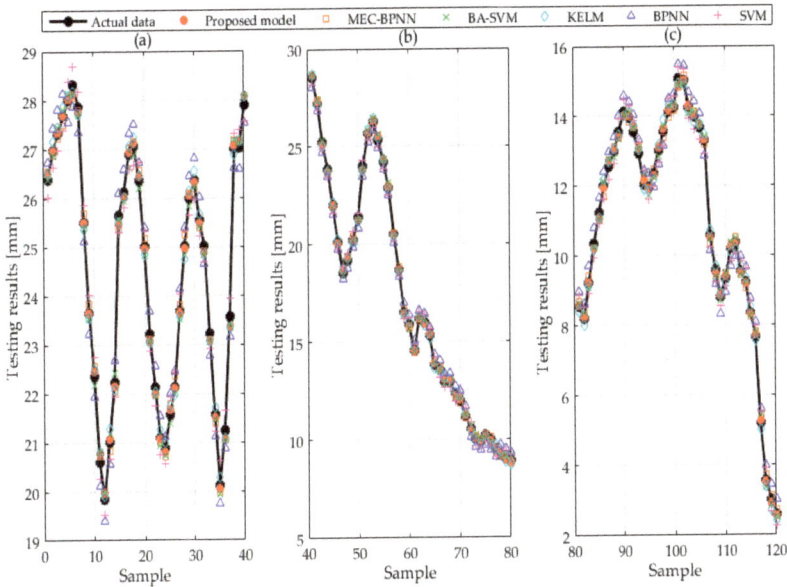

Figure 8. The forecasting values of different methods: (**a**) the forecasting value from 1 to 40 sample point; (**b**) the forecasting value from 41 to 80 sample point; (**c**) the forecasting value from 81 to 120 sample point.

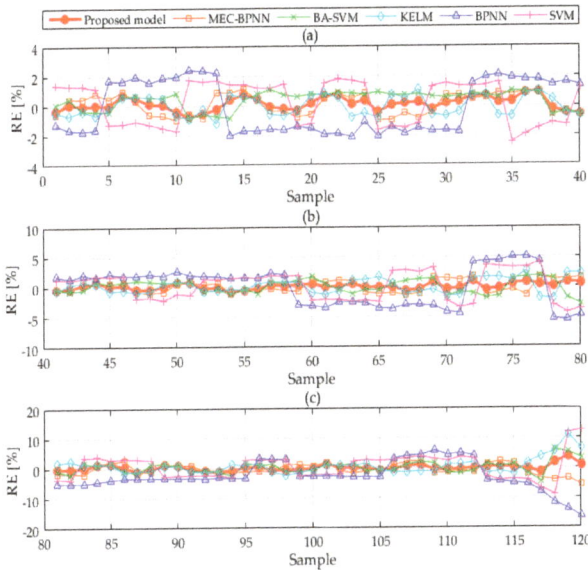

Figure 9. The relative errors curve of each method: (**a**) the relative error from 1 to 40 sample point; (**b**) the relative error from 41 to 80 sample point; (**c**) the relative error from 81 to 120 sample point.

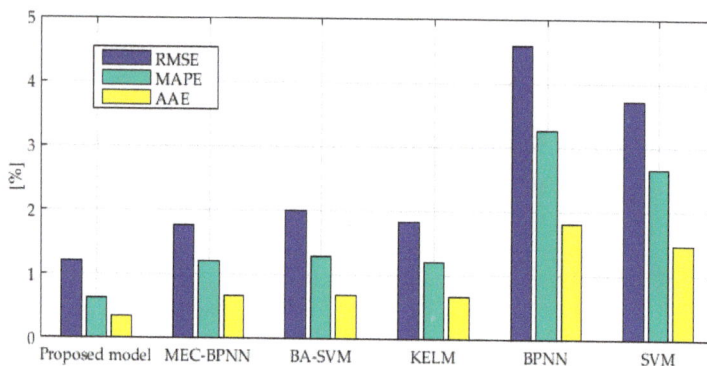

Figure 10. Values of root-mean-square error (RMSE), mean absolute percentage error (MAPE) and average absolute error (AAE).

As is shown in Table 4, the average RMSE values of MEC-BPNN, BA-SVM and KELM are 0.0181, 0.0202 and 0.0186, respectively. In addition, the RMSE standard deviations of MEC-BPNN, BA-SVM and KELM are 0.00130, 0.00136 and 0.00083, respectively. These data illustrate the fact again that the generalization performance of the three modified models has been improved.

As we see from Figure 8 and Table 5, compared with the BPNN and SVM models, the forecasting values of the three improved models are closer to the original values, which shows that the prediction accuracy of the modified model is better. The forecasting value of combination model is between the forecasting values of the three modified models. Although the proposed model is not the most accurate, the change range of the distance between the prediction and the actual value curves is the smallest. Besides, the result of the combination forecast model is closer to the most accurate single model predictive value which further shows that the combination forecasting model greatly improves the stability of prediction under the premise of ensuring accurate prediction.

The relative errors of the BPNN model and the SVM model are at a high level and the fluctuation range is large by observing Figure 9, where the relative forecasting error of the six models are displayed, indicating that the two models' accuracy and robustness are poor. As we can see from Figure 9 and Table 5, the relative errors of MEC-BPNN, BA-SVM and KELM are lower than those of the two basic models with greater volatility, and the value at some points is still large, which shows that its accuracy has been improved while the stability is still not guaranteed. It can be found that the relative errors of the three modified models constantly change their ranking at various points, which can be classified into three cases. For instance, in the first case, the relative errors of MEC-BPNN, BA-SVM and KELM are 1.13%, 1.64% and −1.01%, respectively, at the 60th sample point, where the forecasting accuracy of KELM is the highest. In the second case, the relative errors of MEC-BPNN, BA-SVM and KELM are 1.35%, −3.04% and 2.14%, respectively, at the 80th sample point, where the forecasting accuracy of MEC-BPNN is the highest. In the third case, the relative errors of MEC-BPNN, BA-SVM and KELM are −5.81%, 3.10% and 6.20%, respectively, at the 120th sample point, where the forecasting accuracy of BA-SVM is the highest. However, the relative error curve of the proposed combination model is among the three curves of MEC-BPNN, BA-SVM and KELM and it is close to the most accurate prediction in almost every sample point. In addition, its fluctuation range is also narrower. This indicates that the combination model can obtain both accurate and stable forecasting results.

As is shown in Figure 10, the RMSE, MAPE and AAE of the proposed combination forecasting model are the minimum at 1.20%, 0.63% and 0.34%. This indicates that its whole predictive performance is optimal. By observing these values of the three improved model, it can be found that the whole prediction accuracy of MEC-BPNN is better than KELM's, and KELM's is superior to BA-SVM's. When adopting the VC combined method to solve the problem of dynamic weight distribution to assign weights, MEC-BPNN's weight is the highest, KELM's is the second and BA-SVM's is the minimum. It shows that weight assignation of the proposed combination forecasting method leans toward the most precise model, which enhances the robustness of prediction and also guarantees the accuracy.

By comparing the case with the previous one, it is clear that the weights of the three individual improvement models differ from those of the previous case. The prediction accuracy of BA-SVM is better than that of KELM and its weight is higher than KELM's in the Case 1 according to the calculation results of performance evaluation index. However, the Case 2 is just the opposite. The difference indicates that the VC combined method solved the problem of dynamic weight distribution can adjust the weights according to the prediction accuracy of each individual model. It is so flexible that the overall accuracy of the prediction is improved.

In summary, the paper introduces the MEC-BP model, the BA-SVM model and the KELM model to improve the prediction performance of the individual models. The VC combined method solved the problem of dynamic weight distribution combines these models' advantages, and the weights are flexibly assigned, so the overall instability of the model is reduced and satisfactory prediction results are obtained.

5. Conclusions

In order to obtain better accuracy and stability of icing forecasting, an innovative combination forecasting model using a modified Back Propagation Neural Network-Support Vector Machine-Extreme Learning Machine with Kernel (BPNN-SVM-KELM) based on the variance-covariance (VC) weight determination method is proposed in this paper. First of all, BPNN is optimized by a mind evolutionary algorithm (MEC) to solve the problem that BPNN often falls into local optima and converges slowly. Second, a bat algorithm (BA) is used to optimize SVM, and the problem of choosing SVM key parameters is solved. Third, the kernel function is introduced into ELM to improve the regression forecasting accuracy of the model. Finally, by the dynamic allocation method of VC weights, three improved models of MEC-BPNN, BA-SVM and KELM are combined to obtain the combination forecasting model. In the simulation process, this paper takes into account the strong fluctuation of wind direction data, which will have a negative impact on the accuracy of forecasting. Therefore, according to the influence degree of wind direction on icing thickness, a clustering processing is carried out. Through the simulation of two examples, it is clear that three individual modified models utilize various optimization algorithms to take advantage of the core advantages of the forecasting model, avoiding the defects of the model itself and optimizing the performance of the model. Furthermore, the VC weighted combination method is used to dynamically assign weights, and the forecasting results tend to the best single prediction model. It is proved that the combination method complements the shortcomings of each model and has a strong comprehensive response ability. In summary, the research content of this paper is expected to provide a useful reference for the power sector to deal with icing accidents in advance.

Acknowledgments: This work is supported by the Natural Science Foundation of China (Project No. 71471059), the Fundamental Research Funds for the Central Universities (Project No. 2017XS103) and Wei-Chiang Hong thanks the research grant sponsored by Ministry of Science & Technology, Taiwan (MOST 106-2221-E-161-005-MY2).

Author Contributions: Yi Liang designed this research and processed the data; Dongxiao Niu and Wei-Chiang Hong provided professional guidance. Haichao Wang established the proposed model and wrote this paper; Meng Wang translated and revised this paper.

Conflicts of Interest: The authors declare no conflict of interest.

References

1. Jiang, X.; Xiang, Z.; Zhang, Z.; Hu, J. Predictive model for equivalent ice thickness load on overhead transmission lines based on measured insulator string deviations. *IEEE Trans. Power Deliv.* **2014**, *29*, 1659–1665. [CrossRef]

2. Chang, S.E.; Mcdaniels, T.L.; Mikawoz, J.; Peterson, K. Infrastructure failure interdependencies in extreme events: power outage consequences in the 1998 ice storm. *Nat. Hazards* **2007**, *41*, 337–358. [CrossRef]

3. Hou, H.; Yin, X.; You, D.; Chen, Q.; Tong, G.; Zheng, Y.; Shao, D. Analysis of the Defects of Power Equipment in the 2008 Snow Disaster in Southern China Area. *High Volt. Eng.* **2009**, *35*, 584–590.

4. Lü, Y.; Zhan, Z.; Ma, W. Design and application of online monitoring system for ice-coating on transmission lines. *Power Syst. Technol.* **2010**, *34*, 196–200.

5. Wang, H.L.; Shi-Ming, Y.U.; Fan, Y.M. Transmission line icing monitoring system based on wireless sensor net works. *Manuf. Autom.* **2011**, *17*, 150–153.

6. Luo, J.; Li, L.; Ye, Q.; Hao, Y.; Li, L. Development of Optical Fiber Sensors Based on Brillouin Scattering and FBG for On-Line Monitoring in Overhead Transmission Lines. *J. Lightwave Technol.* **2013**, *31*, 1559–1565. [CrossRef]

7. Imai, I. Studies on ice accretion. *Res. Snow Ice* **1953**, *3*, 35–44.

8. Goodwin, E.J.I.; Mozer, J.D.; Digioia, A.M.J.; Power, B.A. Predicting Ice and Snow Loads for Transmission. Available online: http://www.dtic.mil/docs/citations/ADP001696 (accessed on 11 August 2017).

9. Lenhard, R.W. An indirect method for estimating the weight of glaze on wires. *Bull. Am. Meteorol. Soc.* **1995**, *36*, 1–5.

10. Zhang, Z.; Huang, H.; Jiang, X.; Hu, J.; Sun, C. Analysis of Ice Growth on Different Type Insulators Based on Fluid Dynamics. *Trans. China Electrotech. Soc.* **2012**, *27*, 35–43.

11. Li, P.; Zhao, N.; Zhou, D.; Cao, M.; Li, J.; Shi, X. Multivariable time series prediction for the icing process on overhead power transmission line. *Sci. World J.* **2014**, *2014*, 1–9. [CrossRef] [PubMed]

12. Yang, J.L. Impact on the Extreme Value of Ice Thickness of Conductors from Probability Distribution Models. In Proceedings of the International Conference on Mechanical Engineering and Control Systems, IEEE Computer Society, Washington, DC, USA, 23–25 January 2015.

13. Li, P.; Li, Q.; Cao, M.; Gao, S.; Huang, H. Time Series Prediction for Icing Process of Overhead Power Transmission Line Based on BP Neural Networks. In Proceedings of the 2011 30th Chinese Control Conference (CCC), Yantai, China, 22–24 July 2011.

14. Wang, L.; Luo, Y.; Yao, Y. Analysis of transmission line icing detection and prediction based on neural network. *J. Chongqing Univ. Posts Telecommun.* **2012**, *24*, 254–258.

15. Zheng, Z.; Liu, J. Prediction method of ice thickness on transmission lines based on the combination of GA and BP neural network. *Power Syst. Clean Energy* **2014**, *30*, 27–30.

16. Wang, J.W. Ice thickness prediction model of transmission line based on improved particle swarm algorithm to optimize nrbf neural network. *J. Electr. Power Sci. Technol.* **2012**, *27*, 76–80.

17. Zarnani, A.; Musilek, P.; Shi, X.; Ke, X.; He, H.; Greiner, R. Learning to predict ice accretion on electric power lines. *Eng. Appl. Artif. Intell.* **2012**, *25*, 609–617. [CrossRef]

18. Huang, J.; Yang, H.; Hunan, Y.W. Forecast of line ice-coating degree using circumfluence index & support vector machine method. In Proceedings of the International Conference on Electric Utility Deregulation and Restructuring and Power Technologies, Changsha, China, 26–29 November 2015.

19. Xu, X.; Niu, D.; Wang, P.; Lu, Y.; Xia, H. The weighted support vector machine based on hybrid swarm intelligence optimization for icing prediction of transmission line. *Math. Probl. Eng.* **2015**, *2015*, 1–9. [CrossRef]

20. Ma, T.; Niu, D. Icing Forecasting of High Voltage Transmission Line Using Weighted Least Square Support Vector Machine with Fireworks Algorithm for Feature Selection. *Appl. Sci.* **2016**, *6*, 438. [CrossRef]

21. Ma, T.; Niu, D.; Fu, M. Icing forecasting for power transmission lines based on a wavelet support vector machine optimized by a quantum fireworks algorithm. *Appl. Sci.* **2016**, *6*, 54. [CrossRef]

22. Jie, J.; Zeng, J.; Han, C. An extended mind evolutionary computation model for optimizations. *Appl. Math. Comput.* **2007**, *185*, 1038–1049. [CrossRef]

23. Yang, X.S. A new metaheuristic bat-inspired algorithm. *Comput. Knowl. Technol.* **2010**, *284*, 65–74.

24. Liu, N.; Zhang, Q.; Liu, H. Online Short-Term Load Forecasting Based on ELM with Kernel Algorithm in Micro-Grid Environment. *Trans. China Electrotech. Soc.* **2015**, *30*, 218–224.

25. Wang, D.; Wei, S.; Luo, H.; Yue, C.; Grunder, O. A novel hybrid model for air quality index forecasting based on two-phase decomposition technique and modified extreme learning machine. *Sci. Total Environ.* **2017**, *580*, 719–733. [CrossRef] [PubMed]

Energies **2017**, *10*, 1196

26. Jiang, Y.; Liu, Z.; Luo, H.; Wang, H. Elm indirect prediction method for the remaining life of lithium-ion battery. *J. Electr. Meas. Instrum.* **2016**, *30*, 179–185.

27. Huang, G.B.; Zhu, Q.Y.; Siew, C.K. Extreme learning machine: Theory and applications. *Neurocomputing* **2006**, *70*, 489–501. [CrossRef]

28. Bates, J.M.; Granger, C.W.J. Combination of forecasts. *Oper. Res. Quart.* **1969**, *20*, 451–468. [CrossRef]

29. Liang, Y.; Niu, D.; Cao, Y.; Hong, W.C. Analysis and modeling for china's electricity demand forecasting using a hybrid method based on multiple regression and extreme learning machine: A view from carbon emission. *Energies* **2016**, *9*, 941. [CrossRef]

30. Wang, J.; Hu, J. A robust combination approach for short-term wind speed forecasting and analysis—combination of the arima, elm, svm and lssvm forecasts using a gpr model. *Energy* **2015**, *93*, 41–56. [CrossRef]

31. Hui-Lin, L. Applications of the combination weighted arithmetic averaging operator in the delivery merchandise sale forecast of material resources. *J. Nat. Sci. Heilongjiang Univ.* **2012**, *29*, 22–28.

32. Han, W.; Wang, J.; Zhang, X.H. Application Research of Combined Forecasting Based on Induced Ordered Weighted Averaging Operator. *Manag. Sci. Eng.* **2014**, *8*, 23–26.

33. Zhang, H.; Zhu, Y.J.; Fan, L.L.; Qiao-Ling, W.U. Mid-long term load interval forecasting based on markov modification. *East China Electr. Power* **2013**, *41*, 33–36.

34. Xian, G.M. ε-svr algorithm and its application. *Comput. Eng. Appl.* **2008**, *44*, 40–42.

35. Kohavi, R. A study of cross-validation and bootstrap for accuracy estimation and model selection. In Proceedings of the 14th International Joint Conference on Artificial Intelligence, Montreal, QC, Canada, 20–25 August 1995.

36. Huang, G.B.; Wang, D. H.; Lan, Y. Extreme learning machines: a survey. *Int. J. Mach. Learn. Cybern.* **2011**, *2*, 107–122. [CrossRef]

37. Huang, G.B.; Zhou, H.; Ding, X.; Zhang, R. Extreme learning machine for regression and multiclass classification. *IEEE Trans. Syst. Man Cybern. Part B Cybern.* **2012**, *42*, 513–529. [CrossRef] [PubMed]

38. Huang, X.B.; Ouyang, L.S.; Wang, Y.N.; Li-Cheng, L.I.; Bing, L. Analysis on key influence factor of transmission line icing. *High Volt. Eng.* **2011**, *37*, 1677–1682.

energies

MDPI

Article

Ice Cover Prediction of a Power Grid Transmission Line Based on Two-Stage Data Processing and Adaptive Support Vector Machine Optimized by Genetic Tabu Search

Xiaomin Xu *, Dongxiao Niu, Lihui Zhang, Yongli Wang and Keke Wang

School of Economics and Management, North China Electric Power University, Beijing 102206, China;
niudx@ncepu.edu.cn (D.N.); zlh6699@126.com (L.Z.); wyl_2001_ren@163.com (Y.W.);
15652912329@163.com (K.W.)
* Correspondence: xuxiaomin0701@126.com; Tel.: +86-010-6177-3079

Received: 19 September 2017; Accepted: 8 November 2017; Published: 14 November 2017

Abstract: With the increase in energy demand, extreme climates have gained increasing attention. Ice disasters on transmission lines can cause gap discharge and icing flashover electrical failures, which can lead to mechanical failure of the tower, conductor, and insulators, causing significant harm to people's daily life and work. To address this challenge, an intelligent combinational model is proposed based on improved empirical mode decomposition and support vector machine for short-term forecasting of ice cover thickness. Firstly, in light of the characteristics of ice cover thickness data, fast independent component analysis (FICA) is implemented to smooth the abnormal situation on the curve trend of the original data for prediction. Secondly, ensemble empirical mode decomposition (EEMD) decomposes data after denoising it into different components from high frequency to low frequency, and support vector machine (SVM) is introduced to predict the sequence of different components. Then, some modifications are performed on the standard SVM algorithm to accelerate the convergence speed. Combined with the advantages of genetic algorithm and tabu search, the combination algorithm is introduced to optimize the parameters of support vector machine. To improve the prediction accuracy, the kernel function of the support vector machine is adaptively adopted according to the complexity of different sequences. Finally, prediction results for each component series are added to obtain the overall ice cover thickness. A 220 kV DC transmission line in the Hunan Region is taken as the case study to verify the practicability and effectiveness of the proposed method. Meanwhile, we select SVM optimized by genetic algorithm (GA-SVM) and traditional SVM algorithm for comparison, and use the error function of mean absolute percentage error (MAPE), root mean square error (RMSE) and mean absolute error (MAE) to compare prediction accuracy. Finally, we find that these improvements facilitate the forecasting efficiency and improve the performance of the model. As a result, the proposed model obtains more ideal solutions and has higher accuracy and stronger generalization than other algorithms.

Keywords: ice cover prediction; adaptive support vector machine (ASVM); genetic tabu search (GATS); two-stage data processing; ensemble empirical mode decomposition; fast independent component analysis

1. Introduction

As the terrain and landforms in China are complex and diverse, and the characteristics of micro topography and micro meteorology are extensive, grid transmission lines in these regions are often affected by extreme weather conditions. Ice disasters are one of the most serious natural disasters that affect safe and stable operation of power systems [1]. In recent years, with the increasingly frequent occurrence of extreme weather, icing accidents on transmission lines have occurred more frequently and have attracted the attention of researchers. Transmission lines are important components of power transmission, of which normal and safe operation is an important guarantee to avoid major accidents in the power grid. Severe ice cover will lead to a sharp decline of the mechanical and electrical properties in the transmission line, causing downed transmission line poles, conductor galloping, and broken line accidents. Those accidents will lead to power outages and pose a grave threat to safe and stable operation of the power system [2]. Moreover, the distribution of energy resources in China is uneven, for example, the demand for electricity in the central and eastern regions is vigorous, while the energy resources are mainly distributed in the west. As a result, our country needs to vigorously promote construction of the outgoing channel of the power base to enhance the reliability when transmission lines go through extremely harsh and complex areas of contamination, high altitude, snow, strong acid rain, and fog [3,4].

According to incomplete statistics, since the 1950s, transmission lines in China have suffered from thousands of ice disaster accidents, and the impact of the accidents is increasing [5]. In 2008, the southern part of the country suffered the most severe ice disaster on meteorological record. Disconnections, downed rods, and tripping incidents of large areas resulted in partial grid disaggregation and a large area outage, causing a great loss of more than 1000 billion yuan to the State Grid Corporation [6–8]. The annual maximum thickness of ice cover in some transmission lines shows an increasing trend, as shown in Figure 1. Especially in 2008, the ice cover thickness was the highest during the recent ten years. Along with the promotion of the West–East electricity transmission project and ultra-high voltage transmission project, more and more AC/DC ultra-high voltage transmission lines cross the icing areas. The ice cover becomes one of the main factors influencing safe operation of a power system. Thus, prediction and early warning of the ice cover on transmission line are heated research topics. How to effectively predict the degree of transmission line ice cover thickness has become an important research subject.

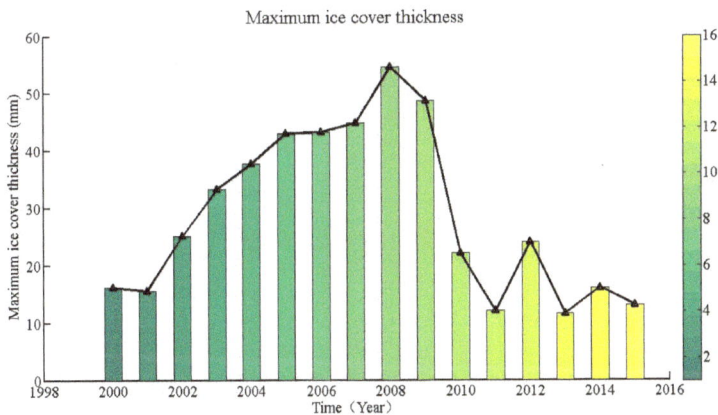

Figure 1. Maximum ice cover thickness data of a transmission line in Hunan Province from 2000 to 2015.

According to different stages of development, physical methods, empirical methods, and intelligent methods are introduced to predict ice cover thickness [9–12]. Some parameters in the physical methods are difficult to measure in real circumstances, which limits the feasibility of these methods. It is very difficult to directly apply physical methods to the forecasting of ice cover on transmission lines [10–12]. Empirical methods do not take the physical mechanism of the real process of icing into consideration. The ice cover model can be directly established through the operation and the experimental summary of the law of the model. The principle of empirical model is simple and clear, and the data are accessible. The most common empirical models are the Lenhard model and Chaine-Skeates model. However, the overall empirical models are rough and cannot be applied to other transmission lines in the geographical environment with large diversity. The generalization ability of the model is not strong. The comparison of the common prediction modeles for the ice cover thickness of transmission lines is shown in Table 1 [13].

Table 1. Comparison of common prediction model for ice cover thickness of transmission line.

Model		Characteristic of Model	Applicable Type
Physics models	Imai model	The model is correct in principle, but the assumption is not constant in practice. Therefore, it may cause a large deviation of the ice cover thickness.	Short term
	Goodwin model	The model assumes that the conductor icing is a circular ice cover and the coefficient is 1, which are not reasonable.	
Empirical models	Growth model	This two models have a large dependence on the regularity of the sample and need large quantity of sample data. Meanwhile, the impact of outliers on the prediction results is large.	Short term & medium term
	Extremum model		Medium term

Intelligent models are popular in the study of ice cover prediction [14–23]. Due to their nonlinear approximation ability, neural networks are widely used in icing prediction [14–17]. Reference [15] used the genetic algorithm to optimize the back propagation (BP) neural network to speed up the convergence rate and reduce the system error, which improves the accuracy of the model. Reference [16] determined the key factors that affect ice cover and used the radial basis function (RBF) neural network for prediction to control the occurrence of error within an acceptable range. However, the neural network algorithm has the disadvantages of local optima, low efficiency, and poor generalization ability, and it is difficult to guarantee the accuracy of the model [18].

Support vector machine (SVM), as one of the most commonly used machine learning algorithms, which can overcome the defects of a neural network. SVM has the advantages of repeated training, fast convergence speed and it can solve real problems with the characteristics of small samples, non-linearity and local extrema. SVM is suitable for ice cover prediction, which is influenced by climate [19–23]. To improve the prediction accuracy, the wavelet method [19], particle swarm optimization algorithm [20], and other algorithms are used to optimize the parameters of SVM. The prediction results are improved to a certain degree. As the single optimization algorithm has some defects of slow convergence rate, local optimum and low accuracy, it does not significantly improve the prediction accuracy. The optimization algorithm combination has been introduced in many fields for prediction. It can complement single algorithms and use their respective advantages to avoid the defects of the algorithm itself. Satisfactory results are achieved in many areas [24–26].

The ice cover of transmission lines is affected by many natural factors and has certain randomness. Therefore, the non-linear and non-stationary characteristics of the collected data of ice thickness are important factors affecting the accuracy of ice thickness predictions. Therefore, in this paper, the original data is denoised firstly, and then decomposed into a certain scale to reduce the non-stationarity of the signal. SVM can make comprehensive consideration of multiple factors on the coefficient, with good non-linear mapping ability and generalization ability. It can also be repeated by training and the convergence speed is fast. In the SVM regression prediction model, the kernel parameter g and the penalty coefficient C have a great influence on the accuracy of the model. In this paper, we use a hybrid algorithm of a genetic arglothm-tabu search (GATS) to optimize the parameters g and C, finding optimal parameters to improve the prediction accuracy of the model. In view of this, this paper presents a hybrid intelligent prediction model. It is applicable to non-linear and non-stationary data signals. Through data denoising and decomposition, the non-stationarity of the data can be better reduced. The hybrid algorithm combines single optimization algorithms, which can make up for the defects of each single algorithm, achieve global search and improve the convergence speed. The combination algorithm is used to optimize the SVM parameters, which can further improve the prediction accuracy. Therefore, the hybrid model proposed in this paper can not only solve the problem of rough and unstable raw signals, but also achieve better prediction results.

The innovations and contributions of this study are further explained as follows:

(a) As a new topic, the current research on the prediction of thickness of ice cover is not often seen. As ice cover on transmission lines will bring many dangers to the safety of power supply, accurate forecasting results are helpful for power grid enterprises to prepare for and control and control their aftermaths in advance.

(b) The method of fast independent component analysis (FICA) has the characteristics of fast convergence and good stability. It can weaken all kinds of interference information while protecting the useful signal. It has a wide application prospect in signal processing field. The FICA method is proposed for the original data to minimize the impact of the extreme conditions on the shock of the original sequence.

(c) Ensemble empirical mode decomposition (EEMD) is a noise-aided data analysis method. This method avoids the difficulty of selecting wavelet bases in wavelet transform. Besides, it inherits the advantages of the empirical mode decomposition (EMD) method and also effectively solves the modal aliasing problem existing in the EMD process. As the thickness of ice cover is greatly affected by climatic factors, the regularity of the original data is not strong, and the sequence has non-stationary characteristics. To better reflect the internal structure of the original sequence, the data after denoising are decomposed by ensembling empirical mode decomposition (EEMD) into a high frequency and low frequency component.

(d) The SVM is based on the principle of structural risk minimization. It can find the best compromise between the complexity and learning ability of the prediction model according to the information of the icing thickness sequence sample to obtain the best generalization ability. EEMD decomposes raw data into high-frequency and low-frequency component sequences with different complexity. In this paper, the optimal model parameters and kernel functions are selected according to the characteristics and complexity of each component, the SVM prediction model suitable for itself are established to improve the accuracy of single prediction model. The support vector machine model used in this paper is called the adaptive support vector machine model (ASVM).

(e) A new hybrid algorithm, which is named of GATS, is presented by combining the genetic algorithm and tabu search. The hybrid algorithm effectively combines the parallel search capability of the genetic algorithm (GA) and the local search capability of the TS algorithm. The combined method can enhance the global search ability and improve the search speed.

(f) The empirical results validate that the proposed model is suitable for ice cover prediction. The model can obtain higher accuracy and satisfactory results. The establishment of the model has important practical significance for the power grid enterprise to effectively confront ice disasters and ensure the safe and reliable operation of the power network [27].

2. Two-Stage Data Pre-Processing Method

2.1. Data De-Noising Processing by Fast Independent Component Analysis (FICA)

Independent component analysis (ICA) is a signal processing method developed in the 1990s. Initially, it was developed for the solution of blind signal separation. Recently, ICA has become a powerful tool for signal processing and data analysis [28,29].

ICA is essentially an optimization algorithm, namely, how to make the separating independent component close to each source signal. The standard ICA problem can be defined as:

Assuming that $x(t) = [x_1(t), x_2(t), \cdots, x_m(t)]^T$ is an M dimensional observation signal vector. It is linearly mixed with N unknown and independent source signals $s(t) = [s_1(t), s_2(t), \cdots, s_n(t)]^T$, where t is a discrete time and its value is 0, 1, 2…. According to [28], the formula can be expressed as:

$$x(t) = As(t) \tag{1}$$

where A is an $m \times n$ dimensional matrix, called the mixed matrix.

The basic idea of ICA is descriped as follows and seen in Figure 2.

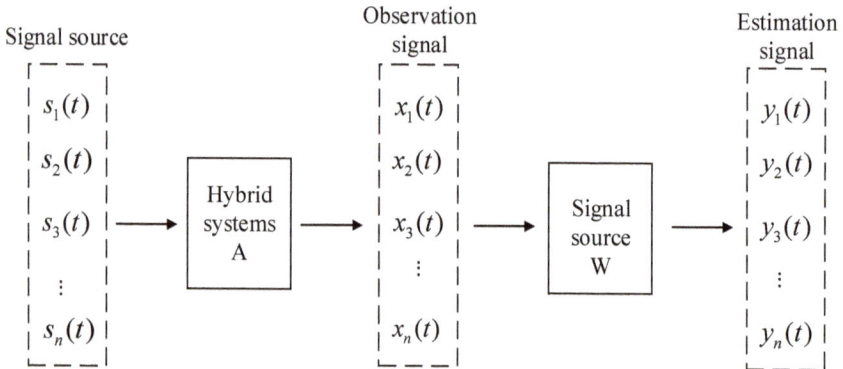

Figure 2. Basic principle diagram of independent component analysis (ICA).

In Figure 2, $s(t)$ is the source signals, A is unknown mixed matrix, $x(t)$ is observed signal, and W is separation matrix. The observed signal $x(t)$ is obtained by mixing the source signal $s(t)$ with the matrix A. In the case of the unknown mixed matrix A and the source signals $s(t)$, the separation matrix W is determined according to only the observed data vector $x(t)$, so that the output signal $y(t) = [y_1(t), y_2(t), \cdots, y_n(t)]^T$ is the estimation of the source signal $s(t)$, and W is an $n \times m$ dimensional matrix. According to [29], the formula can be expressed as:

$$y(t) = Wx(t) = WAs(t) \tag{2}$$

The core of the ICA algorithm is to update W to bring the estimation closer to the source signal. FICA is commonly used for the task. The FICA algorithm is also known as the fixed-point algorithm, uses the Newton iteration method to batch a large number of sampling points of the observed signal. An independent component is separated from the observation signal, which makes the convergence more rapid and stead, and the efficiency of the calculation is improved [30].

2.2. Data Decomposition by Ensemble Empirical Mode Decomposition (EEMD)

Empirical mode decomposition (EMD) is an adaptive signal processing method proposed by Huang in 1998. The method is used for non-linear and non-steady sequences and is directly aimed at data without determining the decomposition basis. EMD is relatively simple and easy to implement. The main idea is to use the Hilbert-Huang transform (HHT) conversion method to transform the non-linear and non-steady time sequence data into a steady sub-time sequence. The method is characterized by the introduction of the concept of the intrinsic mode functions (IMFs). The fluctuation or trend of different scales in the signal is decomposed into a series of data sequence with different feature scales, and each sequence is called an intrinsic mode function. IMF can be linear or non-linear, and each component has practical physical meaning [30]. However, they all have the same number of extreme points and zero crossing points, and there is only one extreme point between two adjacent zero crossing points. Additionally, any two IMF components are independent of each other.

Each IMF is superimposed, and residual component r_n is the source signal. Compared to the method of frequency domain and time domain analysis, EMD is more suitable for the analysis of non-linear and non-steady signals [31].

For the traditional wavelet decomposition algorithm, empirical mode decomposition reduces the influence of human factors on the decomposition results, and has a certain advanced nature. However, in some cases, the algorithm shows the mode mixing phenomenon [32]. Ensemble empirical mode decomposition (EEMD) is a noise-assisted data analysis method aimed at eliminating the disadvantage of the mode mixing of EMD. EEMD includes the white Gaussian noise signals, which are continuous on different time scales, and can eliminate the noise effect by calculating the average of multiple noise processing results. The method solves the EMD mode mixed stack effectively. The basic principle of the EEMD method is as follows:

With the original signal $x(t)$, the EMD algorithm will decompose $x(t)$ into a set of IMF components c_i and residual r, which is [32]:

$$x(t) = \sum_{i=1}^{n} c_i + r \tag{3}$$

EEMD methods add Gaussian white noise to the original signal, using the uniform distribution characteristics of the Gaussian white noise frequency to make the signal in different time scales continuous to eliminate the frequency aliasing phenomenon. Gaussian white noise is subject to the normal distribution of $(0, (\alpha\varepsilon)^2)$, where ε is the standard deviation of the signal and α is the strength parameter of the noise. The specific steps are as follows.

Gaussian white noise $n_i(t)$ with zero mean and constant standard deviation is added to the original signal $x(t)$. According to [33], the formula can be expressed as:

$$x_i(t) = x(t) + k \times n_i(t) \tag{4}$$

$n_i(t)$ is the noise added for the *ith* time, and $x_i(t)$ is the signal after adding the noise the *ith* time. k is proportionality coefficient. It is generally believed that the standard deviation of the white noise added is 0.2 times of the standard deviation of the signal [34].

The following formula is obtained after decomposing noise signals [33]:

$$x_i(t) = \sum_{j=1}^{n} c_{ij} + r_i \tag{5}$$

c_{ij} is the *jth* IMF component after EMD decomposition for the *ith*, and r_i is the remainder after EMD decomposition for the *ith*.

Repeat the above steps for the *Nth* EMD decomposition, calculate the IMF component and the remainder, and obtain the final IMF component c_j and the residual. The formula can be expressed as [34]:

$$c_j = \sum_{i=1}^{N} \frac{c_{ij}}{N} \tag{6}$$

$$r = \sum_{i=1}^{N} \frac{r_i}{N} \tag{7}$$

The final result of the EEMD decomposition is [34]:

$$x(t) = \sum_{j=1}^{n} c_j + r \tag{8}$$

2.3. Ensemble Empirical Mode Decomposition Based on Independent Component Analysis

The formation of ice cover is affected by climatic factors, such as wind direction, wind speed, temperature, humidity, etc. Therefore, collected ice thickness data is non-linear and non-steady, which will be the main factor affecting forecasting results. The effective denoising and decomposition of the ice thickness samples can reduce the non-stationarity of the signals. Independent component analysis belongs to the neural network category. The feature extraction method is not affected by strong background noise or strong interference signals. Fast independent component analysis (FICA) is applied to the signal decomposition to weaken the noise to signal interference, thereby the accuracy of signal decomposition and decomposition efficiency is improved. Ensemble empirical mode decomposition (EEMD) has a great improvement on the traditional empirical mode decomposition method [34]. It effectively solves the problem of mode superposition in traditional empirical mode, which makes the real signal get the maximum reservation. Combining the advantages of the two methods, the paper presents a new feature extraction method based on fast independent component analysis and empirical mode decomposition (FICA-EEMD). Firstly, the signals are separated into statistical independent components by FICA, and then the autocorrelation components of these statistical independent components are analyzed to eliminate the influence of environmental noise [35]. Next, the statistical independent components after denoised are decomposed by EEMD. The same frequency eigenmode function of each statistical independent component is cumulative reconstruction. Finally, extract the intrinsic mode function of ice thickness and constitute a new set of IMFs for the following predictions.

3. Improved Support Vector Machine Prediction Model

3.1. Adaptive Support Vector Machine (ASVM)

The basic idea of SVM is to use the non-linear mapping algorithm to convert the linear undecomposed samples in the low-dimensional space into the high-dimensional feature space, and then it can be divided into linear samples and analyzed by a linear method. SVM regression is used to build a non-linear mapping. The data will be mapped into a high dimensional feature space and a linear regression will be used for analysis [22–24]. Traditional modeling process for support vector machine is seen in [22–24].

An adaptive support vector machine (ASVM) model is proposed in this paper to improve the accuracy of the support vector machine prediction. The selection of the kernel function has a great impact on the veracity of prediction results of the model. By analyzing the complexity of the sample sequence, the kernel functions and parameters of different components are selected to improve the authenticity of the model. The SVM prediction model is established to obtain the final predictive value through component superposition.

3.2. Hybrid Optimization Algorithm of Genetic Algorithm and Tube Search

Genetic algorithm (GA) sets all individuals in the group as objects, and completes the adaptive search of the optimal solution of the problem through biological genetic and evolutionary selection, crossover and mutation mechanism simulation, complete the problem of the optimal solution of the adaptive search. It has the ability of parallel search, and can search for the optimal solution to solve the problem of multi-point departure from a solution space, which can preserve the historical information to a certain extent. It is applicable for the global optimization problem of large-scale arbitrary objective functions. However, GA also has shortcomings of poor local search ability and precocious termination phenomenon. If the algorithm has a small mutation probability, the introduction of new chromosomes is rare. On the contrary, the traditional mutation operator will lead to greater algorithmic randomness and make the search process too blind.

Similarly, due to its flexible memory function and contempt rules, the tube search (TS) algorithm can accept inferior solutions and a strong climbing capabilities in the process of searching. It can jump out of a local optimal solution to search the solution space in other fields, which increase the possibility of obtaining a better global optimal solution. However, the TS algorithm has over-reliance on the initial solution. The convergence rate of the algorithm will be affected by the poor initial solution, and the probability of obtaining the global optimal solution can be reduced due to move from one solution to another during the iterative search.

To improve the computing efficiency, the two algorithms are combined. First, the genetic algorithm is used for global search, which results in individual distribution in the solution space in most regions. Then, the TS search is performed for each individual to improve the quality of the group. The hybrid algorithm effectively combines the parallel search capability of the GA and the local search capability of the TS algorithm. Through the combination of the optimization algorithm, horizontal and vertical can combine to achieve global search. The specific calculation process is shown in Section 3.3. The basic theories and rules of the TS algorithm are found in [36], and the specific GA optimization process is described in [37,38]. We will not repeat those narrations in this paper.

3.3. Modeling Process of the Adaptive Support Vector Machine Model Optimized by Genetic Tabu Search (GATS-ASVM)

Combining the characteristics of different intelligent algorithms, this paper optimizes the parameters of SVM based on the advantages of GA and tabu search (TS). Due to many uncertain natural climatic factors, the data of ice sheet thickness has great volatility and instability. Before forecasting, the original signal was decomposed by ensemble empirical mode decomposition, and the IMF components from high frequency to low frequency are obtained. According to the complexity of each component, different kernel functions of the support vector machine are selected for prediction. To improve the accuracy of the empirical mode decomposition, fast independent component analysis is adopted to perform the data preprocessing. As shown in Figure 2, the concrete forecasting steps include:

Step 1: Data preprocessing by FICA. Collect the real-time ice sheet sequences and use the FICA method based on the negative entropy to denoise the original signals. For $y(t) = Wx(t)$, x is the original signal collected, and W is the solution mixing matrix, which can separate the independent components y in turn. For the multi-independent component decomposed, the useful signal and the noise signal are identified according to the prior knowledge of the signal time and frequency domain. Set the noise signal channels in the independent component to zero and reverse the original signal by the equation of $\hat{x}(t) = W^{-1}y(t)$, where $\hat{x}(t)$ is the signal after denoising.

Step 2: Data decomposition by EEMD. Use the EEMD to decompose the ice cover sequence $\hat{x}(t)$ and obtain the IMF components $c_i(t)$ and the remainder r_n. Add the white noise sequence, which obeys normal distribution of $(0, (\alpha\varepsilon)^2)$. Extract extreme points of the sequence, fit envelope and calculate its mean curve $m(t)$. After $x(t)$ minus $m(t)$, get a new signal $h(t)$. Then take the next step until $h(t)$ satisfies the IMF condition after screening k times. The IMF component $h(t)$ is separated from the original signal, and the residual component $r(t)(r(t) = x(t) - h(t))$ is obtained. The residual component $r(t)$ is used as the new raw data. Repeat the above steps and get the rest of the IMF component and 1 residual component.

Step 3: Normalization and initialization. Normalize the component data and initialize the evolutionary algebra $l = 0$.

Step 4: Genetic tabu operation. Select the individuals based on the selection probability and selection mechanism. Cross-operation based on cross-probability and crossover operator. Perform the genetic tabu mutation operation according to the variation probability and tabu mutation operator.

Step 5: Convergence condition. The new individuals obtained by Step 4 are set to a new generation, the convergence conditions are determined: if the evolution algebra is less than the maximum number of iterations, then let $l = l + 1$ and go back to Step 4; otherwise, terminate the network training, select the optimal individual, and continue to Step 6.

Step 6: Component prediction. Contract the SVM regression model for the IMF components $c_i(t)$ and r_n, select the best parameters and kernel functions, input the forecasting samples and the predicted values of each sequence are obtained.

Step 7: Final prediction. Superimpose each component forecast value to obtain the prediction value of the ice thickness.

In Figure 3, the graphics in the middle show the overall flow, the graph on the left shows the flow of ensemble empirical mode decomposition (EEMD) and the graph on the right shows optimization of support vector machines based on Genetic Tabu Search.

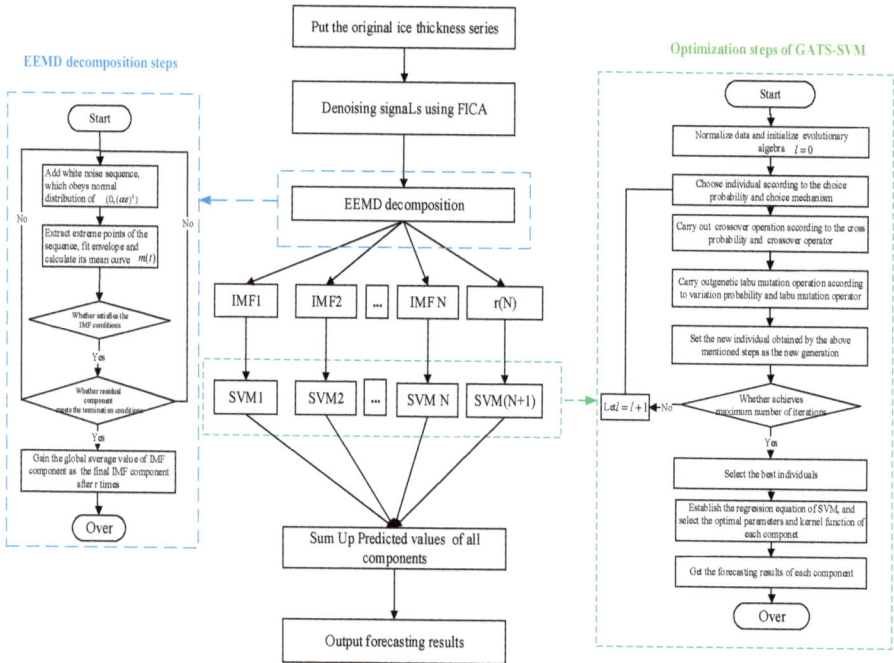

Figure 3. Model structure diagram for ice cover thickness forecasting.

4. Case Study and Results Analysis

In this paper, a 220 kV DC transmission line in Hunan Province is used to verify the proposed model. The transmission line is an important transmission channel in Hunan with a total length of more than 80 km. Line monitoring data include leakage currents, ambient temperature, humidity, wind direction, rainfall and wind speed. Due to the influence of many factors, the time series of ice thickness of transmission line icing is disordered. According to the historical data, the prediction accuracy of the original data is determined, and other forecasting techniques are introduced for comparison.

4.1. Data Preparation

Among the numerous factors influencing the power grid icing and meteorological factors is the most important factor, such as temperature, humidity, wind speed, wind direction and other external climate. A lot of scholars have studied the impact factors preliminary. Reference [39] summarizes the three necessary conditions for the formation of icing by reviewing the results of research of predecessors: Air relative humidity must be above 85%; wind speed should be greater than 1 m/s; the temperature has to reach 0 °C and below.

Analysis of the historical data shows that the most severe seasonal ice cover occurs in December, January, February, and March. On this basis, the main factors considered here are the temperature, relative humidity, and wind speed. Consequently, we choose the typical data for analysis. Per hour icing thickness and meteorological factors from 1 February 2014 to 19 March 2014 for the 220 kV transmission line are selected for the sample data. The raw data trend chart of ice cover thickness, temperature, relative humidity and wind speed are shown in Figure 4.

It can be seen from Figure 4 that the regularity of the ice cover thickness and the influencing factors are weak. There are more noises in the primitive sequence and the trend of the overall data is not strong. In general, the relative humidity fluctuates at 80%, the wind speed is greater than 0, while the temperature is lower than 0 degrees. On this basis, with the decrease of temperature, the increase of wind speed and relative humidity, the ice cover thickness shows an increasing trend. There is a strong correlation between the ice cover thickness and the selected three factors.

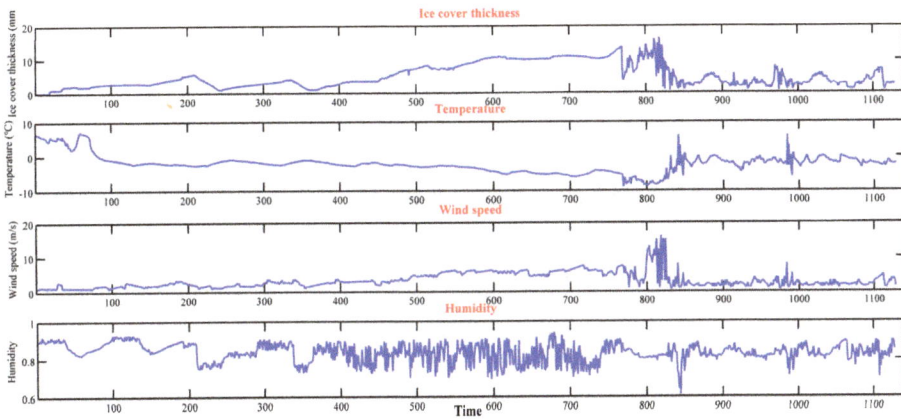

Figure 4. Original data of ice cover thickness, temperature, wind speed and relative humidity from 1 February to 19 March 2014.

The figure above shows that the original sequence of ice thickness is out of order and has some fluctuations. The reprocessing of the original data is needed. Thus, before training samples, we must screen the raw data, adjust the missing data, and remove obvious errors. Finally, we retain 1128 sample points, of which the first 1000 sample points are selected as the training set and the remaining 128 sample points are used as the testing set to prove the validity of the model.

The forecasting model is solved through Matlab on a single core of a 32-bit Lenovo workstation running Windows 7 with 2 dual-core 2.60 GHz CPUs and 4.0 GB of RAM. We extract rules from the past information to forecast the ice cover thickness.

Due to the length limitations of the manuscript, the denoising and decomposition process of the influencing factors in Sections 4.2 and 4.3 are not shown below. We just take ice cover thickness data as an example to show.

4.2. Data Preprocessing and Decomposition

Because the original data have large noise due to the abnormal points, the fast independent component analysis method is used to process with the original data to improve the model's recognition before the decomposition of the EEMD model. Ice thickness data after denoising are decomposed into different sub-time series by the EEMD model, and the IMF component of a series of different scales is generated to achieve the stability sequences of the ice cover.

The EEMD method needs to determine the appropriate value of the strength parameters of noise α and decomposition number n. Reference [40] proved that we can gain greater results when $n = 100$ and the α values range from 0.1 to 0.3. Thus, this paper selects $n = 100$ and $\alpha = 0.2$. EEMD decomposition of the time series of ice cover thickness is obtained by seven intrinsic mode components (from IMF1 to MF7) and a residual sequence. The results are shown in Figure 5. The EEMD decomposition process selects the intrinsic volatility components in the original sequence from high frequency to low frequency. The high-frequency components of high-frequency fluctuations are frequent and chaotic. In some places, the amplitude changes dramatically. The low-frequency components of the periodic law are relatively clear and have strong fluctuation.

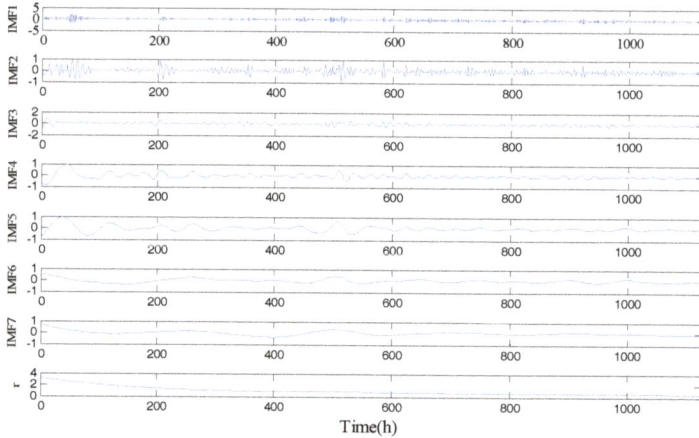

Figure 5. Fast Independent Component Analysis and Ensemble Empirical Mode Decomposition (FICA-EEMD) of the ice thickness.

4.3. Single Prediction

4.3.1. Selection of the Kernel Function

According to the characteristics of each component, the optimal model parameters and kernel functions are selected to establish the SVM model optimized by the genetic tabu search. The selection of the kernel function and its parameters are chosen to improve the accuracy of the sub-models by analyzing the complexity of the sample sequence, as seen in Table 2.

Table 2. Complexity analysis of each component.

Mode	PE Values	Complexity
IMF1	0.99	High level
IMF2	0.782	High level
IMF3	0.573	High level
IMF4	0.441	Low level
IMF5	0.348	Low level
IMF6	0.254	Low level
IMF7	0.219	Low level
Residual	0.155	Low level

To capture the complexity features, the Permutation Entropy (PE) measurement is proposed [41]. It is a method to measure the complexity of time series. The value of permutation entropy represents the stochastic degree of time series. Generally speaking, the smaller the value, the more regular the time series. Otherwise, the time series is of more randomness. From IMF1 to the residue, the PE values gradually decrease from 1.0 to 0.1. Assuming that the threshold is 0.5, the PE values of IMF1 to IMF3 are all above the threshold value, indicating that the three modes have comparatively high-level complexity. In contrast, IMF4 to IMF7 and the residue have relatively low-level complexity because their PE values are all below the threshold. The radial basis kernel function (RBF) with better generalization capability and better processing nonlinear sequence is used for the intrinsic mode components IMF1, IMF2, and IMF3, which have large fluctuation frequency and high complexity; The polynomial kernel function is selected for IMF4, IMF5, IMF6 and IMF7, which have medium- and low-frequency components with the stable change. The residual component is predicted by the linear kernel function [42].

4.3.2. Single Prediction Results

The combined intelligent optimization algorithm of GA and TS is adopted to optimize the parameters g and C to determine an optimal parameter to improve the prediction accuracy of the model. The root mean square error ($RMSE$), mean absolute error (MAE) and mean absolute percentage error ($MAPE$) are set for the prediction effect evaluation of each model. The predictive value and the actual value of the model are, respectively, $\hat{y}_i$ and y_i.

The corresponding formulas are as follows:

$$MAPE = \frac{1}{n}\sum_{i=1}^{n}\left|\frac{\hat{y}_i - y_i}{y_i}\right| \tag{9}$$

$$RMSE = \sqrt{\frac{1}{n}\sum_{i=1}^{n}(\hat{y}_i - y_i)^2} \tag{10}$$

$$MAE = \frac{1}{n}\sum_{i=1}^{n}|\hat{y}_i - y_i| \tag{11}$$

where n represents the sample number.

Set the corresponding components of temperature, humidity and wind speed as input factors, corresponding components of ice thickness sequence as output factors, and put them into the SVM model, we can get the prediction results of each sub-sequence for the testing data as shown in Figure 6. The results of estimated errors in the predictions are given in Figure 7.

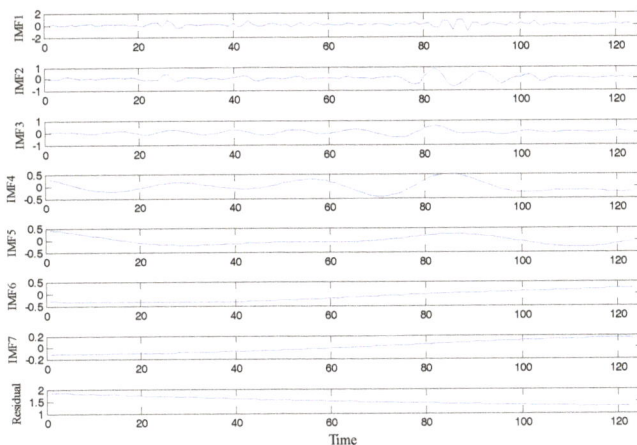

Figure 6. Prediction results of each component for the testing data.

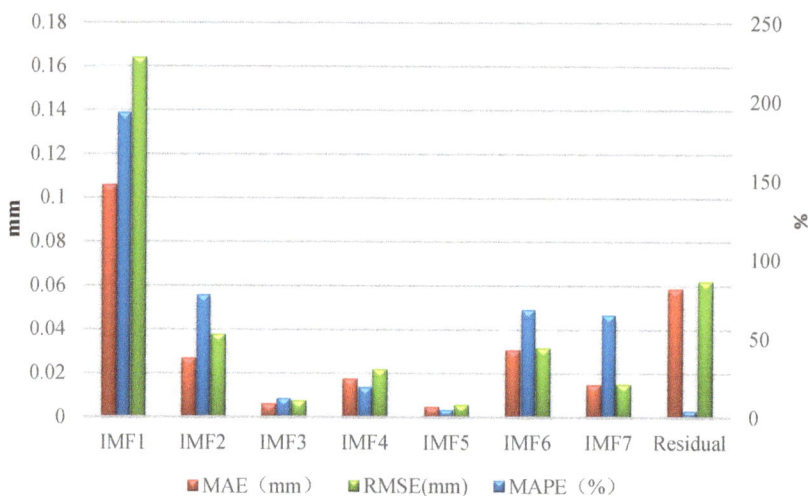

Figure 7. Prediction error of each component for the testing data.

The instantaneous frequency of each IMF component has significant importance at any point. Different IMFs have different meanings, for instance, the low frequency band IMF_n represents the trends of the data, whereas the highest frequency band IMF1 mainly contains noises. Figure 8 shows that the prediction error of IMF_1, which contains the signal of noises, is relatively large. The value of $MAPE$ is close to 200%, and the value of $RMSE$ and MAE is much higher than those of other components. Thus, to improve the overall prediction accuracy and to retain the most important information, we remove the interference term IMF1 to achieve the purpose of the second-time denoising. The prediction results from IMF_2 to IMF_7 and the residual can be added to form the final prediction results.

4.4. Overall Forecasting Results and Error Analysis

Based on the above analysis, the predicted values of the components are superimposed to obtain the predicted values of the thickness of ice cover. The average percentage error $MAPE$ of the forecast value and the actual value are taken as the objective function to search the minimum value of the objective function as the target. The global minimum corresponding to the best fitness function value is the kernel parameter g and penalty factor C of the SVM model when the iteration terminates. The optimized parameters are used in the SVM prediction model to predict the ice thickness.

To investigate the performance of the intelligent model, four algorithms (non-preprocessing adaptive support vector machine optimized by genetic algorithm and tabu search (GATS-ASVM), support vector machine optimized by genetic algorithm (GA-SVM), and standard support vector machine (SVM)) are established for comparison to evaluate the effect of the intelligent model in ice cover thickness prediction. The prediction results of the comparison algorithms are shown in Figure 8.

Figures 8–10 show the results of the four prediction models for forecasting of the thickness of ice cover and the actual measurement of the ice cover thickness. In general, the overall forecasting trends of the four models are close to the real value. The GATS-ASVM algorithm is the closest to the real curve, whereas the standard SVM prediction curves have some deviation.

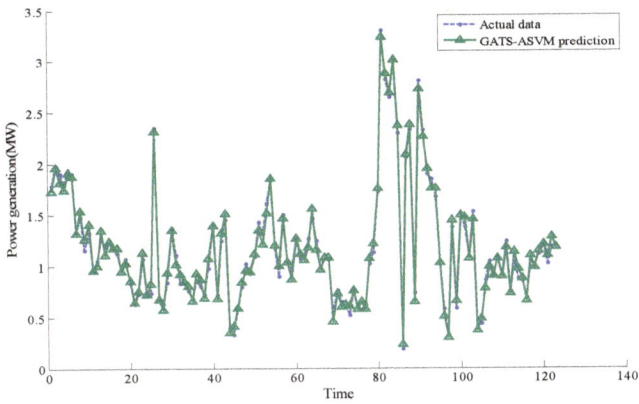

Figure 8. Curves of the proposed model forecasting results and the actual ice thickness.

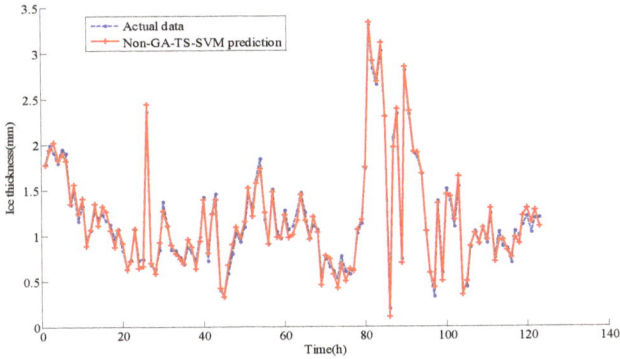

Figure 9. Curves of proposed model forecasting results without data processing. Note: "non- adaptive support vector machine optimized by genetic algorithm and tabu search (GATS-ASVM) prediction" means prediction results of the proposed method without data processing (including fast independent component analysis (FICA) and and ensemble empirical mode decomposition (EEDM)).

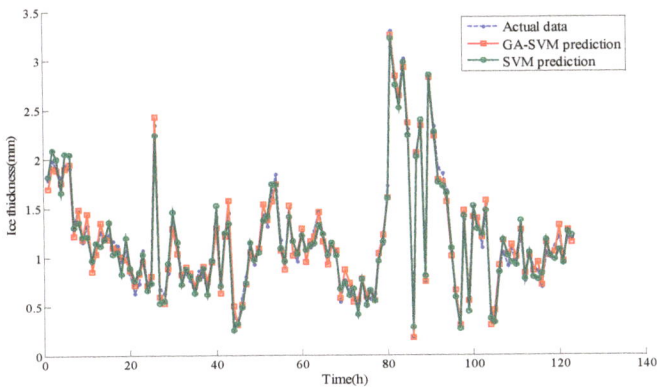

Figure 10. Curves of the comparison model forecasting results.

Various error metrics between the real and forecasting data have been defined to assess the forecasting performance. In our experiments, *MAPE* and *RMSE* are introduced to appraise and compare the different simulation results. Table 3 and Figure 11 show the error distribution of the different models.

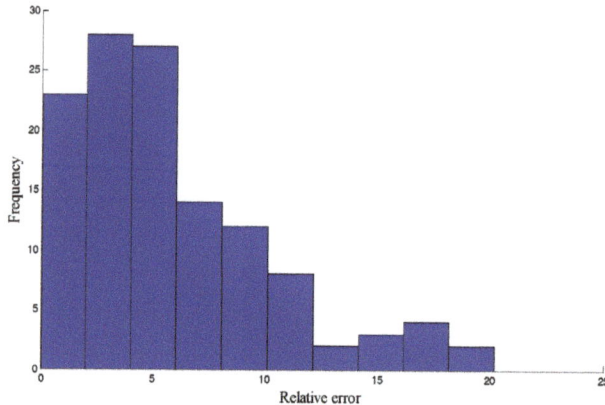

Figure 11. Frequency histogram of the error distribution for the GATS-ASVM algorithm.

Table 3. Error comparison of different algorithms.

Algorithm	MAPE (%)	RMSE (mm)
GATS-ASVM	5.22	1.81
Non-GATS-SVM	6.12	2.5
GA-SVM	6.47	3.51
SVM	7.78	4.68

From Table 3 and Figures 8–11, the following conclusions can be drawn:

(1) The results in Figures 8 and 9 and Table 3 indicate that the model proposed in this paper with the processed data has significant advantages compared to the model with untreated data. Both the error metrics of MAPE and RMSE of the GATS-ASVM with data processing are less than the GATS-ASVM with untreated data because the EEMD method can separate different periods of the fluctuation signal from the original data to make the sub sequence data relatively orderly for higher forecast accuracy. At the same time, FICA denoising is necessary to make the processed data smoother.

(2) Comparing the GATS-ASVM model and GA-SVM model, the former has more advantages. The hybrid optimization algorithm makes up for the defects of the single algorithm, and the adaptive support vector machine takes into account the morphological characteristics of different sub-sequences and selects appropriate kernel functions, which improve the accuracy and the generalization ability of the model and result in better predictions.

(3) From Figures 8–11 and Table 3, the forecasting precision of the combined model is higher than that of the single model. The combined forecasting model uses the complementary advantages of different algorithms to improve the accuracy of the algorithm. The prediction of the single model has large limitations.

(4) Figure 11 and Table 3 show less error in the proposed model. This model has a powerful function in the processing of non-stationary series. Figure 11 shows that the prediction error is controlled within 20%. Most of the errors are distributed between [0, 10%], and the error values greater than 10% are less common. The algorithm has a stable forecasting effect. The error evaluation results

Energies **2017**, *10*, 1862

and the frequency histogram of the error distribution show that the adaptive support vector machine optimized by genetic tabu search based on ensemble empirical mode decomposition has higher prediction accuracy than other algorithms. Compared with the other three methods, the proposed GATS-ASVM method has obvious advantages and can be used for the prediction of the thickness of ice on transmission lines.

5. Conclusions

In this paper, an intelligent model is proposed to predict the thickness of ice cover on a transmission line. The model is useful for the power grid enterprise to effectively control the ice cover. It has great significance for the safe operation of power networks. Through the analysis and summary, this paper reaches the following conclusions:

Firstly, the raw data are processed by fast independent component analysis to remove the abnormal data in the extreme cases and gets great denoising results. The ensemble empirical mode decomposition decomposes data after denoising into different time frequencies from high frequency to low frequency so that the complex signal is decomposed into a finite intrinsic mode function.

Secondly, SVM model optimized by the genetic tabu search algorithm is presented to forecast each sub-sequence function after decomposition. The final icing prediction results are obtained by summing up the total sub-sequence forecasting results. Empirical analysis proves that the model proposed in this paper has strong robustness and generalization ability and can effectively overcome the shortcomings of the premature convergence of the genetic algorithm and the initial value dependence of tabu search. The SVM model is more stable, and the prediction accuracy is improved after optimization of the combined method.

To further improve the accuracy of the prediction, for different sub-sequence functions, the appropriate kernel functions are adaptively selected according to the complexity of the sample sequence to make the model more reliable. After comprehensive analysis of the situation of the transmission line icing on the 220 kV DC transmission line in Hunan Province, a numerical example is given to calculate the cover thickness of the ice cover in this transmission line from 1 February 2014 to 19 March 2014. The simular results of several different combinational algorithms demonstrate that the GATS-ASVM model is ideal for short-term ice cover thickness forecasting for grid transmission lines because it achieved better prediction accuracy.

Acknowledgments: This work is supported by the Natural Science Foundation of China (Project No. 71471059). It was also funded by China Postdoctoral Science Foundation (The sixty-second batch of aid on the surface).

Author Contributions: In this research activity, all the authors were involved in the data analysis and preprocessing phase, simulation, results analysis and discussion, and manuscript preparation. Xiaomin Xu designed this research, wrote and revised this paper; Dongxiao Niu guided the whole idea and framework of the paper and provided a lot of t revised opinions for the paper; Lihui Zhang and Yongli Wang provided professional guidance; Keke Wang collected all the data and revised this paper. All authors have approved the submitted manuscript.

Conflicts of Interest: The authors declare no conflict of interest.

References

1. Li, X.J. Research on Icing Forecasting Model of Transmission Line Based on Data Mining. Master's Thesis, Taiyuan University of Technology, Taiyuan, China, 2016.
2. Li, P.; Zhao, N.; Zhou, D.H.; Cao, M.; Li, J.J.; Shi, X.L. Multivariable Time Series Prediction for the Icing Process on Overhead Power Transmission Line. *Sci. World J.* **2014**, *2014*, 256815. [CrossRef] [PubMed]
3. Li, C.G.; Lv, Y.Z.; Cui, X. The problem of safe operation of power grid in China under conditions of ice and snow disasters. *Power Syst. Technol.* **2008**, *32*, 14–21.
4. Shi, L.L. Assessment of Forest Damage caused by Ice Storm based on MODIS Data—A Case Study of Jiangxi Province, China. *Disaster Adv.* **2013**, *6*, 67–72.
5. Chang, H. Prediction and Experimental Study on Ice Thickness of Overhead Transmission Line Based on Dynamic Tension and Angle. Master's Thesis, Chongqing University, Chongqing, China, 2013.

6.   Liu, H.Z. *DC Transmission Line Icing and Control*; China Electric Power Press: Beijing, China, 2012.

7.   Tian, L. Study on Icing Regularity and Optimal Configuration of Ice Handling Equipment in Hunan UHV Transmission Line. Master's Thesis, Changsha University of Science & Technology, Changsha, China, 2013.

8.   Li, Z.H.; Bai, X.M.; Zhou, Z.G.; Hu, Z.J.; Xu, J.; Li, X.J. The research progress of the method for the icing prevention and control of the power grid. *Power Syst. Technol.* **2008**, *32*, 7–14.

9.   Li, P.; Li, N.; Cao, M.M. Micro-meteorology Features Extraction and Status Assessment for Transmission Line Icing Based on Intelligent Algorithms. *J. Inf. Comput. Sci.* **2010**, *7*, 2043–2052.

10.   Dai, D.; Huang, X.T. Prediction of transmission line icing based on support vector machine model. *High Volt. Technol.* **2013**, *39*, 2822–2828.

11.   Jiang, X.L.; Han, X.B.; Hu, Y.Y.; Yang, Z.Y. The study of Dynamic Wet-Growth Icing model of insulator. *Proc. CSEE* **2017**, *09*, 1–8.

12.   Xu, Q.S.; Lao, J.M.; Hou, W.; Wang, M.L. Real-time monitoring and calculation model of transmission lines unequal ice-coating. *High Volt. Eng.* **2009**, *35*, 2865–2869.

13.   Liu, H.Y.; Zhou, D.; Fu, J.P.; Huang, S.Y. A simple model for predicting glaze loads on wires. *Proc. CSEE* **2001**, *21*, 44–48.

14.   Huang, X.B.; Wang, Y.X.; Zhu, Y.C.; Zheng, X.X.; Li, H.B.; Wang, Y.G. Line icing prediction based on genetic algorithm and fuzzy logic fusion. *High Volt. Eng.* **2016**, *42*, 1228–1235.

15.   Zheng, Z.H.; Liu, J.S. Prediction method of ice thickness on transmission lines based on the combination of GA and BP neural network. *Power Syst. Clean Energy* **2014**, *30*, 27–30.

16.   Liu, R.; Wu, X.D.; Yan, E.M.; Lang, L. Prominent influence factor analysis and RBF ice cover prediction of transmission lines. *Electr. Appl.* **2013**, *32*, 72–75.

17.   Liu, J.; Li, A.J.; Zhao, L.P. Prediction model based on fuzzy and T-S neural network for ice thickness. *Hunan Electr. Power* **2012**, *32*, 1–4.

18.   Wu, Q.D.; Yan, B. Displacement Prediction of Tunnel Surrounding Rock: A Comparison of Support Vector Machine and Artificial Neural Network. *Math. Probl. Eng.* **2014**, *2014*, 351496. [CrossRef]

19.   Ma, T.N.; Niu, D.X.; Fu, M. Icing Forecasting for Power Transmission Lines Based on a Wavelet Support Vector Machine Optimized by a Quantum Fireworks Algorithm. *Appl. Sci.* **2016**, *6*, 54. [CrossRef]

20.   Yin, Z.R.; Su, X.L. Icing thickness forecasting of transmission line based on particle swarm algorithm to optimize SVM. *J. Electr. Power* **2014**, *29*, 6–9.

21.   Zhang, C.; Wei, H.K.; Zhao, J.S.; Liu, T.H.; Zhu, T.T.; Zhang, K.J. Short-term wind speed forecasting using empirical mode decomposition and feature selection. *Renew. Energy* **2016**, *96*, 727–737. [CrossRef]

22.   Wei, T.X.; Ma, G.W.; Huang, W.B. Prediction of runoff based on penalized weighted support vector machine regression model. *J. Hydroelectr. Eng.* **2012**, *31*, 35–38.

23.   Liang, J.J.; Wu, D. Smooth Diagonal Weighted Newton Support Vector Machine. *Math. Probl. Eng.* **2013**, *2013*, 349120. [CrossRef]

24.   Xu, X.M.; Niu, D.X.; Wang, P.; Lu, Y.; Xia, H.C. The weighted support vector machine based on hybrid swarm intelligence optimization for icing prediction of transmission line. *Math. Probl. Eng.* **2015**, *2015*, 798325. [CrossRef]

25.   Khan, A.; Jaffar, M.A. Genetic algorithm and Self Organizing map based fuzzy hybrid intelligent method for color image segmentation. *Appl. Soft Comput.* **2015**, *32*, 300–310. [CrossRef]

26.   Gharehbaghi, S.; Khatibinia, M. Optimal seismic design of reinforced concrete structures under time-history earthquake loads using an intelligent hybrid algorithm. *Earthq. Eng. Eng. Vib.* **2015**, *14*, 97–109. [CrossRef]

27.   Ye, Q.; Wang, M.; Han, J.R. Integrated risk governance in the Yungui Plateau, China: The 2008 ice-snow storm disaster. *J. Alp. Res. Revue Geogr. Alp.* **2012**, *100*, 100–112. [CrossRef]

28.   Sardouie, S.H.; Albera, L.; Shamsollahi, M.B.; Merlet, I. An Efficient Jacobi-Like Deflationary ICA Algorithm: Application to EEG Denoising. *IEEE Signal Process. Lett.* **2015**, *22*, 1198–1202. [CrossRef]

29.   Yao, J.C.; Xiang, Y.; Qian, S.C.; Wang, S.; Wu, S.W. Noise source identification of diesel engine based on variational mode decomposition and robust independent component analysis. *Appl. Acoust.* **2017**, *116*, 184–194. [CrossRef]

30.   Xu, W.L.; Sun, T.; Hu, T.; Hu, T.; Liu, M.H. Huanghua Pear Soluble Solids Contents Vis/NIR Spectroscopy by Analysis of Variables Optimization and FICA. *Spectrosc. Spectr. Anal.* **2014**, *34*, 3253–3256.

Energies **2017**, *10*, 1862

31. Imaouchen, Y.; Kedadouche, M.; Alkama, R.; Thomas, M. A Frequency-Weighted Energy Operator and complementary ensemble empirical mode decomposition for bearing fault detection. *Mech. Syst. Signal Process.* **2017**, *82*, 103–116. [CrossRef]

32. Zeng, P.; Liu, H.X.; Ning, X.B.; Zhuang, J.J.; Zhang, X.G. Study on the energy distribution of the empirical mode decomposition for ECG energy distribution. *Acta Phys. Sin.* **2015**, *64*, 1–8.

33. Wang, H.; Hu, Z.J.; Chen, Z.; Ji, M.L.; He, J.B.; Li, C. A Hybrid Model for Wind Power Forecasting Based on Ensemble Empirical Mode Decomposition and Wavelet Neural Networks. *Trans. China Electrotech. Soc.* **2013**, *28*, 137–144.

34. Fan, L.J. The WPD-EEMD Method and Its Applied Research in the EEG Signals Processing. Master's Thesis, Taiyuan University of Technology, Taiyuan, China, 2010.

35. Chen, J. Study of Denoising Method Based on Wavelet Transform and Independent Component Analysis. Master's Thesis, Fudan University, Shanghai, China, 2010.

36. Qiu, H.; Ming, W.; Zhang, Z.Z. A tabu search algorithm for the multi-period inspector scheduling problem. *Comput. Oper. Res.* **2015**, *59*, 78–93.

37. Bukharon, O.E.; Bogolyubov, D.P. Development of a decision support system based on neural networks and a genetic algorithm. *Expert Syst. Appl.* **2015**, *42*, 6177–6183. [CrossRef]

38. Gan, L.K.; Shek, J.K.H.; Mueller, M.A. Optimised operation of an off-grid hybrid wind-diesel-battery system using genetic algorithm. *Energy Convers. Manag.* **2016**, *126*, 446–462. [CrossRef]

39. Lu, J.Z.; Jiang, Z.L.; Lei, H.C. The Hunan power grid in 2008 ice disaster accident analysis of power systems. *Autom. Electr. Power Syst.* **2008**, *32*, 16–19.

40. Wu, Z.; Huang, N.E. Ensemble empirical mode decomposition: A noise-assisted data analysis method. *Adv. Adapt. Data Anal.* **2009**, *1*, 1–41. [CrossRef]

41. Yu, L.; Wang, Z.S.; Tang, L. A decomposition–ensemble model with data-characteristic-driven reconstruction for crude oil price forecasting. *Appl. Energy* **2015**, *156*, 251–267. [CrossRef]

42. Lin, Y.; Liu, P. Combined Model Based on EMD-SVM for Short-term Ice thickness Prediction. *Proc. CSEE* **2011**, *31*, 102–107.

energies

MDPI

Article

Hybrid Chaotic Quantum Bat Algorithm with SVR in Electric Load Forecasting

Ming-Wei Li [1], Jing Geng [1], Shumei Wang [2] and Wei-Chiang Hong [2,*]

[1] College of Shipbuilding Engineering, Harbin Engineering University, Harbin 150001, Heilongjiang, China; limingwei@hrbeu.edu.cn (M.-W.L.); gengjing@hrbeu.edu.cn (J.G.)

[2] School of Education Intelligent Technology, Jiangsu Normal University/101, Shanghai Rd., Tongshan District, Xuzhou 221116, Jiangsu, China; plum8@163.com

* Correspondence: samuelsonhong@gmail.com, Tel.: +86-516-8350-0307

Received: 9 December 2017; Accepted: 19 December 2017; Published: 19 December 2017

Abstract: Hybridizing evolutionary algorithms with a support vector regression (SVR) model to conduct the electric load forecasting has demonstrated the superiorities in forecasting accuracy improvements. The recently proposed bat algorithm (BA), compared with classical GA and PSO algorithm, has greater potential in forecasting accuracy improvements. However, the original BA still suffers from the embedded drawbacks, including trapping in local optima and premature convergence. Hence, to continue exploring possible improvements of the original BA and to receive more appropriate parameters of an SVR model, this paper applies quantum computing mechanism to empower each bat to possess quantum behavior, then, employs the chaotic mapping function to execute the global chaotic disturbance process, to enlarge bat's search space and to make the bat jump out from the local optima when population is over accumulation. This paper presents a novel load forecasting approach, namely SVRCQBA model, by hybridizing the SVR model with the quantum computing mechanism, chaotic mapping function, and BA, to receive higher forecasting accuracy. The numerical results demonstrate that the proposed SVRCQBA model is superior to other alternative models in terms of forecasting accuracy.

Keywords: support vector regression; chaos theory; quantum behavior; bat algorithm (BA); load forecasting

---

1. Introduction

Electric load forecasting plays an essential role in making optimal action plans for decision makers, such as load unit commitment, energy transfer scheduling, contingency planning load shedding, energy generation, load dispatch, power system operation security, hydrothermal coordination, and so on [1]. Indicated by Bunn and Farmer [2], an 1% increase in electric load forecasting error may lead to a £10 million additional expenditure in operations. Thus, it is important to look for high accurate forecasting models or to develop novel approaches to receive satisfied load forecasting accuracy, which can help decision makers optimize adjust the electricity price/supply and load plan based on the forecasted results, i.e., improve the electricity system operations more efficient, and reduce system operating risks successfully. Unfortunately, affected by several exogenous factors, such as policy, economic production, industrial activities, weather conditions, population, holidays, etc. [3], the electric load data demonstrate seasonality, non-linearity, volatility, randomness and chaos in nature, which increase the difficulty for electric demand forecasting [4].

In the past few decades, lots of electric load forecasting models have been developed to improve load forecasting accuracy. These forecasting methods include two classical types: traditional statistical models and artificial intelligent models. The traditional statistical models are easily to be applied, which include the ARIMA model [5], Kalman filtering/linear quadratic estimation model [6], exponential smoothing model [7], regression model [8], Bayesian estimation model [9], and other time series technologies [10]. However, most of the traditional statistical models are theoretically to deal with the linear relationships among electric loads and other factors; these methods are difficult to well handle the characteristics of non-linearity, volatility, and randomness among historical electric loads and exogenous factors. Thus, they cannot easily receive satisfied electric load forecasting accuracy.

Due to the strong nonlinear fitting ability, various artificial intelligence (AI) based methods have been applied to forecast electric load, to improve the accuracy of load forecasting models since 1980s, such as artificial neural networks (ANNs) [11], expert system-based model [12], and fuzzy inference methodology [13]. To further improve the forecasting performance, these AI methods have been hybridized or combined with each other to obtain new novel forecasting approaches or frameworks, for example, RBF neural network combined with adaptive network-based fuzzy inference system [14], multi-layer perceptron artificial neural network hybridized with knowledge-based feedback tuning fuzzy system (MLPANN) [15], the Bayesian neural network with the hybrid Monte Carlo algorithm [16], fuzzy behavior neural network (WFNN) [17], hybrid artificial bee colony algorithm hybridized with extreme learning machine [18], the random fuzzy variables with ANNs [19], and so on. However, these AI-based approaches still suffer from some embedded drawbacks. The defects of these models include difficulty to set the structural parameters of network [20], time consuming to extract functional approximation, and easily to trapped in local optimal value. More systematic analysis about AI-based models used in load forecasting are shown in references [21].

Support vector machine (SVM) is based on the statistical learning theory and kernel computing techniques, the so-called kernel based neural networks, to effectively deal with small sample size problem, non-linear problem, and high dimensional pattern identification problems. Moreover, it could also be applied to well solve other machine learning problems, such as function approximation, probability density estimation, and so on [22,23]. Rather than by implementing the empirical risk minimization (ERM) principle to minimize the training error, which causes the overfitting problem in the ANNs modeling process, SVM employs the structural risk minimization (SRM) principle to minimize an upper bound on the generalization error, and allow learning any training set without error. Thus, SVMs could theoretically guarantee to achieve the global optimum than ANNs models. In addition, while dealing with the nonlinear problem, SVM firstly maps the data into a higher dimensional space, then, it employs the kernel function to replace the complicate inner product in the high dimensional space. In the other words, it can easily avoid too complex computations with high dimensions, i.e., the so-called dimension disaster problem. This enables SVMs to be a feasible choice for solving a variety of problems in lots of fields which are non-linear in nature. For more detailed mechanisms introduction of SVMs, it is referred to Vaplink [22,23] and Scholkopf and Sloma [24], among others. Along with the introduction of Vapnik's ε-insensitive loss function, SVM also has been extended to solve nonlinear regression estimation problems, which are so-called support vector regression (SVR) [25]. Compared with AI methods, SVR model has the embedded characteristics of small sample learning and generalization ability, which can avoid learning, local minimal point and dimension disaster problem effectively. SVR have been successfully employed to solve forecasting problems in many fields, such as solar irradiation forecasting [26], rainfall/flood hydrological forecasting [27–29], industrial wastewater quality forecasting [30], and so on. Meanwhile, SVR model had also been successfully applied to forecast electric load [31,32]. To improve the forecasting accuracy, Hong and his colleagues propose a series of SVR-based forecasting models via hybridizing with different evolutionary algorithms [33–36]. Based on Hong's series research results, well determining parameters of an SVR model is critical to improve the forecasting performance. Henceforth, Hong and his successors have employed chaotic mapping functions (including logistic function and cat mapping function) to enrich diversity of population over the whole space, and also have applied cloud theory to execute the three parameters selection carefully to receive significant improvements in terms of forecasting accuracy.

Bat algorithm [37] is a new swarm intelligent optimization proposed by Yang in 2010. It is originated from the simulation of bat's prey detection and obstacle avoidance by sonar. This algorithm is a simulation technology based on iteration. The population is initialized randomly, then the optimal resolution is searched through iteration, finally the local new resolutions are found around the optimal resolution by random flying, hence, the local search is strengthened. Compared with other algorithms, BA has the advantages of parallelism, quick convergence, distribution and less parameter adjusted. It has been proved that BA is superior to PSO in terms of convergent rate and stability [38]. Nowadays, BA is widely applied in natural science, such as PFSP dispatch problem [39], K-means clustering optimization [40], engineering optimization [41], and multi-objective optimization [42], etc. Comparing with other evolutionary algorithms, such as, PSO and GA, BA has greater improving potential. However, similar to those optimization algorithms which are based on population iterative searching mechanism, standard BA also suffers from slow convergent rate in the later searching period, weak local search ability and premature convergence tendency [41].

On the other hand, quantum computing technique is an important research hotspot in the field of intelligent computing. The principle of qubit and superposition of states in quantum computing is used. The units are represented by qubit coding, and the revolution is updated by quantum gate, which expands its ergodic ability in solution space. Recently, it has received some hot attention that quantum computing concepts could be theoretically hybridized with those evolutionary algorithms to improve their searching performances. Huang [43] proposes an SVR-based forecasting model by hybridizing the quantum computing concepts and the cat mapping function with the PSO algorithm into an SVR model, namely SVRCQPSO forecasting model, and receives satisfied forecasting accurate levels. Lee and Lin [44,45] also hybridize the quantum computing concepts and cat mapping function with tabu search algorithm and genetic algorithm to propose SVRCQTS and SVRCQGA models, respectively, and also receive higher forecasting accuracy. Li et al. [46] also applied quantum non-gate to realize quantum mutation to avoid premature convergence. Their experiments on classical complicated functions also reveal that the improved algorithm could effectively avoid local optimal solutions. However, due to the population diversity decline along with iterative time increasing, the BA and QBA still suffers from the very problem that trapping into local optima and premature convergence.

Considering the core drawback of the BA and QBA, i.e., trapping into local optima, causing unsatisfied forecasting accuracy, this paper would continue to explore the feasibility of hybridizing quantum computing concepts with BA, to overcome the premature problem of BA, eventually, to determine more suitable parameter combination of an SVR model. Therefore, this paper employs quantum computing concepts to empower each bat to expand the search space during the searching processes of BA; in the meanwhile, also applies the chaotic mapping function to execute global perturbation operation to help the bats jump from the local optima when the diversity of the population is poor; then, receive more suitable parameter combination of an SVR model. Finally, a new load forecasting model, via hybridizing cat mapping function, quantum computing concepts and BA with an SVR model, namely SVRCQBA model, is proposed. Furthermore, the forecasting results of SVRCQBA model are used to compare with that of other alternatives proposed by Huang [43] and Lee and Lin [44,45] to test its superiority in terms of forecasting accuracy. The main innovative contribution of this paper is continuing to hybridize the SVR model with the quantum computing mechanism, chaotic mapping theory and evolutionary algorithms, to well explore the load forecasting model with higher accurate levels.

The remainder of this article is organized as follows. The basic formulation of an SVR model, the proposed CQBA and the implementation details of the proposed SVRCQBA model are illustrated in Section 2. Section 3 presents a numerical example and achieves the compared analysis among the proposed model and published alternative models. Finally, Section 4 concludes this paper.

2. Methodology of SVRCQBA Model

2.1. Support Vector Regression (SVR) Model

The brief ideas of SVR are demonstrated. A non-linear mapping function, $\varphi(x)$, is defined to map the input data set, $\{(x_i, y_i)\}_{i=1}^{N}$, into a high dimensional feature space. Then, there theoretically exists a linear function, f, to formulate the non-linear relationships between input data and output data. The linear function, f, is the so-called the SVR function, and is shown as Equation (1),

$$f(\mathbf{x}) = \mathbf{w}^T \varphi(\mathbf{x}) + b \tag{1}$$

where $f(\mathbf{x})$ represents the forecasting values; $\varphi(\mathbf{x})$ is the feature mapping function, non-linearly mapping the input space, $\mathbf{x}$, into the feature space; the coefficients, $\mathbf{w}$ and b, are determined by minimizing the empirical risk, as shown in Equation (2),

$$R_{emp}(f) = \frac{1}{N} \sum_{i=1}^{N} L_\varepsilon(y_i, \mathbf{w}^T \varphi(\mathbf{x}_i) + b) \tag{2}$$

where $L_\varepsilon(\mathbf{y}, f(\mathbf{x}))$ is the ε-insensitive loss function as shown in Equation (3),

$$L_\varepsilon(\mathbf{y}, f(\mathbf{x})) = \begin{cases} |f(\mathbf{x}) - \mathbf{y}| - \varepsilon & if \ |f(\mathbf{x}) - \mathbf{y}| \geq \varepsilon \\ 0 & otherwise \end{cases} \tag{3}$$

In addition, $L_\varepsilon(\mathbf{y}, f(\mathbf{x}))$ is used to look for an optimum hyper plane in the feature space, to maximize the distance separating the training data into two subsets. Thus, the SVR focuses on looking for the optimum hyper plane, and minimizing the training errors between the training data and the ε-insensitive loss function.

Therefore, the SVR modeling problem could be illustrated as minimizing the overall errors, shown in Equation (4),

$$\min_{w,b,\xi^*,\xi} R_\varepsilon(w, \xi^*, \xi) = \frac{1}{2} \mathbf{w}^T \mathbf{w} + C \sum_{i=1}^{N} (\xi_i^* + \xi_i) \tag{4}$$

with the constraints,

$$\mathbf{y}_i - \mathbf{w}^T \varphi(\mathbf{x}_i) - b \leq \varepsilon + \xi_i^*,$$
$$-\mathbf{y}_i - \mathbf{w}^T \varphi(\mathbf{x}_i) - b \leq \varepsilon + \xi_i,$$
$$\xi_i^* \geq 0$$
$$\xi_i \geq 0$$
$$i = 1, 2, \ldots, N$$

The first term of Equation (4), representing the concept that maximizes the distance within two separated training data, is used to penalize large weights, in the meanwhile, to maintain the flatness of $f(\mathbf{x})$. The second term penalizes training errors via the ε-insensitive loss function. C is a parameter to trade off of $f(\mathbf{x})$ and $\mathbf{y}$. Training errors under ε are denoted as ξ_i^*, whereas training errors above ε are denoted as ξ_i.

After solving the quadratic optimization problem with inequality constraints, the parameter vector $\mathbf{w}$ in Equation (1) is computed as Equation (5),

$$\mathbf{w} = \sum_{i=1}^{N} (\alpha_i^* - \alpha_i) \varphi(\mathbf{x}_i) \tag{5}$$

where α_i^*, α_i are computed and named as Lagrangian multipliers. Finally, the SVR regression function is obtained as Equation (6) in the dual space,

$$f(\mathbf{x}) = \sum_{i=1}^{N}(\alpha_i^* - \alpha_i)K(\mathbf{x}_i, \mathbf{x}) + b \qquad (6)$$

where $K(\mathbf{x}_i, \mathbf{x}_j)$ is the so-called kernel function, and its value could be calculated via the inner product of two vectors, $\mathbf{x}_i$ and $\mathbf{x}_j$, in the feature space, $\varphi(\mathbf{x}_i)$ and $\varphi(\mathbf{x}_j)$, respectively, i.e., $K(\mathbf{x}_i, \mathbf{x}_j) = \varphi(\mathbf{x}_i) \cdot \varphi(\mathbf{x}_j)$. Any function that satisfies Mercer's condition [25] could be used as the kernel function.

The most famous kernel functions are the Gaussian RBF with a width of σ, and the polynomial kernel with an order of d and constants a_1 and a_2, as shown in Equations (7) and (8), respectively. If the value of σ is large enough, the RBF kernel function would approximate to the linear kernel (i.e., polynomial with an order of 1). In addition, the Gaussian RBF kernel function is not only easier to implement, but also capable to non-linearly map the data into the higher dimensional space, thus, it is suitable to deal with non-linear problems. Therefore, the Gaussian RBF kernel function (Equation (7)) is used in this paper.

$$K(\mathbf{x}_i, \mathbf{x}_j) = e^{-\frac{||\mathbf{x}_i - \mathbf{x}_j||^2}{2\sigma^2}} \qquad (7)$$

$$K(\mathbf{x}_i, \mathbf{x}_j) = (a_1\mathbf{x}_i\mathbf{x}_j + a_2)^d \qquad (8)$$

The selection of the three parameters, σ, C, and ε of an SVR model influence the accuracy of forecasting. For parameter, ε, it represents the parameter of the ε-insensitive loss function. It controls the width of insensitive area (i.e., low noise of the data set) from data set, thus, it determines the amount of support vectors. If ε is too large, the amount of support vectors would be few, thus, the forecasting model would become relative simple and with low accuracy; on the contrary, if ε is very small, the regression accuracy could be enhanced, however, the forecasting model would become relatively complicate and with low general adoptions.

For parameter, C, it represents the penalty for those data outside the ε-tube. It determines the complexity and stability of the forecasting model. If C is very small, the penalty is mall, i.e., the training errors are large; on the contrary, if C is too large, the learning accuracy would also be enhanced, however, the forecasting model would be with low general adoptions. In addition, the values of C would also affect the fatness of the forecasting model, i.e., the arrangements of outliers. For a suitable C, it could deal with the disturbance of these outliers, and hence, it could guarantee the stability of the forecasting model. Therefore, the suitable parameter determination of C and ε, it could receive more accurate and more stable forecasting model.

For parameter, σ, it not only represents the basic capability of the Gaussian RBF kernel function to deal with nonlinear relationships among data, but also reflects the correlations among support vectors. For example, if σ is very small, the correlation among those support vectors is weak, then, the process of machine learning is relatively complex, i.e., it cannot guarantee to receive general adoptions; on the contrary, if σ is too large, the correlation among those support vectors is too strong to receive sufficient accuracy. Therefore, in the modeling processes, if σ is approximating smaller, it is suggested set a larger value of C.

Based on the above analysis of these three parameters, the complexity and general adoptions of an SVR model are determined by these three parameters and their interactions. Therefore, too look for a novel algorithm to optimize the parameter combination is an important issue to improve the forecasting accuracy of an SVR model.

2.2. Chaotic Quantum Bat Algorithm (CQBA)

2.2.1. Bat Algorithm (BA)

Bats detect preys and avoid obstacles with sonar. According to echolocation in acoustic theory, bats judge preys' size through adjusting phonation frequency. By the variation of echolocation, bats would detect the distance, direction, velocity, size, etc. of objects, which guarantees bats' accurate flying and hunting [47]. While searching for preys, they change the volume, $A(i)$, and emission velocity, $R(i)$, of impulse automatically. During the prey-searching period, the ultrasonic volume that they send out is high, while the emission velocity is relatively low. Once the prey is locked, the impulse volume turns down and emission velocity increases with the distance between bat and prey being shortened.

The bat algorithm is a meta heuristic algorithm for intelligent search. The theory is as followings, (1) Bat's position and velocity are initialized, and are treated as the solution in problem space; (2) The optimal fitness function value of the problem is calculated; (3) The volume and velocity of bat units are adjusted, and are transformed towards optimal unit; (4) The optimal solution is finally received. The bat algorithm involves global search and local search.

In global search, suppose that the search space is with d dimensions, at the time, t, the ith bat has its position, x_i^t, and velocity, v_i^t. At the time, $t + 1$, its position, x_i^{t+1}, and velocity, v_i^{t+1}, are updated as Equations (9) and (10), respectively,

$$x_i^{t+1} = x_i^t + v_i^{t+1} \tag{9}$$

$$v_i^{t+1} = v_i^t + (x_i^t - x_*)F_i \tag{10}$$

where x_* is the current global optimal solution; F_i is the sonic wave frequency, as shown in Equation (11),

$$F_i = F_{min} + (F_{max} - F_{min})\beta \tag{11}$$

where $\beta \in [0, 1]$ is a random number; F_{max} and F_{min} are respectively the sonic wave max frequency and min frequency of the ith bat at this moment. In the process of practice, according to the scope that this problem needs to search, the initialization of each bat is assigned one random frequency following uniform distribution in $[F_{min}, F_{max}]$.

In local search, once a solution is selected in the current global optimal solution, each bat would produce new alternative solution in the mode of random walk according to Equation (12),

$$x_{new}(i) = x_{old} + \lambda A^t \tag{12}$$

where x_{old} is a solution randomly chosen in current optimal disaggregation; A^t is the average of volume in current bat population; λ is a D dimensional vector in $[-1, 1]$.

The bat's velocity and position update steps are similar to that in standard PSO. In PSO, F_i actually dominates the moving range and space of the particle swarm. To a certain degree, BA could be treated as the balance and combination between standard PSO and augmented local search. The balance is dominated by impulse volume, $A(i)$, and impulse emission rate, $R(i)$. When the bat locks the prey, the volume, $A(i)$, is reduced and the emission rate, $R(i)$, is increased. The impulse volume, $A(i)$, and impulse emission rate, $R(i)$, are updated as Equations (13) and (14), respectively,

$$A^{t+1}(i) = \gamma A^t(i) \tag{13}$$

$$R^{t+1}(i) = R^0(i)\left[1 - e^{-\delta t}\right] \tag{14}$$

where, $0 < \gamma < 1$, $\delta > 0$, are both constants. It is obviously that as $t \to \infty$, then, $A^t(i) \to 0$ and $R^t(i) = R^0(i)$. In the practice process, $\gamma = \delta = 0.95$.

2.2.2. Quantum Computing for BA

a.   Quantum Bat Population Initialization

In quantum bat algorithm, the probability amplitude of qubit is applied as the code of bat in current position. Considering the randomness of code in population initialization, the coding program of the bat B^i in this paper is given as Equation (15),

$$B^i = \begin{bmatrix} \cos\theta_1^i & \cos\theta_2^i & \cdots & \cos\theta_j^i & \cdots & \cos\theta_d^i \\ \sin\theta_1^i & \sin\theta_2^i & \cdots & \sin\theta_j^i & \cdots & \sin\theta_d^i \end{bmatrix} \tag{15}$$

where, $\theta_j^i = 2\pi \times rand(\cdot)$, $rand(\cdot)$ is the random number in $(0,1)$; $i = 1, 2, \ldots, N$; $j = 1, 2, \ldots, d$; d is the space dimensionality.

Thus, it can be seen that each bat occupies 2 positions in the ergodic space. The probability amplitudes of each corresponding to the quantum state of $|0\rangle$ and $|1\rangle$ are defined as Equations (16) and (17), respectively. For convenience, B_c^i is called cosinusoidal position, B_s^i is called sinusoidal position.

$$B_c^i = \left(\cos\theta_1^i, \cos\theta_2^i, \ldots, \cos\theta_j^i, \ldots, \cos\theta_d^i \right) \tag{16}$$

$$B_s^i = \left(\sin\theta_1^i, \sin\theta_2^i, \ldots, \sin\theta_j^i, \ldots, \sin\theta_d^i \right) \tag{17}$$

b.   Quantum Bat Global Search and Local Search

In QBA, the move of bat's position is actualized by quantum revolving gate. Thus, in standard BA, the update of bat's moving velocity transforms into the update of quantum revolving gate, the update of bat's position transforms into the update of bat's qubit probability amplitude. The optimal positions of the current population are set as Equations (18) (for quantum state of $|0\rangle$) and (19) (for quantum state of $|1\rangle$), respectively,

$$B_c^g = \left(\cos\theta_1^g, \cos\theta_2^g, \ldots, \cos\theta_d^g \right) \tag{18}$$

$$B_s^g = \left(\sin\theta_1^g, \sin\theta_2^g, \ldots, \sin\theta_d^g \right) \tag{19}$$

Based on the assumption above, the update rule of bats' state is as followings.

In global search, the update rule of the qubit probability amplitude increment of bat B^i is as Equation (20),

$$\Delta\theta_j^i(t+1) = \Delta\theta_j^i(t) + F_i\Delta\theta_g \tag{20}$$

where $\Delta\theta_g$ is defined as Equation (21),

$$\Delta\theta_g = \begin{cases} 2\pi + \theta_j^g - \theta_j^i, & \theta_j^g - \theta_j^i < -\pi \\ \theta_j^g - \theta_j^i, & -\pi \leq \theta_j^g - \theta_j^i \leq \pi \\ \theta_j^g - \theta_j^i - 2\pi, & \theta_j^g - \theta_j^i > \pi \end{cases} \tag{21}$$

In local search, the update rule of the qubit probability amplitude corresponding to the current optimal phase increment of bat B^i is defined as Equation (22),

$$\Delta\theta_j^i(t+1) = e^{-\frac{\omega \cdot gen}{gen\_max} \cdot average(A) \cdot \rho} \tag{22}$$

where, ω is constant; *gen* is the current iteration number; *gen_max* is the maximal iteration number; *average(A)* is the average of current amplitude of each bat; ρ is the random integer in $[-1, 1]$.

c.   Quantum bat location updating

Based on quantum revolving gate, the quantum probability amplitude is updated as Equation (23),

$$
\begin{bmatrix} \cos\left(\theta_j^i(t+1)\right) \\ \sin\left(\theta_j^i(t+1)\right) \end{bmatrix} = \begin{bmatrix} \cos\left(\Delta\theta_j^i(t+1)\right) & -\sin\left(\Delta\theta_j^i(t+1)\right) \\ \sin\left(\Delta\theta_j^i(t+1)\right) & \cos\left(\Delta\theta_j^i(t+1)\right) \end{bmatrix} \times \begin{bmatrix} \cos\left(\theta_j^i(t)\right) \\ \sin\left(\theta_j^i(t)\right) \end{bmatrix}
$$

$$
= \begin{bmatrix} \cos\left(\theta_j^i(t) + \Delta\theta_j^i(t+1)\right) \\ \sin\left(\theta_j^i(t) + \Delta\theta_j^i(t+1)\right) \end{bmatrix} \tag{23}
$$

The two new updated positions (for the quantum state of $|0\rangle$ and $|1\rangle$) of bat B^i are shown as Equations (24) and (25), respectively,

$$
P_c^i(t+1) = \left(\cos(\theta_1^i(t) + \Delta\theta_1^i(t+1)), \ldots, \cos\left(\theta_d^i(t) + \Delta\theta_d^i(t+1)\right)\right) \tag{24}
$$

$$
P_s^i(t+1) = \left(\sin(\theta_1^i(t) + \Delta\theta_1^i(t+1)), \ldots, \sin\left(\theta_d^i(t) + \Delta\theta_d^i(t+1)\right)\right) \tag{25}
$$

It demonstrates that quantum revolving gate actualizes the simultaneous movements of bat's two positions by updating qubit phase which depicts the bat's position. Thus, under the condition of unchanging total population size, the qubit encoding can enhance ergodicity, which helps improving the efficiency of the algorithm.

2.2.3. Chaotic Quantum Global Perturbation

As a bionic evolutionary algorithm, with the increasing number of iterations, the diversity of the population will decline, which leads to premature convergence during optimization processes. As mentioned, the chaotic variable can be used to maintain diversity of the population to avoid premature convergence. Many scholars have published papers using improved chaotic algorithm [48,49]. Authors also have used cat map to the improve GA and PSO algorithm [50,51], the results of numerical experiments show that the searching ability of new GA and PSO improved by chaos is enhanced. Hence, in this paper, the cat mapping function is employed to be the global chaotic perturbation strategy (GCPS), i.e., the so-called CQBA, based on the QBA to adopt GCPS while suffering from premature convergence problem in the iterative searching processes.

The two-dimensional cat mapping function is shown as Equation (26),

$$
\begin{cases} y^{t+1} = frac(y^t + z^t) \\ z^{t+1} = frac(y^t + 2z^t) \end{cases} \tag{26}
$$

where *frac* function is employed for the fractional parts of a real number y by subtracting an appropriate integer.

The global chaotic perturbation strategy (GCPS) is illustrated as followings.

(1)   **Generate $\frac{N}{2}$ chaotic disturbance bats.** For each $Bat^i(i = 1, 2, \ldots, N)$, apply Equation (26) to generate d random numbers, z_j ($j = 1, 2, \ldots, d$). Then, the Equations (27) and (28) are used to map these numbers, z_j, into y_j (with valued from -1 to 1). Set y_j as the qubit (with quantum state, $|0\rangle$) amplitude, $\cos\theta_j^i$, of Bat^i.

$$
\frac{z_j - 0}{1 - 0} = \frac{y_j - (-1)}{1 - (-1)} \tag{27}
$$

$$
\cos\theta_j^i = y_j = 2z_j - 1 \tag{28}
$$

(2) **Determine the $\frac{N}{2}$ bats with better fitness**. Calculate fitness value of each bat from current QBA, and arrange these bats to be a sequence in the order of fitness values. Then, select the bats with the $\frac{N}{2}$th ranking ahead in the fitness values.

(3) **Form the new CQBA population**. Mix the $\frac{N}{2}$ chaotic perturbation bats with the $\frac{N}{2}$ bats which are with better fitness selected from current QBA, and form a new population that contains new N bats, and named it as CQBA population.

(4) **Complete global chaotic perturbation**. After obtaining the new CQBA population, take the new CQBA population as the new population of QBA, and continue to execute the QBA process.

2.2.4. Implementation Steps of CQBA

The procedure of the hybrid CQBA with an SVR model is detailed as followings and the associate flowchart is provided as Figure 1.

Step 1   **Parameter Setting**. Initialize the population size, N; maximal iteration, *gen_max*; expected criteria, ϑ; pulse emission rate, $R(i)$; maximum and minimum of emission frequencies, F_{max} and F_{min}, respectively.

Step 2   **Population Initialization of Quantum Bats**. According to quantum bat population initialization strategy, initialize quantum bat population randomly.

Step 3   **Evaluate Fitness**. Evaluate the objective fitness by employing the coding information of quantum bats. Each probability amplitude of qubit is corresponding to an optimization variable in solution space. Assumed that the jth qubit of the bat B^i is $\begin{bmatrix} \eta_j^i \\ \zeta_j^i \end{bmatrix}$, the element's value of the qubit is between the interval, $[-1, 1]$; the solution space variable corresponding to that is $\begin{bmatrix} (X_j^i)_c \\ (X_j^i)_s \end{bmatrix}$, set the element's value be between the interval, $[a_j, b_j]$. Then, the solution could be calculated by the equal proportion relationship (i.e., Equations (29) and (30)),

$$\frac{(X_j^i)_c - a_j}{b_j - a_j} = \frac{\eta_j^i - (-1)}{1 - (-1)} \tag{29}$$

$$\frac{(X_j^i)_s - a_j}{b_j - a_j} = \frac{\zeta_j^i - (-1)}{1 - (-1)} \tag{30}$$

Eventually, the solution $\begin{bmatrix} (X_j^i)_c \\ (X_j^i)_s \end{bmatrix}$ is obtained as shown in Equations (31) and (32).

$$(X_j^i)_c = \frac{1}{2}\left[b_j\left(1 + \eta_j^i\right) + a_j\left(1 - \eta_j^i\right) \right] \tag{31}$$

$$(X_j^i)_s = \frac{1}{2}\left[b_j\left(1 + \zeta_j^i\right) + a_j\left(1 - \zeta_j^i\right) \right] \tag{32}$$

Each bat corresponds to 2 solutions of the optimal problem, the probability amplitude η_j^i of the quantum state of $|0\rangle$ corresponds to $(X_j^i)_c$; the probability amplitude ζ_j^i of the quantum state of $|1\rangle$ corresponds to $(X_j^i)_s$, where $i = 1, 2, \ldots, N; j = 1, 2, \ldots, d$.

After the transformation of solution space, the parameter combination (σ, C, ε) for each bat is obtained. The forecasting values could also be received, then, the forecasting error is calculated as the fitness value for each bat by the mean absolute percentage error (MAPE), as shown in Equation (33).

$$\text{MAPE} = \frac{1}{N} \sum_{i=1}^{N} \left| \frac{f_i(x) - \hat{f}_i(x)}{f_i(x)} \right| \times 100\% \tag{33}$$

where N is the total number of data; $f_i(x)$ is the actual load value at point i; $\hat{f}_i(x)$ is the forecasted load value at point i.

Step 4   **Quantum Global Search.** According to quantum bat global search strategy, employ Equations (20) and (23) to implement the global search process of quantum bats, update the optimal location and fitness of the population.

Step 5   **Quantum Local Search.** This step considers two situations to implement quantum local search.

Step 5.1   **If** $rand(\cdot) > R(i)$, use Equations (22) and (23), around the optimal bat of the current population, to implement quantum local search, and obtain the new position; else, go to Step 6.

Step 5.2   **If** $rand(\cdot) < A(i)$ and the new position is superior to the original position, then, update the bat's position, and employ Equations (13) and (14) to update $A(i)$ and $R(i)$, respectively, go to Step 5.3; else, go to Step 6.

Step 5.3   Update the optimal location and fitness of the population. Go to Step 6.

Step 6   **Premature Convergence Test.** To improve the global disturbance efficiency, set the expected criteria ϑ, when the population aggregation degree is higher, the global chaotic disturbance for population should be executed once. The mean square error (MSE), as shown in Equation (34), is used to evaluate the premature convergence status,

$$\text{MSE} = \frac{1}{N} \sum_{i=1}^{N} \left(\frac{f_i(x) - f_{avg}(x)}{f(x)} \right)^2 \tag{34}$$

where, N is the number of forecasting samples, $f_i(x)$ is the actual value of the ith period; $f_{avg}(x)$ is average objective value of the current status; $f(x)$ can be obtained by Equation (35),

$$f(x) = \max\left\{ 1, \max_{\forall i \in N}\{|f_i(x) - f_{avg}(x)|\} \right\} \tag{35}$$

If the value of MSE is less than δ, the individual aggregation degree of population is higher, it can be seen that premature convergence appears, go to Step 7, else go to Step 8.

Step 7   **Chaotic Global Perturbation.** Based on cat mapping, i.e., the GCPS as illustrated Section 2.2.1, generate $\frac{N}{2}$ chaotic perturbation bats, sort bats obtained from QBA according to fitness values, and select the $\frac{N}{2}$th bats with better fitness. Then, form the new population which includes the $\frac{N}{2}$ chaotic perturbation bats and the $\frac{N}{2}$ bats with better fitness selected from current QBA. After forming the new population, the QBA is implemented continually.

Step 8   **Stop Criteria.** If the number of search steps is greater than a given maximum search step, *gen_max*, then, the coded information of the best bat among the current population is determined as parameters (σ, C, ε) of an SVR model; otherwise, go back to Step 4 and continue searching the next generation.

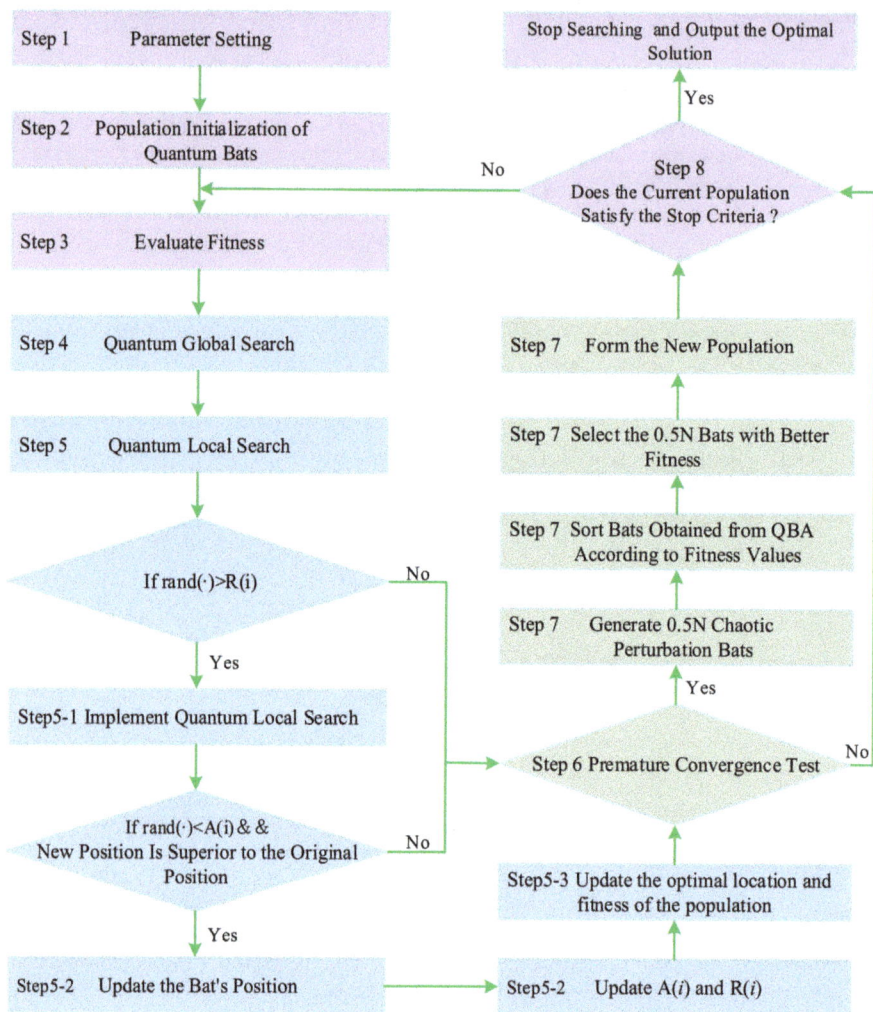

Figure 1. Chaotic quantum bat algorithm flowchart.

3. Experimental Examples

3.1. Data Set of Numerical Examples

To compare the performances of the proposed SVRCQBA model and other hybrid chaotic quantum SVR-based models, this paper uses the hourly load data provided in 2014 Global Energy Forecasting Competition [52]. The load data totally contains 744-h load values, i.e., from 00:00 1 December 2011 to 00:00 1 January 2012. To be based on the same comparison conditions, the data set is divided based on the same means as shown in the previous papers [43–45]. Therefore, the load data are also divided into three sub-sets, the training set with 552-h load values (i.e., from 01:00 1 December 2011 to 00:00 24 December 2011), the validation set with 96-h load values (i.e., from 01:00 24 December 2011 to 00:00 28 December 2011), and the testing set al.so with 96-h load values (i.e., from 01:00 28 December 2011 to 00:00 1 January 2012).

The rolling-based procedure, proposed by Hong [32], is employed to help CQBA searching suitable parameter's value of an SVR model in the training process. In the course of specific training, the training set is further divided into two subsets, namely the fed-in and the fed-out, respectively. Firstly, for each pair of parameters (σ, C, ε) determined by CQBA, the preceding n load values are used as the fed-in vector; then, one-step-ahead forecasted load is computed by the SVR model, i.e., the ($n + 1$)th forecasted load. Secondly, the next n load data, i.e., from 2nd to ($n + 1$)th load values, are set as the new fed-in vector and similarly the second one-step-ahead forecasted load is received, namely the ($n + 2$)th forecasted load. Repeat this procedure until the 552nd forecasted load is computed. The training error can be simultaneously calculated in each iteration, and the validation error would be also calculated.

The adjusted parameter combination only with the smallest validation and testing errors will be selected as the most appropriate parameter combination. Special emphasis is that the testing data set is never used in parameter search and model training, it is only employed for examining the forecasting accurate level. Eventually, the 96 h load data are forecasted by the SVRCQBA model.

3.2. The SVRCQBA Load Forecasting Model

3.2.1. Parameters Setting in CQBA Algorithm

Experiences have indicated that the parameter setting of a model would affect significantly the forecasting accuracy. The parameters of CQBA for the experimental example are set as follows: The population size, N, is set to be 200; the maximal iteration, *gen_max*, is set as 1000; expected criteria, δ, is set to 0.01; the minimal and maximal values of the pulse frequencies, F_{min} and F_{max} are set as -1 and 1, respectively.

In the process of parameter optimization, for the SVR model, the feasible regions of three parameters are set practically, $\sigma \in [0, 10]$, $\varepsilon \in [0, 100]$, and $C \in [0, 3 \times 10^3]$. Considering that the influence of iterative time would affect performances of models, and, to ensure the reliability of forecasting results, the optimization time of each algorithm is set at the same as far as possible.

3.2.2. Forecasting Accuracy Evaluation Index

This article selects the MAPE mentioned above (Equation (33)), the root mean square error (RMSE), and the mean absolute error (MAE) as performance criteria to test the forecasting performance of each model. The RMSE and MAE are calculated by Equations (36) and (37), respectively,

$$\text{RMSE} = \sqrt{\frac{\sum_{i=1}^{N}\left(f_i(x) - \hat{f}_i(x)\right)^2}{N}} \tag{36}$$

$$\text{MAE} = \frac{1}{N}\sum_{i=1}^{N}|f_i(x) - \hat{f}_i(x)| \tag{37}$$

where N is the total number of data; $f_i(x)$ is the actual load value at point i; $\hat{f}_i(x)$ is the forecasted load value at point i.

3.2.3. Forecasting Performance Improvement Tests

To ensure the forecasting performance improvement of the proposed model is significant, it is essential to conduct some statistical test. Based on Diebold and Mariano's [53] and Derrac et al. [54] suggestions, Wilcoxon signed-rank test [55] is conducted in this paper. The Wilcoxon signed-rank test is used to detect the significance of a difference in the central tendency of two data series when the size is equal. Let d_i be the difference between the forecasting errors from any two compared forecasting models on ith forecasting value. The differences would be ranked based on their absolute values; if the differences are tied, the use of average ranks for dealing with ties is recommended, for example, if two differences are tied in the assignation of ranks 1 and 2, assign rank 1.5 to both differences. Let R^+ be the sum of ranks that the first model outperforms the second, on the contrary, R^- the sum of ranks that the second model outperforms the first. If ranks of $d_i = 0$, then, exclude the compared and reduce sample size. The statistic W is represented as Equation (38),

$$W = \min\{R^+, R^-\}$$ (38)

If W is smaller than or equal to the value of Wilcoxon distribution under n degrees of freedom, then, the null hypothesis of performance equality from two compared forecasting models is rejected; this implies that the proposed forecasting model outperforms the other alternative. Furthermore, along with the comparing size increasing, the sampling distribution of W converges to a normal distribution, thus, the associate p-value could also be calculated.

3.2.4. Forecasting Results and Analysis

Considering the GEFCOM 2014 load data set is also used for analysis in references [43–45], therefore, those proposed models are also employed to compare with the proposed model. These alternative models include, SVRBA, SVRQBA, SVRCQBA, SVRQPSO (SVR with chaotic particle swarm optimization algorithm) [43], SVRCQPSO (SVR with chaotic quantum particle swarm optimization algorithm) [43], SVRQTS (SVR with quantum tabu search algorithm) [44], SVRCQTS (SVR with chaotic quantum tabu search algorithm) [44], SVRQGA (SVR with quantum genetic algorithm) [45], SVRCQGA (SVR with chaotic quantum genetic algorithm) [45].

The parameter combinations of SVR are eventually determined by the BA, QBA, CQBA, QTS, CQTS, QPSO, CQPSO, QGA, and CQGA, respectively. The details of the most appropriate parameters of all employed compared models for GEFCOM 2014 data set are shown in Table 1. It is clearly to learn about that the proposed SVRCQBA model receives the smallest forecasting accuracy, and computation time savings.

Table 1. Parameters combination of SVR determined by CQBA and other algorithms.

Optimization Algorithms	Parameters			MAPE of Testing (%)	Computation Time (Seconds)
	σ	C	ε		
SVRQPSO [43]	9.000	42.000	0.180	1.960	635.73
SVRCQPSO [43]	19.000	35.000	0.820	1.290	986.46
SVRQTS [44]	25.000	67.000	0.090	1.890	489.67
SVRCQTS [44]	12.000	26.000	0.320	1.320	858.34
SVRQGA [45]	5.000	79.000	0.380	1.750	942.82
SVRCQGA [45]	6.000	54.000	0.620	1.170	1327.24
SVRBA	8.000	37.000	0.750	3.160	326.87
SVRQBA	13.000	61.000	0.560	1.744	549.68
SVRCQBA,	11.000	76.000	0.670	1.098	889.36

Based on the parameters combiation of each compared SVR-based model, use the training data set to conduct the training work, and receive the well trained SVR model. These trained models are further employed to forecast the load. The forecasting comparison curves of nine models mentioned above and actual values are shown as in Figure 2. Table 2 illustrates the forecasting accurate indexes for the proposed SVRCQBA and other alternative compared models.

Figure 2 clearly demonstrates that the proposed SVRCQBA model achieves results closer to the actual load values than other alternative compared models. In Table 2, the MAPE, RMSE and MAE of the proposed SVRCQBA model are 1.0982%, 1.4835, and 1.4372, respectively, which are smaller than that of other eight compared models. It also indicates that the proposed SVRCQBA model provides very contributions of improvements in terms of load forecasting accuracy. The concrete analysis results are as follows.

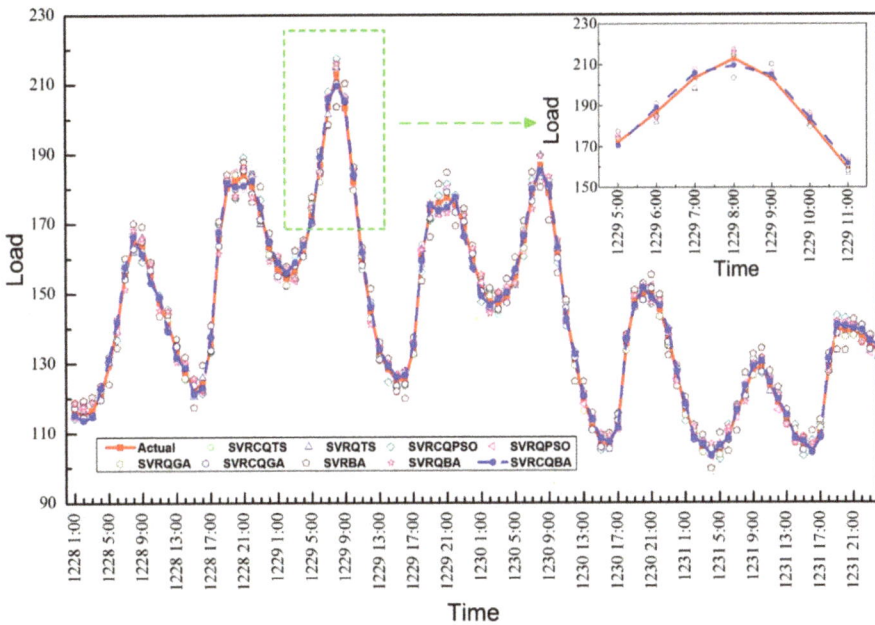

Figure 2. Forecasting values of SVRCQBA and other alternative compared models.

Table 2. Forecasting indexes of SVRCQBA and other alternative compared models.

Indexes	SVRQPSO [43]	SVRCQPSO [43]	SVRQTS [44]	SVRCQTS [44]	SVRQGA [45]	SVRCQGA [45]
MAPE (%)	1.9600	1.3200	1.8900	1.2900	1.7500	1.1700
RMSE	2.9358	1.9909	2.8507	1.9257	1.6584	1.4927
MAE	2.8090	1.8993	2.7181	1.8474	1.6174	1.4522

Indexes	SVRBA		SVRQBA		SVRCQBA	
MAPE (%)	3.1600		1.7442		1.0982	
RMSE	4.7312		2.5992		1.4835	
MAE	4.5234		2.4968		1.4372	

For forecasting performance comparison between SVRQBA and SVRBA models, the values of RMSE, MAPE and MAE for the SVRQBA model are smaller than that of the SVRBA model. It demonstrates that empowering the bats to have quantum behaviors, i.e., using quantum revolving gate (Equation (23)) in the BA to let any bats have comprehensive flying direction choices, which is an appropriate method to improve the solution, then, to improve the forecasting accuracy while the BA is hybridized with an SVR model. For example, in Table 2, the introduction of the quantum computing mechanism changes the forecasting performances (MAPE = 3.1600%, RMSE = 4.7312, MAE = 4.5234) of SVRBA model to much better performances (MAPE = 1.7442%, RMSE = 2.5992, MAE = 2.4968) of SVRQBA model. Employing the quantum revolving gate could improve almost 1.5% (=3.1600% − 1.7442%) forecasting accuracy in terms of MAPE, which plays the critical role in forecasting accuracy improvement contributions. Therefore, it is important to look for any more advanced quantum gates to empower more selection choices for any bats in the searching processes.

Meanwhile, comparing the RMSE, MAPE, MAE of SVRCQBA model with that of SVRQBA model, the forecasting accuracy of SVRCQBA model is superior to that of SVRQBA model. It reveals that the CQBA determines more appropriate parameters combination for an SVR model by introducing cat mapping function, which has a critical role in looking for an improved solution when the QBA algorithm are trapped in local optima or requires a long time to solve the problem of interest.

For example, as shown in Table 2, searching parameters for an SVR model by CQBA instead of by QBA is excellently to shift the performances (MAPE = 1.7442%, RMSE = 2.5992, MAE = 2.4968) of the SVRQBA model to much better performances (MAPE = 1.0982%, RMSE = 1.4835, MAE = 1.4372) of the SVRCQBA model. Applying cat mapping function could improve almost 0.7% (=1.7442% − 1.0982%) forecasting accuracy in terms of MAPE, which also reveals the very contributions in forecasting accuracy improvement. Therefore, it is also an interesting issue to use other novel chaotic mapping functions to effectively enrich the diversity of population while searching iterations reach to a large scale.

In addition, the forecasting indexes results in Table 2 also illustrate that employing the CQPSO, CQTS, and CQGA, it could receive the best solution, (σ, C, ε) = (19.000, 35.000, 0.820), (σ, C, ε) = (12.000, 26.000, 0.320), and (σ, C, ε) = (6.000, 54.000, 0.620), with forecasting error, (MAPE = 1.3200%, RMSE = 1.9909, MAE = 1.8993), (MAPE = 1.2900%, RMSE = 1.9257, MAE = 1.8474), and (MAPE = 1.1700%, RMSE = 1.4927, MAE = 1.4522), respectively. As mentioned above that it is superior to classical TS, PSO, and GA algorithms. However, the solution still could be further improved by the CQBA algorithm to (σ, C, ε) = (11.000, 76.000, 0.670) with more accurate forecasting performance, (MAPE = 1.0982%, RMSE = 1.4835, MAE = 1.4372). It illustrates that hybridizing the cat mapping function and quantum computing mechanism with BA to select suitable parameters combination of an SVR model is a more powerful approach to receive satisfied the forecasting accuracy. Therefore, hybridizing CQBA with an SVR model could only improve at most 0.22% (=1.3200% − 1.0982%) forecasting accuracy in terms of MAPE, which also reveals the selection of advanced evolutionary algorithms could also contribute to forecasting accuracy improvements, however, along with the mature development of evolutionary algorithms, the contributions seem to be minor. Therefore, it should be a valuable remark that hybridizing other optimization approaches (such as chaotic mapping functions, quantum computing mechanism, cloud theory, and so on) to targeted overcome some embedded drawbacks of existed evolutionary algorithms is with much contributions to forecasting accuracy improvements. Based on the remark, it indicates that hybridizing novel optimization techniques with novel evolutionary algorithms could be the most important research tendency in the SVR-based load forecasting work.

Finally to ensure the significant contribution in terms of forecasting accuracy improvement for the proposed SVRQBA and SVRCQBA models the Wilcoxon signed-rank test is then implemented. In this paper the test is conducted under two significant levels α = 0.025 and α = 0.005 by one-tail test. The test results are demonstrated in Table 3 which indicate that the proposed SVRCQBA model has received significant forecasting performance than other alternative compared models.

Table 3. Results of Wilcoxon signed-rank test.

Compared Models	Wilcoxon Signed-Rank Test		
	α = 0.025; W = 2328	α = 0.005; W = 2328	*p*-Value
SVRCQBA vs. SVRQPSO	1087 [T]	1087 [T]	0.00220 **
SVRCQBA vs. SVRCQPSO	1184 [T]	1184 [T]	0.00156 **
SVRCQBA vs. SVRQTS	1123 [T]	1123 [T]	0.00143 **
SVRCQBA vs. SVRCQTS	1246 [T]	1246 [T]	0.00234 **
SVRCQBA vs. SVRQGA	1207 [T]	1207 [T]	0.00183 **
SVRCQBA vs. SVRCQGA	1358 [T]	1358 [T]	0.00578 *
SVRCQBA vs. SVRBA	874 [T]	874 [T]	0.00278 **
SVRCQBA vs. SVRQBA	1796 [T]	1796 [T]	0.00614 *

[T] Denotes that the SVRCQGA model significantly outperforms the other alternative compared models; * represents that the test has rejected the null hypothesis under α = 0.025; ** represents that the test has rejected the null hypothesis under α = 0.005.

4. Conclusions

This paper proposes an electric demand forecasting by hybridizing SVR model with the cat mapping function quantum computing mechanism and the BA. The experimental results illustrate that the proposed model demonstrates significant forecasting performance than other hybrid chaotic quantum evolutionary algorithm SVR-based forecasting models in the literature. This paper continues to extend the search space with the limitations from conventional Newtonian dynamics by using quantum computing mechanism and to enhance ergodicity of population and to enrich the diversity of the searching space by using cat mapping function. Consequently, quantum computing mechanism is applied to endow bits to act as quantum behaviors hence to extend the searching space of BA and eventually to improve forecasting accuracy. Cat mapping function is further used to avoid premature convergence while the QBA is modeling and also contribute to accurate forecasting performances.

This paper also provides some important conclusions and indicates some valuable research directions for future research. Firstly, empowering the bats to have quantum behaviors by using quantum revolving gate could contribute most accuracy improvements. Therefore, in the future the successive researchers could consider constructing an n-dimensional quantum gate where n is the dimensions of employed data set i.e., for each bat in the modeling process it has n probability amplitudes instead of only one amplitude. Based on this new design it is expected to look for more abundant search results via those bats with n probability amplitudes.

Secondly applying chaotic mapping functions could also improve forecasting accuracy. Therefore, in the future any approaches which could enrich the diversity of population during modeling process are deserved to employ to receive more satisfied forecasting accuracy such as other novel chaotic mapping functions or novel design of mutation or crossover operations and so on.

Finally, only hybridizing different evolutionary algorithms could contribute minor forecasting accuracy improvements. Therefore, hybridizing different novel optimization techniques with novel evolutionary algorithms could contribute most in terms of forecasting accuracy improvements and would be the most important research tendency in the SVR-based load forecasting work in the future.

Acknowledgments: The work is supported by the following project grants National Natural Science Foundation of China (51509056); Heilongjiang Province Natural Science Fund (E2017028); Fundamental Research Funds for the Central Universities (HEUCF170101); Open Fund of the State Key Laboratory of Coastal and Offshore Engineering (LP1610); Heilongjiang Sanjiang Project Administration Scientific Research and Experiments (SGZL/KY-08); and Jiangsu Distinguished Professor Project by Jiangsu Provincial Department of Education.

Author Contributions: Ming-Wei Li and Wei-Chiang Hong conceived and designed the experiments; Jing Geng and Shumei Wang performed the experiments; Ming-Wei Li and Wei-Chiang Hong analyzed the data and wrote the paper.

Conflicts of Interest: The authors declare no conflict of interest.

References

1. Xiao, L.; Wang, J.; Hou, R.; Wu, J. A combined model based on data pre-analysis and weight coefficients optimization for electrical load forecasting. *Energy* **2015**, *82*, 524–549. [CrossRef]
2. Bunn, D.W.; Farmer, E.D. Comparative models for electrical load forecasting. *Int. J. Forecast.* **1986**, *2*, 241–242.
3. Fan, G.; Peng, L.-L.; Hong, W.-C.; Sun, F. Electric load forecasting by the SVR model with differential empirical mode decomposition and auto regression. *Neurocomputing* **2016**, *173*, 958–970. [CrossRef]
4. Wang, J.; Wang, J.; Li, Y.; Zhu, S.; Zhao, J. Techniques of applying wavelet de-noising into a combined model for short-term load forecasting. *Int. J. Electr. Power Energy Syst.* **2014**, *62*, 816–824. [CrossRef]
5. Pappas, S.S.; Ekonomou, L.; Karampelas, P.; Karamousantas, D.C.; Katsikas, S.K.; Chatzarakis, G.E.; Skafidas, P.D. Electricity demand load forecasting of the Hellenic power system using an ARMA model. *Electr. Power Syst. Res.* **2010**, *80*, 256–264. [CrossRef]
6. Zhang, M.; Bao, H.; Yan, L.; Cao, J.; Du, J. Research on processing of short-term historical data of daily load based on Kalman filter. *Power Syst. Technol.* **2003**, *9*, 39–42.

7.   Maçaira, P.M.; Souza, R.C.; Oliveira, F.L.C. Modelling and forecasting the residential electricity consumption in Brazil with pegels exponential smoothing techniques. *Procedia Comput. Sci.* **2015**, *55*, 328–335. [CrossRef]

8.   Dudek, G. Pattern-based local linear regression models for short-term load forecasting. *Electr. Power Syst. Res.* **2016**, *130*, 139–147. [CrossRef]

9.   Zhang, W.; Yang, J. Forecasting natural gas consumption in China by Bayesian model averaging. *Energy Rep.* **2015**, *1*, 216–220. [CrossRef]

10.   Li, H.Z.; Guo, S.; Li, C.J.; Sun, J.Q. A hybrid annual power load forecasting model based on generalized regression neural network with fruit fly optimization algorithm. *Knowl.-Based Syst.* **2013**, *37*, 378–387. [CrossRef]

11.   Ertugrul, Ö.F. Forecasting electricity load by a novel recurrent extreme learning machines approach. *Int. J. Electr. Power Energy Syst.* **2016**, *78*, 429–435. [CrossRef]

12.   Bennett, C.J.; Stewart, R.A.; Lu, J.W. Forecasting low voltage distribution network demand profiles using a pattern recognition based expert system. *Energy* **2014**, *67*, 200–212. [CrossRef]

13.   Akdemir, B.; Çetinkaya, N. Long-term load forecasting based on adaptive neural fuzzy inference system using real energy data. *Energy Procedia* **2012**, *14*, 794–799. [CrossRef]

14.   Hooshmand, R.-A.; Amooshahi, H.; Parastegari, M. A hybrid intelligent algorithm based short-term load forecasting approach. *Int. J. Electr. Power Energy Syst.* **2013**, *45*, 313–324. [CrossRef]

15.   Mahmoud, T.S.; Habibi, D.; Hassan, M.Y.; Bass, O. Modelling self-optimised short term load forecasting for medium voltage loads using tunning fuzzy systems and artificial neural networks. *Energy Convers. Manag.* **2015**, *106*, 1396–1408. [CrossRef]

16.   Niu, D.X.; Shi, H.; Wu, D.D. Short-term load forecasting using Bayesian neural networks learned by hybrid Monte Carlo algorithm. *Appl. Soft Comput.* **2012**, *12*, 1822–1827. [CrossRef]

17.   Hanmandlu, M.; Chauhan, B.K. Load forecasting using hybrid models. *IEEE Trans. Power Syst.* **2011**, *26*, 20–29. [CrossRef]

18.   Li, S.; Wang, P.; Goel, L. Short-term load forecasting by wavelet transform and evolutionary extreme learning machine. *Electr. Power Syst. Res.* **2015**, *122*, 96–103. [CrossRef]

19.   Lou, C.W.; Dong, M.C. A novel random fuzzy neural networks for tackling uncertainties of electric load forecasting. *Int. J. Electr. Power Energy Syst.* **2015**, *73*, 34–44. [CrossRef]

20.   Suykens, J.A.K.; Vandewalle, J.; De Moor, B. Optimal control by least squares support vector machines. *Neural Netw.* **2001**, *14*, 23–35. [CrossRef]

21.   Sankar, R.; Sapankevych, N.I. Time series prediction using support vector machines: A survey. *IEEE Comput. Intell. Mag.* **2009**, *4*, 24–38.

22.   Vapnik, V. *The Nature of Statistical Learning Theory*, 2nd ed.; Springer: New York, NY, USA, 2000; ISBN 978-0-387-98780-4.

23.   Vapnik, V. *Statistical Learning Theory*; Wiley: New York, NY, USA, 1998; ISBN 978-0-471-03003-4.

24.   Scholkopf, B.; Smola, A.J. *Learning with Kernels: Support Vector Machines, Regularization, Optimization, and Beyond*; The MIT Press: Cambridge, MA, USA, 2002; ISBN 978-0-262-19475-4.

25.   Vapnik, V.; Golowich, S.; Smola, A. Support vector machine for function approximation, regression estimation, and signal processing. *Adv. Neural Inf. Process. Syst.* **1996**, *9*, 281–287.

26.   Antonanzas, J.; Urraca, R.; Martinez-De-Pison, F.J.; Antonanzas-Torres, F. Solar irradiation mapping with exogenous data from support vector regression machines estimations. *Energy Convers. Manag.* **2015**, *100*, 380–390. [CrossRef]

27.   Yu, P.S.; Chen, S.T.; Chang, I.F. Support vector regression for real-time flood stage forecasting. *J. Hydrol.* **2006**, *328*, 704–716. [CrossRef]

28.   Pai, P.-F.; Hong, W.-C. A recurrent support vector regression model in rainfall forecasting. *Hydrol. Process.* **2007**, *21*, 819–827. [CrossRef]

29.   Granata, F.; Gargano, R.; de Marinis, G. Support vector regression for rainfall-runoff modeling in urban drainage: A comparison with the EPA's storm water management model. *Water* **2016**, *8*, 69. [CrossRef]

30.   Granata, F.; Papirio, S.; Esposito, G.; Gargano, R.; de Marinis, G. Machine learning algorithms for the forecasting of wastewater quality indicators. *Water* **2017**, *9*, 105. [CrossRef]

31.   Kavaklioglu, K. Modeling and prediction of Turkey's electricity consumption using Support Vector Regression. *Appl. Energy* **2011**, *88*, 368–375. [CrossRef]

32. Hong, W.C. Electric load forecasting by seasonal recurrent SVR (support vector regression) with chaotic artificial bee colony algorithm. *Energy* **2011**, *36*, 5568–5578. [CrossRef]

33. Hong, W.-C.; Dong, Y.; Zhang, W.; Chen, L.-Y.; Panigrahi, B.K. Cyclic electric load forecasting by seasonal SVR with chaotic genetic algorithm. *Int. J. Electr. Power Energy Syst.* **2013**, *44*, 604–614. [CrossRef]

34. Ju, F.-Y.; Hong, W.-C. Application of seasonal SVR with chaotic gravitational search algorithm in electricity forecasting. *Appl. Math. Model.* **2013**, *37*, 9643–9651. [CrossRef]

35. Geng, J.; Huang, M.-L.; Li, M.-W.; Hong, W.-C. Hybridization of seasonal chaotic cloud simulated annealing algorithm in a SVR-based load forecasting model. *Neurocomputing* **2015**, *151*, 1362–1373. [CrossRef]

36. Peng, L.-L.; Fan, G.-F.; Huang, M.-L.; Hong, W.-C. Hybridizing DEMD and quantum PSO with SVR in electric load forecasting. *Energies* **2016**, *9*, 221. [CrossRef]

37. Yang, X.-S. A new metaheuristic bat-inspired algorithm. In *Nature Inspired Cooperative Strategies for Optimization*; González, J.R., Pelta, D.A., Cruz, C., Terrazas, G., Krasnogor, N., Eds.; Springer: Berlin/Heidelberg, Germany, 2010; Volume 284, pp. 65–74. ISBN 978-3-642-12537-9.

38. Yang, X.-S. *Nature Inspired Meta-heuristic Algorithms*, 2nd ed.; Luniver Press: Frome, UK, 2010; pp. 97–104. ISBN 978-1-905986-28-6.

39. Sheng, X.-H.; Ye, C.-M. Application of bat algorithm to permutation flow-shop scheduling problem. *Ind. Eng. J.* **2013**, *16*, 119–124.

40. Komarasamy, G.; Wahi, A. An optimized k-means clustering technique using bat algorithm. *Eur. J. Sci. Res.* **2012**, *84*, 263–273.

41. Yang, X.-S.; Gandomi, A.H. Bat algorithm: A novel approach for global engineering optimization. *Eng. Comput.* **2012**, *29*, 464–483. [CrossRef]

42. Yang, X.-S. Bat algorithm for multi-objective optimization. *Int. J. Bio-Inspired Comput.* **2011**, *3*, 267–274. [CrossRef]

43. Huang, M.-L. Hybridization of chaotic quantum particle swarm optimization with SVR in electric demand forecasting. *Energies* **2016**, *9*, 426. [CrossRef]

44. Lee, C.-W.; Lin, B.-Y. Application of hybrid quantum tabu search with support vector regression (SVR) for load forecasting. *Energies* **2016**, *9*, 873. [CrossRef]

45. Lee, C.-W.; Lin, B.-Y. Applications of the chaotic quantum genetic algorithm with support vector regression in load forecasting. *Energies* **2017**, *10*, 1832. [CrossRef]

46. Li, Z.-Y.; Ma, L.; Zhang, H.-Z. Quantum bat algorithm for function optimization. *J. Syst. Manag.* **2014**, *23*, 717–722.

47. Moss, C.F.; Sinha, S.R. Neurobiology of echolocation in bats. *Curr. Opin. Neurobiol.* **2003**, *13*, 751–758. [CrossRef] [PubMed]

48. Yuan, X.; Wang, P.; Yuan, Y.; Huang, Y.; Zhang, X. A new quantum inspired chaotic artificial bee colony algorithm for optimal power flow problem. *Energy Convers. Manag.* **2015**, *100*, 1–9. [CrossRef]

49. Peng, A.N. Particle swarm optimization algorithm based on chaotic theory and adaptive inertia weight. *J. Nanoelectron. Optoelectron.* **2017**, *12*, 404–408. [CrossRef]

50. Li, M.-W.; Geng, J.; Hong, W.-C.; Chen, Z.-Y. A novel approach based on the Gauss-vSVR with a new hybrid evolutionary algorithm and input vector decision method for port throughput forecasting. *Neural Comput. Appl.* **2017**, *28*, S621–S640. [CrossRef]

51. Li, M.-W.; Hong, W.-C.; Geng, J.; Wang, J. Berth and quay crane coordinated scheduling using chaos cloud particle swarm optimization algorithm. *Neural Comput. Appl.* **2017**, *28*, 3163–3182. [CrossRef]

52. Global Energy Forecasting Competition. 2014. Available online: http://www.drhongtao.com/gefcom/ (accessed on 28 November 2017).

53. Diebold, F.X.; Mariano, R.S. Comparing predictive accuracy. *J. Bus. Econ. Stat.* **1995**, *13*, 134–144.

54. Derrac, J.; García, S.; Molina, D.; Herrera, F. A practical tutorial on the use of nonparametric statistical tests as a methodology for comparing evolutionary and swarm intelligence algorithms. *Swarm Evol. Comput.* **2011**, *1*, 3–18. [CrossRef]

55. Wilcoxon, F. Individual comparisons by ranking methods. *Biom. Bull.* **1945**, *1*, 80–83. [CrossRef]

energies

MDPI

Article

A New Hybrid Prediction Method of Ultra-Short-Term Wind Power Forecasting Based on EEMD-PE and LSSVM Optimized by the GSA

Peng Lu [1], Lin Ye [1,*], Bohao Sun [2], Cihang Zhang [1], Yongning Zhao [1] and Jingzhu Teng [1]

[1] College of Information and Electrical Engineering, China Agricultural University, Beijing 100083, China;
    lupeng@cau.edu.cn (P.L.); zhangch@cau.edu.cn (C.Z.); zyn@cau.edu.cn (Y.Z.); tengjingzhu@cau.edu.cn (T.Z.)
[2] China Electric Power Research Institute, Haidian District, Beijing 100192, China; hobson_choice@126.com
* Correspondence: yelin@cau.edu.cn; Tel.: +86-010-6273-7842

Received: 31 December 2017; Accepted: 9 February 2018; Published: 21 March 2018

Abstract: Wind power time series data always exhibits nonlinear and non-stationary features, making it very difficult to accurately predict. In this paper, a novel hybrid wind power time series prediction model, based on ensemble empirical mode decomposition-permutation entropy (EEMD-PE), the least squares support vector machine model (LSSVM), and gravitational search algorithm (GSA), is proposed to improve accuracy of ultra-short-term wind power forecasting. To process the data, original wind power series were decomposed by EEMD-PE techniques into a number of subsequences with obvious complexity differences. Then, a new heuristic GSA algorithm was utilized to optimize the parameters of the LSSVM. The optimized model was developed for wind power forecasting and improved regression prediction accuracy. The proposed model was validated with practical wind power generation data from the Hebei province, China. A comprehensive error metric analysis was carried out to compare the performance of our method with other approaches. The results showed that the proposed model enhanced forecasting performance compared to other benchmark models.

Keywords: wind power prediction; ensemble empirical mode decomposition-permutation entropy (EEMD-PE); least squares support vector machine (LSSVM); heuristic algorithm

1. Introduction

As a clean renewable energy, wind energy is regarded as a good alternative to deal with environmental problems and energy crises [1,2]. According to a report published by the World Wind Energy Association (WWEA), worldwide wind capacity reached 54 GW by the end of 2017, with a growth rate of 11.8% [3]. The total installed capacity is reported in Figure 1. The intermittent nature of wind power generation has posed a big challenge for maximizing the utilization of the wind power industry [4]. It is of practical significance to optimize the wind power prediction algorithm and make it more suitable for the operation and wind conditions of a specific wind farm.

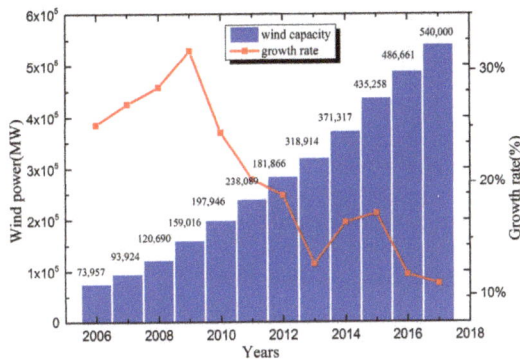

Figure 1. Total installed wind power from 2005 to 2017 worldwide.

Wind power forecasting is difficult to achieve due to its intermittency and stochastic fluctuation, which brings great challenges to power system operation and control [5,6]. Over the past few decades, a large amount of research has been devoted to the development of effective and reliable wind speed/power forecasting methods, models, and tools [7]. Generally, these methods can be broadly divided into four major models [8–11]: (a) physical models, (b) statistical models, (c) hybrid models, and (d) spatial correlation models. Physical models take into account parameters such as topography, temperature, and pressure, at the location of the wind farm, which are often utilized in short-term forecasting 6–72 h ahead and long-term forecasting with multiple weather variables. Typical physical wind power forecasting systems, among others, are the Prediktor system, developed by the Risoe National Laboratory in Denmark [12]; Previento, developed by the University of Oldenburg in Germany [13]; and eWind, developed by AWS True Wind Inc. (New York, NY, USA) [14]. Statistical models are built based on historical power/speed time series data, which establishes a functional relationship between historical data and forecast data [15]. These models analyze the relationship between various explanatory variables and online measurements. The well-known pure statistical models are the autoregressive (AR) model [16], autoregressive moving average (ARMA) model [17], autoregressive integrated moving average (ARIMA) model [18], seasonal autoregressive integrated moving average (SARIMA) model [19], and the autoregressive integrated moving average with exogenous variables (ARMAX) model [20]. However, statistical models based on the assumption that linear structures exist among time series data cannot capture non-linear patterns very well.

Individual models lack the ability to deal with big data and fail to capture the majority of the complex characteristics of the original wind power series data [21]. To make use of the advantages of statistical and physical models, a number of hybrid models with data pre-processing techniques, error post-processing techniques, and parameter selection and optimization techniques, have been proposed.

Data pre-processing techniques involve analyzing and processing data make original time series into multiple sequences or matrices, which have more obvious characteristics. Therefore, to some extent, pre-processing techniques can improve forecasting accuracy. Wavelet decomposition [22–26] and empirical mode decomposition [27–29] are the prevailing data pre-processing techniques, which can analyze the original wind power series in time and frequency domains. De et al. [25] compared the hybrid artificial neural networks (ANN) method. Case studies show that wavelet decomposition (WD)-LSSVM performs better than WD-ANN. Zhang et al. [29] used variational mode decomposition (VMD) to process original wind power series, then established a novel combined model based on machine learning methods. Zhao et al. [30] analyzed the characteristics of the outliers caused by wind curtailment, then, a data-driven outlier elimination approach, combining quartile method and density-based clustering method was proposed; however, variational mode decomposition (VMD) is prone to mode mixing problems. Niu et al. [31] used empirical mode decomposition (EMD) to decompose original wind speed data, then, a novel hybrid forecasting model based on the general regression neural network (GRNN) method, optimized by the fruit fly optimization algorithm (FOA), was proposed. Ye et al. [32] discussed EMD, EEMD, complementary ensemble empirical mode decomposition (CEEMD), and complete empirical mode decomposition with adaptive noise (CEEMDAN). Their results showed that the proposed CEEMDAN-support vector regression (CEEMDAN-SVR) model out-performed the other models.

Error post-processing (EP-P) techniques use estimated error, which is obtained from a forecasting model, to correct final forecasting results. Huang et al. [33] proposed a new, real-time decomposition model based on the feature selection and error correction of wind speed forecasting, which improved prediction accuracy. Platon et al. [34] used an advanced technique to estimate surface wind gusts, then, combined dynamic and statistical techniques into the wind power forecasting model. Liang et al. [35] improved wind speed forecasting performance using a correlation analysis method to analyze multi-step forecast errors and proposed a novel hybrid wind speed prediction model based on error forecast correction. Federica et al. [36] employed a principal component analysis (PCA), combined with post-processing, to reduce computational costs and forecast errors. Li et al. [37] proposed a new combined approach based on Extreme Learning Machine (ELM) and an error correction model, which improved prediction accuracy over a short-term time scale (6–72 h).

Parameter selection and optimization techniques can improve prediction accuracy and reduce prediction time through the training model. Xiao et al. [38] employed a new hybrid prediction model based on a modified bat algorithm (BA) with the conjugate gradient (CG) method to multi-step wind speed prediction, which optimized the initial weights of the neural networks. Wang et al. [39] proposed a novel combined forecasting model based on a multi-objective bat algorithm (MOBA), multi-step-ahead wind speed forecasting. Huang et al. [40] proposed a novel forecasting model, using a quantum particle swarm optimization (PSO) algorithm, to receive higher forecast accuracy levels. Chang et al. [41] compared the persistence method, the back propagation artificial neural network (BP) model, and radial basis neural network (RBF) model. Case studies showed that the proposed forecasting method was more accurate and reliable than the other three models. The clonal selection algorithm (CSA) [42], gravitational search algorithm (GSA) [43], particle swarm optimization (PSO) [44,45], simplified swarm optimization (SSO) [46], and cuckoo search algorithm (CS) [47], among others, are the prevailing methods to optimize the parameters of wind power/speed forecasting models.

Spatial correlation models characterize the relationship between the wind power or speed of a target wind farm and a reference wind farm at different spatial locations. Zhou et al. [48] proposed a spatial and temporal correlation model and it was found that this model could improve ultra-short-term wind power forecasting accuracy. Tascikaraoglu et al. [49] proposed a novel method, which first utilized a Wavelet Transform (WT) method to decompose the wind speed data into more stationary components and then used a spatio-temporal model on each of the subseries to incorporate both the temporal and spatial information for wind speed forecasting. Ye et al. [50] analyzed uncertainty and dependence in wind power output, and employed a physical spatio-temporal correlation model. They found that this method outperformed statistical models.

In this paper, a novel combine model is proposed based on ensemble empirical mode decomposition, permutation entropy, least squares support vector machine, and gravitational search algorithm for ultra-short wind power forecasting. To investigate the effectiveness of the model, the proposed method will be thoroughly tested and benchmarked on real wind power data from Hebei, China. The main contributions of this research will be as follows:

(1) Using pre-processing techniques to deal with the complex wind power time series.

The ensemble empirical mode decomposition-permutation entropy will be used to analyze the original wind power series, by which the original wind power time series will be translated into some new, relatively stable subsequences. Ensemble empirical mode decomposition can decompose original wind power time series into a series of intrinsic mode functions (IMF) with different characteristic scales; however, it fails to capture weak changes in time series. Permutation entropy will be used to reconstitute subsequences by similar principles, which can promote weak time signals.

(2)   Employing the LSSVM forecasting model, optimized by GSA.

LSSVM will be employed as the basic forecasting model, due to the features of regression for wind power prediction. To improve the forecasting accuracy and stability of LSSVM directly, the hyper-parameters of LSSVM will, firstly, be optimized by GSA to obtain the best hyper-parameters.

(3)   Using comprehensive error metrics to assess the performance of the proposed model.

The error indicators, in this paper, will include the normalized mean absolute of errors (NMAE), normalized root mean square error (NRMSE), and Pearson correlation coefficient (R). In this paper, we will also introduce two improvement percentage error indexes, $\tilde{\zeta}_{NMAE(\%)}$ and $\tilde{\zeta}_{NRMSE(\%)}$.

The remainder of the paper will be organized as follows: the details of the proposed hybrid model based on EMD-PE-LSSVM-GSA for wind power forecasting will be illustrated in Section 2. Forecasting performance evaluation indicators will be described in Section 3. Experimental examples will be presented in Section 4. The resulting analysis and forecasting performance of the proposed method, compared with other methods, will be given in Section 5. Finally, conclusions will be given in Section 6.

2. Proposed Methodology

The approaches used, including ensemble empirical mode decomposition, permutation entropy, the least squares support vector machine model, and gravity search algorithm, are described in this section. The EEMD-PE-LSSVM-GSA wind power prediction process is shown in Figure 2.

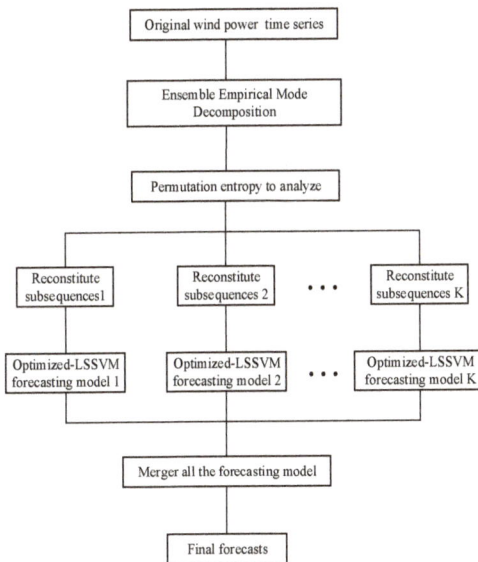

Figure 2. The procedure of the new proposed prediction model.

2.1. Ensemble Empirical Mode Decomposition (EEMD)

EMD is frequently subject to a mode mixing problem, where a portion of the IMF may have properties that are quite similar to adjacent IMFs. EEMD is based on EMD and is an algorithm-based method of processing signals, which can be used to developed to solve the mode mixing problem [51]. White noise is added to the wind power time series at different scales. In order to solve the EMD mode mixing problem, a detailed explication is given in [52], as follows:

(1) Add white noise series to the original wind power series:

$$x_{new,i}(t) = x(t) + w_i(t) \tag{1}$$

where $x(t)$ is the original wind power series, and $w_i(t)$ is the white noise series. Then, find the corresponding EMD components.

(2) Find the local maxima and minima of $x_{new,i}(t)$.
(3) Find the upper envelope $x_{new,iu}(t)$ and lower envelope $x_{new,il}(t)$.
(4) Calculate the mean of the wind power time series with white noise and the difference between $x_{new,iu}(t)$ and $x_{new,il}(t)$.

$$m(i) = \frac{x_{new,iu}(t) + x_{new,il}(t)}{2} \tag{2}$$

$$d_i(i) = x_{new,i}(t) - m(i) \tag{3}$$

(5) Repeat Steps 1–3 with $d(i)$ instead of $x_{new,i}(t)$, until $m(i) \leq \delta$ (where δ is the acceptable error). Then, take $c_i^1(i) = d_i(i)$ as the first EMD component of $x_{new,i}(t)$, and the residual is as follows:

$$w_i^1 = x_{new,i}(t) - c_i^1(i) \tag{4}$$

(6) The wind power time series $x(t)$ can be decomposed as follows:

$$x(t) = \sum_{i=1}^{n} c_i^m + w_i^n \tag{5}$$

where c_i^m represents the IMFs, and w_i^n is the final residue.

2.2. Permutation Entropy (PE)

In the case of nonlinear analysis, the complexity of the signal can be effectively determined according to its entropy values [53], such as scale entropy, sample entropy, and multi-scale entropy. Permutation entropy is widely used in sequence complexity and nonlinear analysis because of its high robustness, efficiency, and simplicity.

This method's motivation is the classification of the complex system. The larger the permutation entropy value, the higher the time series randomness of the sequence and the more likely another pattern will occur. Conversely, the smaller the permutation entropy value, the lower the time series randomness of the subsequence and the less likely another pattern will occur. The algorithm implementation process of PE is given below.

For L time series samples, $\{x(i), i = 1, 2, 3, \ldots, K\}$, the time series are reconstructed by m-dimension phase space.

$$X(i) = [x(i), x(i + \tau), \ldots, x(i + (m - 1)\tau)] \tag{6}$$

where m is the embed dimension of the wind power time series, and τ is the delay time.

$$[x(i + (j_1 - 1)\tau) \leq x(i + (j_2 - 1)\tau) \leq, \ldots, \leq x(i + (j_m - 1)\tau)] \tag{7}$$

where $j_1, j_2, \ldots, j_m$ represents the index number of the column in which each element in the reconstruction vector resides. Each vector can be mapped to a set of symbols.

$$S(g) = [j_1, j_2, \ldots, j_m] \tag{8}$$

where $g = 1, 2, \ldots, k, k \leq m!$

We calculated the probability of occurrence for each symbol sequence, $P_1, P_2, \ldots, P_k, \sum_{l=1}^{k} P_l$, where:

$$\sum_{l=1}^{k} P_l = 1 \tag{9}$$

In the form of Shannon entropy, the permutation entropy of the wind power time series can be expressed as:

$$H_p(m) = -\sum_{l=1}^{k} P_l \ln P_l \tag{10}$$

When $P_l = \frac{1}{m!}$, $H_p(m)$ reaches the maximum $\ln(m!)$, the standardized processing can be achieved by:

$$H_p = \frac{H_p(m)}{\ln(m!)} \tag{11}$$

Permutation entropy values were used to evaluate the complexity of each IMFs signal, and the adjacent entropy values were used to reconstitute IMFs into new subsequences (RS).

2.3. Least Squares Support Vector Machine (LSSVM)

The support vector machine (SVM) is an effective machine algorithm for data classification and regression [54]. SVM can overcome data over-fitting problems and improve generalization performance by minimizing structural risk instead of empirical risk. The standard SVM uses nonnegative errors in the cost function and inequality constraints, while the LSSVM uses square errors and equality constraints. Therefore, LSSVM is a variation of the standard SVM.

Considering the wind power training dataset, (x_i, y_j), $i = 1, \ldots, NL$, NL is the number of training datasets, $x_i \in R^d$ is the input vector, $y_i \in R$ is the corresponding output, and d is the dimension of x_i. The optimal decision function can be constructed by mapping the input space into the high-dimension feature space as follows:

$$f(x) = \omega^T \varphi(x) + b \tag{12}$$

where $\varphi(x)$ is the nonlinear function, ω is the weight, and b is the bias.

$$R = \frac{1}{2} \|\omega\|^2 + \gamma R_e \tag{13}$$

where $\|\omega\|$ is the model complexity, γ is the regularization parameter to balance the complex degree and approximation accuracy of the model, and R_e is the empirical risk function. The objective function of LSSVM can be framed:

$$\min Z(\omega, \xi) = \frac{1}{2} \|\omega\|^2 + \gamma \sum_{i=1}^{t} \xi_i^2 \tag{14}$$

$$s.t. y_i = \omega \varphi(x_i) + \xi_i + b \quad i = 1, 2, \ldots, N \tag{15}$$

$$L(\omega, b, \xi, \lambda) = \frac{1}{2} \|\omega\|^2 + \gamma \sum_{i=1}^{t} \xi_i^2 - \sum_{i=1}^{t} \lambda_i (\omega \varphi(x_i) + \xi_i + b - y_i) \tag{16}$$

where $\lambda_i(1, 2, \ldots, N)$ represents the Lagrange multipliers.

Based on the Karush-Kuhn-Tucker (KKT) conditions, Equation (15) is given by:

$$\begin{cases} \omega - \sum_{i=1}^{t} \lambda_i \xi_i^2 = 0 \\ \sum_{i=1}^{t} \lambda_i = 0 \\ \lambda_i - \lambda_i \xi_i = 0 \\ \omega\varphi(x_i) + \xi_i + b - y_i = 0 \end{cases} \tag{17}$$

Based on Equation (15) the following expression can be derived:

$$\begin{bmatrix} 0 & I^T \\ I & J + I/\gamma \end{bmatrix} \begin{bmatrix} b \\ \lambda \end{bmatrix} = \begin{bmatrix} 0 \\ y \end{bmatrix} \tag{18}$$

where $I = [1, 1, \ldots, 1]^T$ is a $t \times 1$ dimensional vector, $\lambda = [\lambda_1, \lambda_2, \ldots, \lambda_t]^T$ is the coefficient matrix, $y = [y_1, y_2, \ldots, y_t]^T$ is the output matrix, $K(x_i, y_j) = \varphi(x_i)^T \varphi(x_j)$, and K is the kernel function on the basis of Mercer's condition. The regression function of the LSSVM model can be described as:

$$f(x) = \sum_{i=1}^{t} \lambda_i K(x_i, y_i) + b \tag{19}$$

The radial basis function is selected as the kernel function, which is given as follows:

$$K(x_i, y_j) = \exp\left[\frac{-\|x_i - y_i\|^2}{2\sigma^2} \right] \tag{20}$$

where σ is the kernel parameter.

2.4. Gravitational Search Algorithm (GSA)

The GSA was first proposed in 2009 [55] and is a population optimization algorithm based on the law of gravity and Newton's second law. The algorithm searches for the optimal solution by moving the particle position of the population. That is, as the algorithm iterates, the particles move continuously in the search space by the gravitation between them.

Assuming that the optimization problem can be given in (14) and (15), the particle's position is the solution. The position of particle i is defined as:

$$X_i = (x_i^1, \ldots, x_i^d, \ldots, x_i^n) \tag{21}$$

Step 1: Initialize the speed and position of random particles and calculate the fitness of each particle.

Step 2: Calculate the gravitational constant $G(t)$ and the inertia mass of each particle:

$$G(t) = G_0 \times e^{-\alpha/T} \tag{22}$$

$$m_i(t) = \frac{fit_i(t) - worst(t)}{best(t) - worst(t)} \tag{23}$$

$$M_i(t) = \frac{m_i(t)}{\sum_{j=1}^{N} m_j(t)} \tag{24}$$

where G_0 is the initial gravitational constant, α is the decay rate, T is the maximum generation, and

$$best(t) = \min_{j \in (1, \ldots, n)} fit_j(t), worst(t) = \max_{j \in (1, \ldots, n)} fit_j(t).$$

Step 3: Calculate the resultant particle force, which can be given as:

$$F_{ij}^d(t) = G(t)\frac{M_{pi}(t) \times M_{aj}(t)}{R_{ij}(t) + \varepsilon}(x_j^d(t) - x_i^d(t)) \tag{25}$$

where $F_{ij}^d(t)$ is gravitation with the particles i and j, with dimension, d, at the t generation; $M_{pi}(t)$ is the passive gravitational mass related to particle i; ε is the small constant; $x_i^d(t)$ and $x_j^d(t)$ is the position of dimension, d, of particles i and j at the t generation; $R_{ij}(t) = \|x_i(t), x_i(t)\|_2$ is the Euclidean distance between particles i and j.

Step 4: Calculate accelerated speed. According to Newton's second law, the acceleration is obtained as follows:

$$a_i^d(t) = \frac{F_i^d(t)}{M_{ii}(t)} \tag{26}$$

Step 5: Update speed and position:

$$v_i^d(t+1) = rand \cdot v_i^d(t) + a_i^d(t) \tag{27}$$

$$x_i^d(t+1) = x_i^d(t) + v_i^d(t+1) \tag{28}$$

where $rand_j$ is a random number with a uniform distribution [0,1].

Step 6: Check the termination condition. Terminate the optimization if the stopping criteria requirements are met, and, if not, repeat the procedure from step 2 to 5 until the termination condition requirements are met.

2.5. The Proposed Method for Wind Power Forecasting

The flowchart of the proposed hybrid model based on EEMD-PE-LSSVM-GSA is illustrated in

Stage 1: EEMD process

To build an effective prediction model, the features of the original wind power datasets must be fully analyzed and considered. EEMD techniques can be used to decompose the original wind power time series, $x(t)$, into new, relatively stable subsequences, $x_i(t)$, $(i = 1, 2, 3, \ldots)$.

Stage 2: PE process

PE techniques can be used to analyze the intrinsic mode signals, $x_i(t)$, $(i = 1, 2, 3, \ldots)$, and reconstitute subsequences by combination stacking, to give reconstituted subsequences, $RS = [x_{1j}, x_{2j}, \ldots, x_{nj}] + \cdots + [x_{1m}, x_{2m}, \ldots, x_{nm}]$, $(j, m < i = 1, 2, \ldots)$.

Stage 3: Optimize parameters in the LSSVM process

The LSSVM forecasting model can be employed to forecast the reconstituted subsequences, $RS = [x_{1j}, x_{2j}, \ldots, x_{nj}] + \cdots + [x_{1m}, x_{2m}, \ldots, x_{nm}]$, $(j, m < i = 1, 2, \ldots)$, and the RBF kernel functions can be chosen to initialize the LSSVM.

(1): Initialize: Setting the parameters of GSA, the particle number is L, gravitational constant is G_0, attenuation rate is α, and the dimensions of GSA are d.

(2): Calculate: Calculate the fitness function $F_{fitness}$ as follows:

$$F_{fitness} = \sqrt{\frac{1}{N}\sum_{i=1}^{N}(x_i - \hat{x}_i)^2} \tag{29}$$

where x_i is the real wind power value, $\hat{x}_i$ is the forecasting value, and N is the number of samples.

(3): Update: The states are updated as follows:

$$\begin{cases} x_i^d(t+1) = x_i^d(t) + v_i^d(t+1) \\ v_i^d(t+1) = rand_i \times v_i^d(t) + a_i^d(t) \end{cases} \tag{30}$$

where $x_i^d(t)$ is the position of the particle, $v_i^d(t)$ is the speed of search, and $rand_i$ is a uniform random variable, with a value in the range of [0, 1].

(4): Selection: If the iteration reaches its maximum, or the $F_{fitness}$ reaches its minimum, the best hyper parameters (σ and γ) and corresponding kernel parameters can be found.

(5): Validation: Output wind power prediction values for every new subsequence. The wind power forecasting errors, in terms of different criteria, are computed to validate the method. The results are compared with that of other methods. The best parameters of the optimized model will be obtained.

Stage 4: Hybrid process

Combine all the reconstituted subsequences of forecasting results and output the final forecasting results.

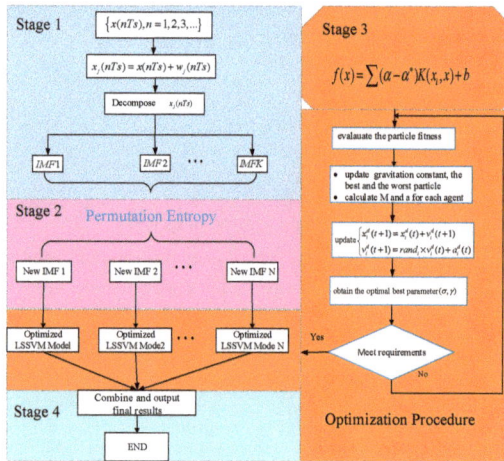

Figure 3. The overall framework of the proposed model.

3. Performance Criterion

In this paper, the error indexes include the normalized mean absolute of errors (NMAE), which reflects actual prediction error, normalized root mean square errors (NRMSE), which reflects large forecasting deviations, and the Pearson correlation coefficient. They are defined, respectively, as follows:

$$\text{NMAE} = \frac{1}{N}\sum_{i=1}^{N}\frac{|x_i - \hat{x}_i|}{P_{Inst}} \times 100\% \tag{31}$$

$$\text{NRMSE} = \frac{1}{P_{Inst}}\sqrt{\frac{1}{N}\sum_{i=1}^{N}(x_i - \hat{x}_i)^2} \times 100\% \tag{32}$$

$$R = \frac{N\sum x_i\hat{x}_i - \sum x_i\sum \hat{x}_i}{\sqrt{\sum x_i^2 - (\sum x_i)^2}\sqrt{\sum \hat{x}_i^2 - (\sum \hat{x}_i)^2}} \tag{33}$$

where x_i is the actual wind power value, $\hat{x}_i$ is the forecasting value, P_{Inst} is the installed wind power capacity, and N is the number of samples.

Additionally, this paper introduces two percentage error indexes, which are defined as follows:

$$\zeta_{\text{NMAE}(\%)} = \frac{\text{NMAE}_2 - \text{NMAE}_1}{\text{NMAE}_1} \times 100\% \tag{34}$$

$$\xi_{NRMSE(\%)} = \frac{NRMSE_2 - NRMSE_1}{NRMSE_1} \times 100\% \tag{35}$$

where a negative value of $\xi_{NMAE(\%)}$ means model 2 decreases $|\xi_{NMAE}|\%$ NMAE value relative to model 1, and a positive value of $\xi_{NRMSE(\%)}$ means model 2 increases $|\xi_{NRMSE}|\%$ NRMSE value relative to model 1.

4. Experimental Examples

4.1. Dataset Description

In this paper, a total of 5760 samples were collected from a wind farm in Hebei, China. Considering the influence of seasonal factors, the whole dataset was divided into four parts, Datasets A, B, C, and D, which were independently used to verify the effectiveness of the proposed method. Dataset A was from 1–15 January 2016, Dataset B was from 1–15 April, Dataset C was from 1–15 July, and Dataset D was from 1–13 October. Wind power generation data were 15 min averaged values. The forecasting methods were applied over very short time horizons, of up to 4 steps (i.e., 1 h) ahead, with each step being 15 min. The samples are shown in Figure 4.

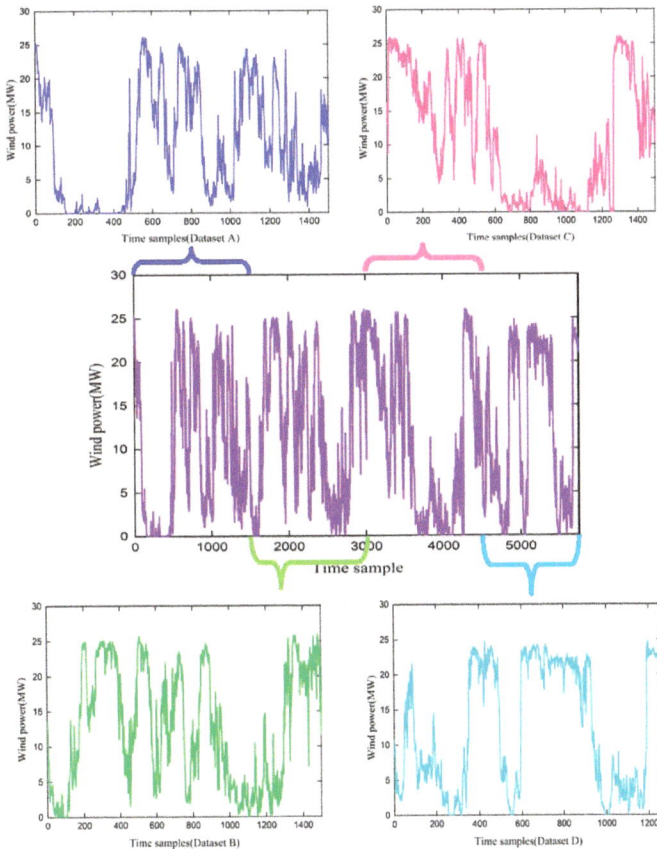

Figure 4. Time samples from Hebei, China.

4.2. Data Processing

EEMD-PE was used to analyze the wind power series, by which the wind power time series could be translated into new, relatively stable subsequences. EEMD was also used to decompose the wind power time series into a series of IMFs with different characteristic scales. Then, PE was used to analyze the IMFs.

There were two important EEMD parameters: the number of the ensemble and the amplitude of the added white noise. In this experiment, 200 groups of white noise signals were added, with a standard deviation was 0.2. There were twelve independent IMF compositions. Decomposition results are shown in Figure 5.

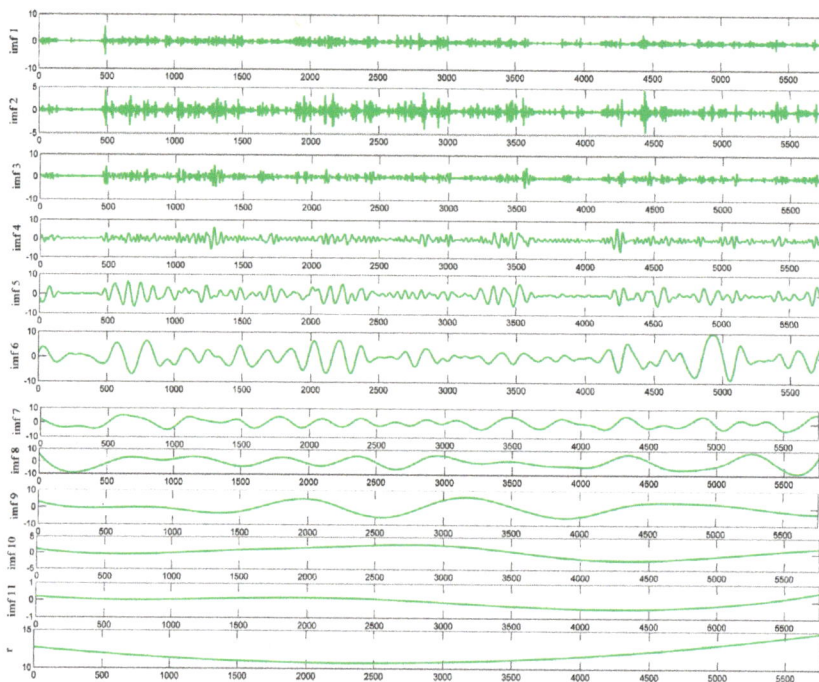

Figure 5. Decomposition by EEMD of wind power series.

The PE parameters (m and τ) have an impact on simulation time and prediction accuracy; if the parameter m is too large, prediction accuracy will be reduced, and if the parameter τ is too small, the simulation time will become longer. Therefore, we discuss the processing of the two parameters.

A number of different parameter values were chosen to forecast the series. It is known that the higher the embedded dimensions, the more complex the structure will be and the more modeling time will be spent. Considering the forecast time and model complexity, $m = 3, 4, 5, 6, 7$ and $\tau = 0.1, 0.5, 1$ models are discussed for (1–4)-step-ahead wind power forecasting, and the errors are shown in Table 1.

In Table 1, the NRMSE of $m = 3$ and $\tau = 1$ are the smallest among all parameters; then, error increases slowly with the dimension of embedding. Performance evaluation, using NMAE and NRMSE, shows that $m = 3$ and $\tau = 1$ is better than other parameters. When $m = 7$ and $\tau = 1$, the NRMSE of the testing sample is 3.0773, which is the largest level of error, and the modeling time is 10.9929. Comparative analysis shows that increasing the embedded dimension and computation time impacts on the prediction at different time scales. Considering prediction accuracy and simulation time, $m = 3$ and $\tau = 1$ are selected to predict wind power.

Table 1. The NRMSE and NMAE of the testing sample with different values of m and τ.

τ_i		$\tau=0.1$			$\tau=0.5$			$\tau=1$	
Indicator	NRMSE	NMAE	Modeling Time (s)	NRMSE	NMAE	Modeling Time (s)	NRMSE	NMAE	Modeling Time (s)
3	2.9778	1.9901	9.9635	2.9813	1.9601	10.1566	2.9738	1.9601	10.7045
4	2.9886	1.9783	10.3289	3.0112	1.9605	10.2562	3.0685	1.9978	10.0467
5	3.0137	1.9644	10.5294	2.9867	1.9384	10.3086	3.0334	1.9824	11.7248
6	3.0152	1.9735	10.4951	2.9863	1.9583	10.3605	2.9744	1.9463	10.5937
7	3.0548	2.0172	11.0421	3.1121	1.9858	11.6372	3.3363	2.0956	11.8948
average	3.0100	1.9847	10.4718	3.0078	1.9606	10.5434	3.0773	1.9964	10.9929

Wind power time series data has nonlinear and non-stationary features. It can be seen from Figure 5 that there are a lot of IMF components after decomposition. If the LSSVM model is used to build each component respectively, the computing time will increase significantly. PE technology can be used to evaluate complexity of each IMF signal.

In order to forecast ultra-short-term wind power effectively, this paper used PE technology to analyze the complexity features of each IMF component. The PE values of all IMFs are shown in Figure 6. In Figure 6, the IMF component frequency decreases from high to low, and the PE value also decreases, which verifies that the PE theory is effective. The PE value indicates the stochastic degree of the time series, where a smaller PE value means more regular time series, and a larger PE value means more randomness. To reduce the computing complexity of the proposed method, according to the PE values, the IMFs were classified and merged to reconstituted subsequences. From IMF 1 to IMF 11 and residue (r), the PE values gradually decreased from 1.7916 to 0.6320. IMF 1 was assigned to RS I, since it had the highest frequency. IMF 2 and IMF 3 PE value differences were about 0.2~0.3, so they could be set as RS II. IMF 4 and IMF 5 PE value differences were about 0.1, and, thus, could be set as RS III. IMF 5~IMF 11+r PE value differences were about 0.02~0.06, and were set as RS IV. The reconstituted subsequences processed by PE are shown in Figure 7.

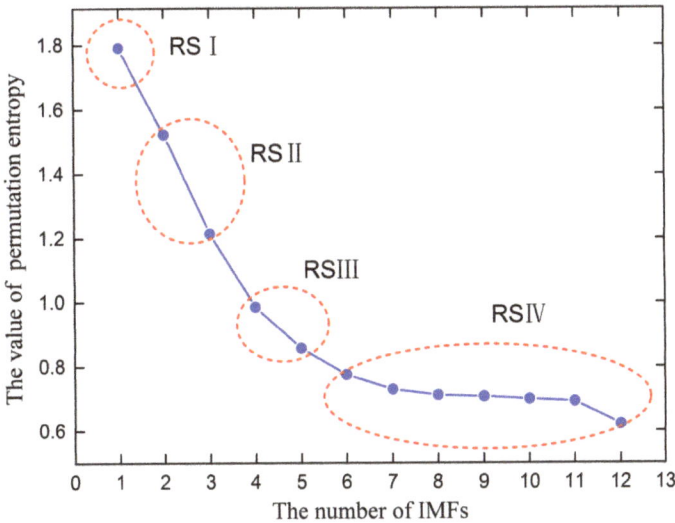

Figure 6. The permutation entropy values of IMFs.

4.3. Parameter and Training Dataset Settings

4.3.1. The Parameter Setting of the Forecasting Models

The simulation was done on a Windows 7 PC with a 64-bit, 2.20 GHz Intel Core i3 2330M CPU, and 6 GB of memory. The wind power forecasting experiments were employed in MATLAB R2014a. The experimental parameters [43] are shown in Table 2.

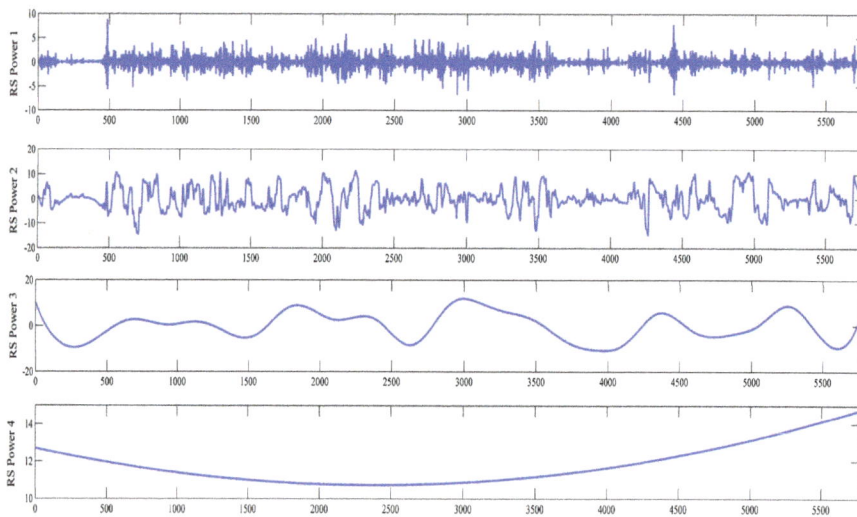

Figure 7. Decomposition results by EEMD-PE.

Table 2. Setting the experimental parameters.

Model	Experimental Parameters	Default Value
GSA	particle number	30
	maximum evolutionary generation number	30
	gravitational constant	100
	attenuation rate	10
	range of the test function	[0.01, 100]
	dimension of the test function	2
LSSVM	value range of kernel parameter c	[0.01, 10]
	value range of parameter γ	[0.1, 1200]
RBFNN	training precision	0.0001
	neuron number of the input layer	1
	neuron number of the hidden layer	3
	neuron number of the output layer	1

The optimal parameters, which result from using RBF kernel functions in the LSSVM model, is shown in Table 3.

Table 3. Optimal kernel function parameters.

Types of Kernel Function	Penalty Factor	Kernel Function Parameters
RBF	$\gamma = 101.628$	$\sigma = 0.1184$

4.3.2. Length of Training Datasets

The length of training datasets is an important factor affecting prediction accuracy. The 1-step-ahead NRMSE of different forecasting methods, with training Datasets A and B, are presented in Figure 8.

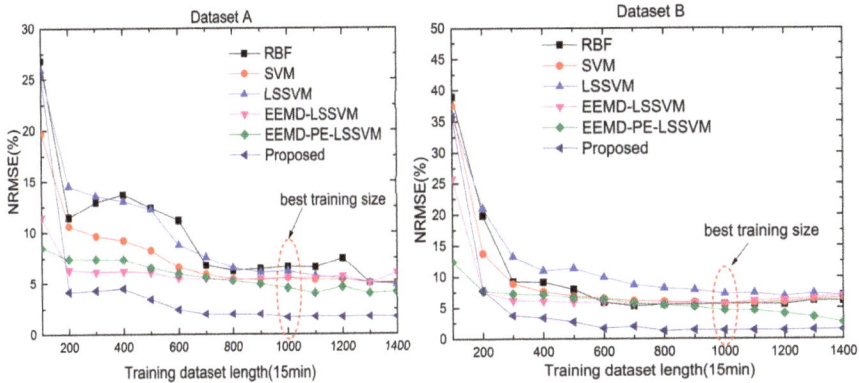

Figure 8. The NRMSE of 1-step-ahead forecasting by different models with two training datasets.

In Figure 8, for Datasets A and B, the NRMSE values for six models tended to decrease with the length of the training dataset.

For Dataset A, the NRMSE of each method varied irregularly when the length of the training dataset increased from 100 to 700. The values in this interval could not be selected as the length of the training dataset. In the range of 700–1400, the trend of NRMSE became flat. The proposed model was most insensitive to training dataset length, and the NRMSE of the proposed approach remained almost unchanged as the dataset length was greater than 700, which shows the proposed model was a simple, but powerful forecasting method.

For Dataset B, the NRMSE for each method varied irregularly when the length of training dataset increased from 100 to 600. Similarly, the values in this interval could not be selected as the length of the training dataset. In the range of 600–1400, the trend of NRMSE became flat, and the other five methods remained unchanged after 1000. However, the EEMD-PE-LSSVM model kept decreasing at 1000.

Taking into account the sensitivity of each method to the data, 1000 data points, as the length of the training set, was appropriate.

5. Results and Discussion

The proposed hybrid model was employed to forecast ultra-short-term wind power, and the corresponding results from the proposed model and other contrast models are discussed in the following section.

5.1. Experiment 1: The Comparison Results of the Proposed Model and Other Models

Ultra-short-term wind power for 1-step, 2-step, 3-step and 4-step-ahead prediction was implemented for Datasets A, B, C, and D. Results from the analyses will be clearly demonstrated in Tables 4–7 to reveal the effectiveness of each model.

To further verify the applicability, performance, and superiority of the proposed hybrid model, the wind power data from Datasets A, B, C, and D were employed for modeling, with five alternative forecasting models (the SVM model, RBF model, LSSVM model, EEMD-LSSVM model, and EEMD-PE-LSSVM model) that were compared with the proposed hybrid model. The results are shown in Figures 9–12 and Tables 4–7.

In Figures 9–12, it can be seen that the error of prediction results for the proposed model was much smaller than the SVM model, RBF model, LSSVM model, EEMD-LSSVM model, and EEMD-PE-LSSVM, which implies that the proposed method performs much better than other five models. The prediction results of the EEMD-PE-LSSVM model lagged behind the proposed model for all forecasting time horizons. The forecasting results of the RBF model were the worst compared to the other models.

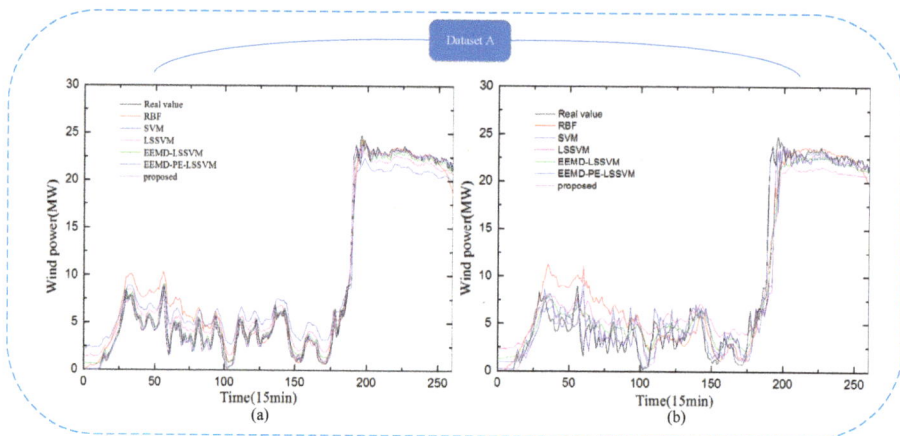

Figure 9. Comparison between the forecast and real values for Dataset A. (**a**) 1-step-ahead result; (**b**) 4-step-ahead result.

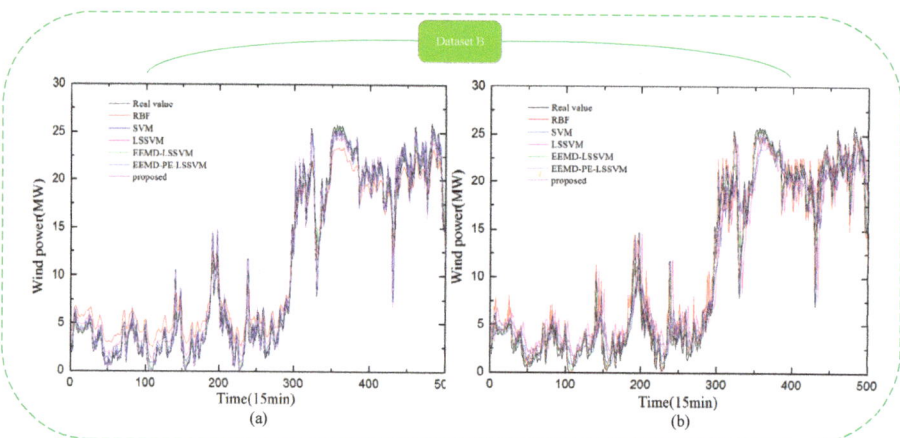

Figure 10. Comparison between the forecast and real values for Dataset B. (**a**) 1-step-ahead result; (**b**) 4-step-ahead result.

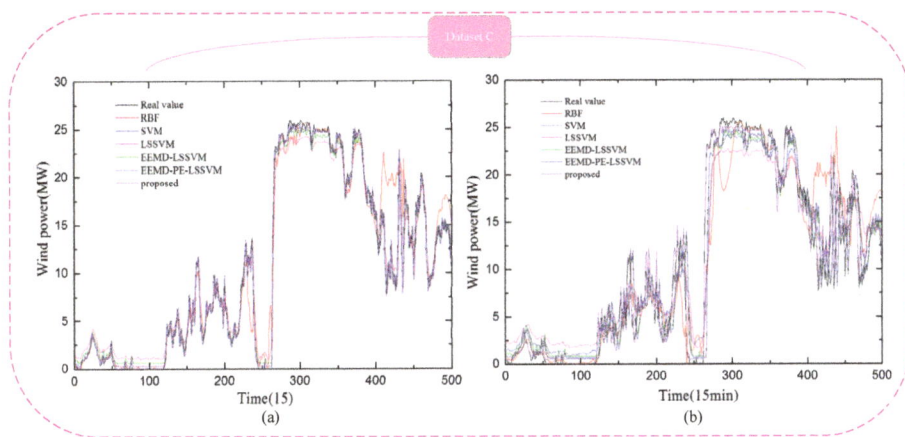

Figure 11. Comparison between the forecast and real values for Dataset C. (**a**) 1-step-ahead result; (**b**) 4-step-ahead result.

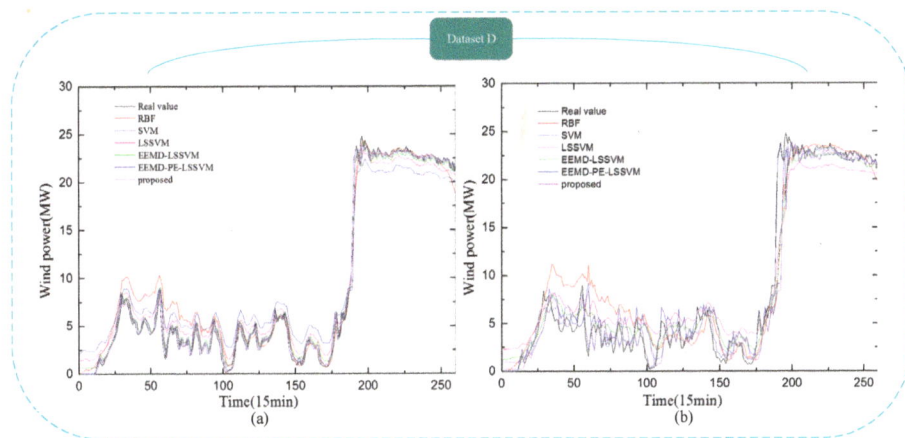

Figure 12. Comparison between the forecast and real values for Dataset D. (**a**) 1-step-ahead result; (**b**) 4-step-ahead result.

In order to test the accuracy of wind power forecasting, NRMSE, NMAE, R, $\zeta_{\text{NMAE}(\%)}$, and $\zeta_{\text{NRMSE}(\%)}$ were used in this paper, for Datasets A, B, C, and D with 1-step-ahead and 4-steps-ahead. Detailed numerical analysis is given in Tables 4–7.

It can be observed from Tables 4–7 that the NRMSE and NMAE values of the proposed method were the lowest and the R values were the highest, compared with other models for (1–4)-step-ahead prediction during the entire evaluation period, which demonstrates the superior performance of the proposed method. For all the forecasting horizons investigated in this paper, the proposed method always reached the minimum values of NRMSE and NMAE and the maximum value of R. This indicates that the proposed model significantly outperforms benchmark models. Thus, the proposed model is an effective tool for wind power forecasting.

Table 4. The values of NRMSE, NMAE and R for six models in Dataset A.

Indicator	NRMSE (%)				NMAE (%)				R (%)			
Time horizon	1-step	2-step	3-step	4-step	1-step	2-step	3-step	4-step	1-step	2-step	3-step	4-step
RBF	6.6432	8.3647	9.0717	10.7424	4.3353	5.6122	6.5253	7.6562	90.67	86.57	81.63	75.78
SVM	5.5528	7.1953	8.6525	10.1627	4.3245	5.5247	6.6472	7.7864	94.87	90.72	85.89	80.20
LSSVM	6.2582	7.8868	9.1352	10.1816	5.0344	6.2646	7.1246	7.8225	93.65	85.53	77.73	71.21
EEMD-LSSVM	5.5714	8.4635	10.5156	12.0235	4.1735	6.3327	7.9133	8.9266	95.7	91.46	86.3	82.24
EEMD-PE-LSSVM	4.5363	6.9125	8.5431	9.7176	3.3664	4.3626	5.9448	8.9611	96.75	92.32	88.9	84.35
Proposed	4.1884	6.7376	7.3766	9.0113	3.1867	4.1885	5.3326	8.6643	99.46	98.96	97.99	97.68

Table 5. The values of NRMSE, NMAE and R for six models in Dataset B.

Indicator	NRMSE (%)				NMAE (%)				R (%)			
Time horizon	1-step	2-step	3-step	4-step	1-step	2-step	3-step	4-step	1-step	2-step	3-step	4-step
RBF	5.1226	6.4646	6.9672	7.5973	3.4726	4.5874	4.9797	5.3514	96.05	96.19	94.46	93.44
SVM	5.6434	6.5854	7.3128	8.3364	4.5374	5.1553	5.6744	6.1772	94.47	96.30	95.09	93.90
LSSVM	7.2826	8.3536	8.7153	8.9557	6.5156	7.2438	7.4639	7.5716	97.9	96.32	95.3	94.61
EEMD-LSSVM	5.6447	8.1175	9.3537	10.2974	4.0737	5.7973	6.7243	7.5343	96.62	92.88	90.56	88.58
EEMD-PE-LSSVM	4.5356	6.6436	7.8475	8.7764	3.8383	5.7538	6.8929	7.9927	98.25	96.22	94.76	93.36
Proposed	4.3235	6.3164	7.5626	8.9962	2.9927	4.5847	5.9927	6.3338	99.81	99.49	99.42	99.26

Table 6. The values of NRMSE, NMAE and R for six models in Dataset C.

Indicator	NRMSE (%)				NMAE (%)				R (%)			
Time horizon	1-step	2-step	3-step	4-step	1-step	2-step	3-step	4-step	1-step	2-step	3-step	4-step
RBF	9.4631	9.8434	9.4552	12.1662	5.3172	5.4662	5.6846	7.7175	92.05	87.07	87.95	83.39
SVM	4.9958	6.2820	7.2430	8.3250	4.0549	4.7668	5.4573	6.1938	98.32	96.71	95.30	95.30
LSSVM	6.7291	7.8755	8.5629	9.4239	5.8439	6.5553	6.9874	7.4729	98.09	95.54	94.24	92.74
EEMD-LSSVM	4.3984	6.9439	8.8273	10.2339	3.1895	4.5649	5.6227	6.4657	98.08	95.00	91.85	89.00
EEMD-PE-LSSVM	3.4629	5.6243	7.1929	8.3749	2.8136	4.2865	5.4144	6.2728	99.04	97.44	95.77	94.25
Proposed	3.9255	5.2738	6.2736	8.1817	2.5687	4.0292	5.2535	6.1093	99.92	99.85	99.49	99.26

Table 7. The values of NRMSE, NMAE and R for six models in Dataset D.

Indicator	NRMSE (%)				NMAE (%)				R (%)			
Time horizon	1-step	2-step	3-step	4-step	1-step	2-step	3-step	4-step	1-step	2-step	3-step	4-step
RBF	5.7482	6.6848	6.2486	9.1483	4.0141	4.6735	4.0979	6.2376	97.03	96.58	95.21	91.00
SVM	4.4814	5.1492	5.7937	6.8674	3.7321	3.9372	4.1503	4.6592	99.03	98.11	96.95	95.49
LSSVM	6.6836	7.1152	7.5974	8.2797	6.1482	6.2483	6.4237	6.7349	99.2	98.15	97.18	96.11
EEMD-LSSVM	3.2975	5.6603	6.3875	7.7385	2.2579	3.3714	4.1872	4.8933	98.89	96.95	95.03	93.25
EEMD-PE-LSSVM	3.1542	5.2385	6.1969	7.5508	1.9632	3.1385	3.9653	4.6587	99.45	98.46	97.44	96.48
Proposed	2.8432	5.1474	6.0947	6.3749	1.1869	4.9532	4.3978	5.9485	99.93	99.8	99.59	99.20

5.2. Experiment II: Comparison Results of Improvement Percentage Error Indexes and Modeling Time

In order to compare performance differences between the combined model and other benchmark models, $\zeta_{NMAE(\%)}$ and $\zeta_{NRMSE(\%)}$ were utilized in this study. By using this type of criterion, the improvement percentage values of the proposed model and benchmark models are given in Figure 13.

The histogram of $\zeta_{NMAE(\%)}$ and $\zeta_{NRMSE(\%)}$ for all models regarding Datasets A, B, C, and D for (1–4)-step-ahead wind power forecasting are shown in Figure 13. For Dataset A, except for the 1- and 4-step-ahead forecasting, the $\zeta_{NRMSE(\%)}$ values of the proposed method compared with EEMD-PE-LSSVM were positive, which shows that the performance of the proposed model was worse than the EEMD-PE-LSSVM model. In the 1-step-ahead forecasting, the negative value of $\zeta_{NRMSE(\%)}$ for the proposed method, compared with the RBF model, was minimal, which indicates that the forecasting performance of the proposed model was powerful. In the 2-, 3- and 4-step-ahead forecasting, the proposed method, compared with the EEMD-LSSVM model, performed best.

In (1–3)-step-ahead forecasting, the proposed method had a positive value of $\zeta_{NMAE(\%)}$ compared with EEMD-PE-LSSVM, which shows that the proposed model was worse than the EEMD-PE-LSSVM model. In the 4-step ahead forecasting, the proposed model was slightly worse than the RBF, SVM, and LSSVM models.

There were similar results for Datasets B, C, and D, which shows that the proposed model was effective for ultra-short-term wind power prediction.

As demonstrated in Figure 13, we can derive the following conclusions: (a) heuristic algorithms have good optimization capabilities in wind power forecasting; (b) hybrid models obtain better performance compared with individual and other combined models without optimization; and (c) the proposed model performed the best among all of the studied models.

The simulation time of the 4-step-ahead wind power forecasting, with regard to Datasets A, B, C, and D, for all methods, is given in Table 8. Although the simulation time of the proposed method had higher time consumption than the other prediction models, it achieved the best prediction accuracy, and this simulation time is acceptable in practical implementation.

Table 8. The simulation time for all methods.

Approaches	Proposed	EEMD-PE-LSSVM	EEMD-LSSVM	LSSVM	SVM	RBF
Dataset A	158.4554	130.6064	129.5019	30.4655	12.5090	14.5357
Dataset B	158.1473	130.3876	129.4410	30.3430	11.6306	13.1059
Dataset C	159.9275	132.2656	129.4538	28.4974	12.6486	14.9028
Dataset D	154.4916	127.4774	126.4261	27.3058	11.4950	12.6474
Average	157.75545	130.18425	128.7057	29.152925	12.0708	13.79795

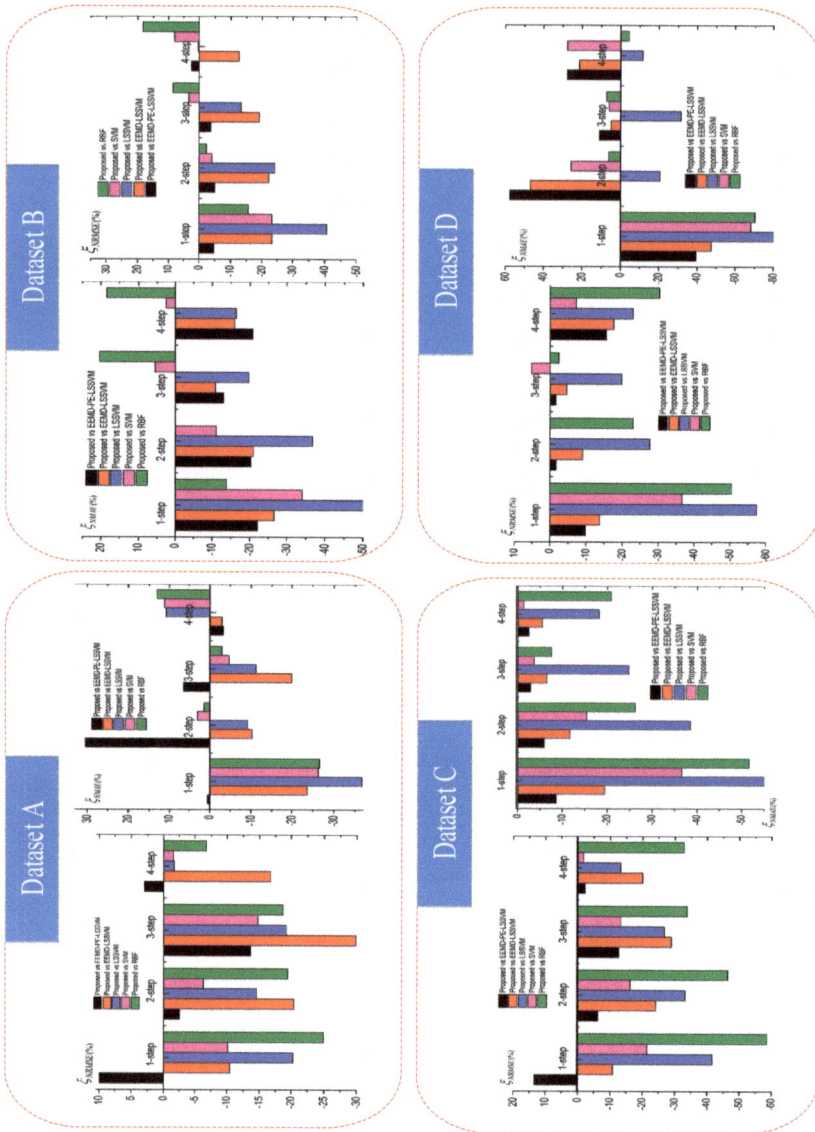

Figure 13. Histograms of NRMSE and NMAE for all models using Datasets A, B, C, and D, with (1–4)-step-ahead.

6. Conclusions

A new hybrid prediction method of ultra-short-term wind power forecasting, based on EEMD-PE and LSSVM optimization by GSA, was proposed in this paper. The EEMD method was used to decompose raw wind power data time series into a series of IMFs with different scales to solve the mode mixing problem. To effectively reduce the computational complexity of the combined forecasting method, PE was introduced into the complexity assessment of each IMF component, based on the PE value, then the IMF components were recombined to generate new subsequences with significant differences in complexity. The GSA model was utilized to optimize the parameters of LSSVM, which avoided the choice of parameters; then, the optimized model was used in wind power forecasting, which improved regression prediction accuracy. For a fair, clear, comparative study, the proposed method was tested on a practical wind farm (in Hebei, China) and compared with several other models, including the EEMD-PE-LSSVM, EEMD-LSSVM, LSSVM, SVM, and RBF models. The results of the experiments indicated that the proposed model satisfactorily forecasted ultra-short-term wind power for the different datasets.

Acknowledgments: This work was supported by the National Natural Science Foundation of China (Project No. 51477174, 51677188, and 51711530227), National Key Research and Development Program of China (Project No. 2017YFB0902200), Project of State Grids Corporation of China (Project No. 5201011600TS), and the Open Fund of State Key Laboratory of Operation and Control of Renewable Energy and Storage Systems (Project No. TSTE-00833-2016).

Author Contributions: Peng Lu designed the research and wrote the paper; Lin Ye provided professional guidance; Bohao Sun, Jingzhu Teng, Cihang Zhang, and Yongning Zhao translated and revised this paper.

Conflicts of Interest: The authors declare no conflict of interest.

Abbreviations

$x(t)$	the original wind power series	NRMSE	the normalized root mean square error
$w_i(t)$	the white noise series	NMAE	the normalized mean absolute error
$x_{new,i}(t)$	the wind power series with white noise	R	pearson correlation coefficient
N	the total number of data	EMD	empirical mode decomposition
$S(g)$	a set of symbols	EEMD	ensemble empirical mode decomposition
P_k	the probability of each symbol sequence	VMD	variational mode decomposition
$H_p(m)$	the permutation entropy	WT	wavelet Transform
RS	reconstitute subsequences	WD	wavelet decomposition
$\varphi(x)$	is the nonlinear function	SVM	support vector machine
ω	the weight	LSSVM	least squares support vector machine
b	the bias	RBF	radial basis neural network
R_e	the empirical risk function	GA	genetic algorithm
$G(t)$	the gravitational constant	SSO	simplified swarm optimization
G_0	the initial gravitational constant	CSA	clonal selection algorithm
α	the decay rate	PSO	particle swarm optimization
T	the maximum generation	FFA	firefly optimization algorithm
$a_i^d(t)$	acceleration	GSA	gravitational search algorithm
v_i^d	the speed of particle ith in dimension	MOBA	multi-objective bat algorithm
x_i^d	the position of particle ith in dimension	BP	back propagation artificial neural network
x_i	the actual wind power value	BA	bat algorithm
$\hat{x}_i$	the forecasting value		

References

1. Zhao, Y.; Ye, L.; Li, Z.; Song, X.; Lang, Y.; Su, J. A novel bidirectional mechanism based on time series model for wind power forecasting. *Appl. Energy* **2016**, *177*, 793–803. [CrossRef]
2. Lydia, M.; Kumar, S.S.; Selvakumar, A.I.; Kumar, G.E.P. A comprehensive review on wind turbine power curve modeling techniques. *Renew. Sustain. Energy Rev.* **2014**, *30*, 452–460. [CrossRef]

3. World Wind Market Has Reached 486 GW from Where 54 GW Has Been Installed Last Year. Available online: http://www.wwindea.org/11961-2/ (accessed on 23 January 2018).

4. Ren, G.; Liu, J.; Wan, J.; Guo, Y.; Yu, D.; Yan, J. Overview of wind power intermittency: Impacts, measurements, and mitigation solutions. *Appl. Energy* **2017**, *204*, 47–65. [CrossRef]

5. Liu, Y.Q.; Sun, Y.; Infield, D.; Zhao, Y.; Han, S.; Yan, J. A hybrid forecasting method for wind power ramp based on orthogonal test and support vector machine (ot-svm). *IEEE Trans. Sustain. Energy* **2017**, *8*, 451–457. [CrossRef]

6. Jiang, Y.; Chen, X.Y.; Yu, K.; Liao, Y.C. Short-term wind power forecasting using hybrid method based on enhanced boosting algorithm. *J. Mod. Power Syst. Clean Energy* **2017**, *5*, 126–133. [CrossRef]

7. Alencar, D.B.D.; Affonso, C.D.M.; Oliveira, R.L.D.; Rodríguez, J.M.; Leite, J.; Filho, J.R. Different models for forecasting wind power generation: Case study. *Energies* **2017**, *10*, 1976.

8. Tascikaraoglu, A.; Uzunoglu, M. A review of combined approaches for prediction of short-term wind speed and power. *Renew. Sustain. Energy Rev.* **2014**, *34*, 243–254. [CrossRef]

9. Dong, L.; Wang, L.J.; Khahro, S.F.; Gao, S.; Liao, X.Z. Wind power day-ahead prediction with cluster analysis of nwp. *Renew. Sustain. Energy Rev.* **2016**, *60*, 1206–1212. [CrossRef]

10. Chang, W.Y. A literature review of wind forecasting methods. *J. Power Energy Eng.* **2014**, *2*, 161–168. [CrossRef]

11. Genikomsakis, K.N.; Lopez, S.; Dallas, P.I.; Ioakimidis, C.S. Simulation of wind-battery microgrid based on short-term wind power forecasting. *Appl. Sci.* **2017**, *7*, 1142. [CrossRef]

12. Landberg, L.; Watson, S.J. Short-term prediction of local wind conditions. *Bound.-Layer Meteorol.* **1994**, *70*, 171–195. [CrossRef]

13. Focken, U.; Lange, M.; Waldl, H.-P. Previento-a wind power prediction system with an innovative upscaling algorithm. In Proceedings of the European Wind Energy Conference, Copenhagen, Denmark, 2–6 July 2001.

14. Development and Testing of Improved Statistical Wind Power Forecasting Methods. Available online: digital.library.unt.edu/ark:/67531/metadc829494/ (accessed on 23 January 2018).

15. Aggarwal, S.K. Wind power forecasting: A review of statistical models. *Int. J. Energy Sci.* **2013**, *3*, 1–10.

16. Lydia, M.; Kumar, S.S.; Selvakumar, A.I.; Kumar, G.E.P. Linear and non-linear autoregressive models for short-term wind speed forecasting. *Energy Convers. Manag.* **2016**, *112*, 115–124. [CrossRef]

17. Jiang, W.; Yan, Z.; Feng, D.H.; Hu, Z. Wind speed forecasting using autoregressive moving average/generalized autoregressive conditional heteroscedasticity model. *Eur. Trans. Electr. Power* **2012**, *22*, 662–673. [CrossRef]

18. Abdelaziz, A.Y.; Rahman, M.A.; El-Khayat, M.M.; Hakim, M.A. Short term wind power forecasting using autoregressive integrated moving average approach. *J. Energy Power Eng.* **2013**, *7*, 2089.

19. Fang, T.; Lahdelma, R. Evaluation of a multiple linear regression model and sarima model in forecasting heat demand for district heating system. *Appl. Energy* **2016**, *179*, 544–552. [CrossRef]

20. Maggina, A. *Market-Based Accounting Research (Mbar) Models: A Test of Arimax Modeling*; Springer: New York, NY, USA, 2015; pp. 279–298.

21. Costa, A.; Crespo, A.; Navarro, J.; Lizcano, G.; Madsen, H.; Feitosa, E. A review on the young history of the wind power short-term prediction. *Renew. Sustain. Energy Rev.* **2008**, *12*, 1725–1744. [CrossRef]

22. Kiplangat, D.C.; Asokan, K.; Kumar, K.S. Improved week-ahead predictions of wind speed using simple linear models with wavelet decomposition. *Renew. Energy* **2016**, *93*, 38–44. [CrossRef]

23. Liu, H.; Tian, H.Q.; Li, Y.F. Comparison of new hybrid feemd-mlp, feemd-anfis, wavelet packet-mlp and wavelet packet-anfis for wind speed predictions. *Energy Convers. Manag.* **2015**, *89*, 1–11. [CrossRef]

24. Liu, D.; Niu, D.; Wang, H.; Fan, L. Short-term wind speed forecasting using wavelet transform and support vector machines optimized by genetic algorithm. *Renew. Energy* **2014**, *62*, 592–597. [CrossRef]

25. Giorgi, M.G.D.; Campilongo, S.; Ficarella, A.; Congedo, P.M. Comparison between wind power prediction models based on wavelet decomposition with least-squares support vector machine (LS-SVM) and artificial neural network (ANN). *Energies* **2014**, *7*, 5251–5272. [CrossRef]

26. Jyothi, M.N.; Rao, P.R. Very-short term wind power forecasting through adaptive wavelet neural network. In Proceedings of the 2016 Biennial International Conference on Power and Energy Systems: Towards Sustainable Energy (PESTSE), Bangalore, India, 21–23 January 2016; pp. 1–6.

27. Guo, Z.; Zhao, W.; Lu, H.; Wang, J. Multi-step forecasting for wind speed using a modified emd-based artificial neural network model. *Renew. Energy* **2012**, *37*, 241–249. [CrossRef]

28. Fan, G.F.; Peng, L.L.; Zhao, X.; Hong, W.C. Applications of hybrid emd with pso and ga for an svr-based load forecasting model. *Energies* **2017**, *10*, 1713. [CrossRef]
29. Zhang, Y.; Liu, K.; Qin, L.; An, X. Deterministic and probabilistic interval prediction for short-term wind power generation based on variational mode decomposition and machine learning methods. *Energy Convers. Manag.* **2016**, *112*, 208–219. [CrossRef]
30. Zhao, Y.; Ye, L.; Wang, W.; Sun, H.; Ju, Y.; Tang, Y. Data-driven correction approach to refine power curve of wind farm under wind curtailment. *IEEE Trans. Sustain. Energy* **2018**, *9*, 95–105. [CrossRef]
31. Niu, D.; Liang, Y.; Hong, W.-C. Wind speed forecasting based on emd and grnn optimized by foa. *Energies* **2017**, *10*, 2001. [CrossRef]
32. Ren, Y.; Suganthan, P.N.; Srikanth, N. A comparative study of empirical mode decomposition-based short-term wind speed forecasting methods. *IEEE Trans. Sustain. Energy* **2014**, *6*, 236–244. [CrossRef]
33. Jiang, Y.; Huang, G. Short-term wind speed prediction: Hybrid of ensemble empirical mode decomposition, feature selection and error correction. *Energy Convers. Manag.* **2017**, *144*, 340–350. [CrossRef]
34. Patlakas, P.; Drakaki, E.; Galanis, G.; Spyrou, C.; Kallos, G. *Wind Gust Estimation by Combining a Numerical Weather Prediction Model and Statistical Post-Processing*; EGU: Munich, Germany, 2017.
35. Liang, Z.; Liang, J.; Wang, C.; Dong, X.; Miao, X. Short-term wind power combined forecasting based on error forecast correction. *Energy Convers. Manag.* **2016**, *119*, 215–226. [CrossRef]
36. Davò, F.; Alessandrini, S.; Sperati, S.; Monache, L.D.; Airoldi, D.; Vespucci, M.T. Post-processing techniques and principal component analysis for regional wind power and solar irradiance forecasting. *Sol. Energy* **2016**, *134*, 327–338. [CrossRef]
37. Li, Z.; Ye, L.; Zhao, Y.; Song, X.; Teng, J.; Jin, J. Short-term wind power prediction based on extreme learning machine with error correction. *Prot. Control Mod. Power Syst.* **2016**, *1*. [CrossRef]
38. Xiao, L.; Qian, F.; Shao, W. Multi-step wind speed forecasting based on a hybrid forecasting architecture and an improved bat algorithm. *Energy Convers. Manag.* **2017**, *143*, 410–430. [CrossRef]
39. Wang, J.; Heng, J.; Xiao, L.; Wang, C. Research and application of a combined model based on multi-objective optimization for multi-step ahead wind speed forecasting. *Energy* **2017**, *125*, 591–613. [CrossRef]
40. Huang, M.L. Hybridization of chaotic quantum particle swarm optimization with svr in electric demand forecasting. *Energies* **2016**, *9*, 426. [CrossRef]
41. Chang, W.Y. Short-term wind power forecasting using the enhanced particle swarm optimization based hybrid method. *Energies* **2013**, *6*, 4879–4896. [CrossRef]
42. Chitsaz, H.; Amjady, N.; Zareipour, H. Wind power forecast using wavelet neural network trained by improved clonal selection algorithm. *Energy Convers. Manag.* **2015**, *89*, 588–598. [CrossRef]
43. Yuan, X.; Chen, C.; Yuan, Y.; Huang, Y.; Tan, Q. Short-term wind power prediction based on LSSVM–GSA model. *Energy Convers. Manag.* **2015**, *101*, 393–401. [CrossRef]
44. Osório, G.J.; Matias, J.C.O.; Catalão, J.P.S. Short-term wind power forecasting using adaptive neuro-fuzzy inference system combined with evolutionary particle swarm optimization, wavelet transform and mutual information. *Renew. Energy* **2015**, *75*, 301–307. [CrossRef]
45. Dong, Z.; Yang, D.; Reindl, T.; Walsh, W.M. A novel hybrid approach based on self-organizing maps, support vector regression and particle swarm optimization to forecast solar irradiance. *Energy* **2015**, *82*, 570–577. [CrossRef]
46. Yeh, W.C.; Yeh, Y.M.; Chang, P.C.; Ke, Y.C.; Chung, V. Forecasting wind power in the mai liao wind farm based on the multi-layer perceptron artificial neural network model with improved simplified swarm optimization. *Int. J. Electr. Power Energy Syst.* **2014**, *55*, 741–748. [CrossRef]
47. Xiao, L.; Shao, W.; Yu, M.; Ma, J.; Jin, C. Research and application of a hybrid wavelet neural network model with the improved cuckoo search algorithm for electrical power system forecasting. *Appl. Energy* **2017**, *198*, 203–222. [CrossRef]
48. Zhou, H.; Xue, Y.; Guo, J.; Chen, J. Ultra-short-term wind speed forecasting method based on spatial and temporal correlation models. *J. Eng.* **2017**, *2017*, 1071–1075.
49. Tascikaraoglu, A.; Sanandaji, B.M.; Poolla, K.; Varaiya, P. Exploiting sparsity of interconnections in spatio-temporal wind speed forecasting using wavelet transform. *Appl. Energy* **2016**, *165*, 735–747. [CrossRef]
50. Ye, L.; Zhao, Y.; Zeng, C.; Zhang, C. Short-term wind power prediction based on spatial model. *Renew. Energy* **2017**, *101*, 1067–1074. [CrossRef]

51. Huang, N.E.; Shen, Z.; Long, S.R.; Wu, M.C.; Shih, H.H.; Zheng, Q.; Yen, N.C.; Chi, C.T.; Liu, H.H. The empirical mode decomposition and the hilbert spectrum for nonlinear and non-stationary time series analysis. *Proc. Math. Phys. Eng. Sci.* **1998**, *454*, 903–995. [CrossRef]

52. Wu, Z.; Huang, N.E. Ensemble empirical mode decomposition: A noise-assisted data analysis method. *Adv. Adapt. Data Anal.* **2009**, *1*, 1–41. [CrossRef]

53. Bandt, C.; Pompe, B. Permutation entropy: A natural complexity measure for time series. *Phys. Rev. Lett.* **2002**, *88*, 174102. [CrossRef] [PubMed]

54. Wu, Q.; Peng, C. Wind power grid connected capacity prediction using lssvm optimized by the bat algorithm. *Energies* **2015**, *8*, 14346–14360. [CrossRef]

55. Xing, B.; Gao, W.J. *Gravitational Search Algorithm*; Springer International Publishing: Berlin, Germanuy, 2014; pp. 355–364.

Article

Towards Cost and Comfort Based Hybrid Optimization for Residential Load Scheduling in a Smart Grid

Nadeem Javaid [1,*], Fahim Ahmed [1], Ibrar Ullah [1,2], Samia Abid [1], Wadood Abdul [3], Atif Alamri [3] and Ahmad S. Almogren [3]

1   COMSATS Institute of Information Technology, Islamabad 44000, Pakistan;
    fahim.bsee1214@gmail.com (F.A.); sammia.abid@gmail.com (S.A.)
2   University of Engineering and Technology Peshawar, Bannu 28100, Pakistan;
    ibrarullah@uetpeshawar.edu.pk
3   Pervasive and Mobile Computing, College of Computer and Information Sciences, King Saud University,
    Riyadh 11633, Saudi Arabia; aabdulwaheed@ksu.edu.sa (W.A.); ahalmogren@ksu.edu.sa (A.A.);
    atif@ksu.edu.sa (A.A.)
*   Correspondence: nadeemjavaidqau@gmail.com; Tel.: +92-300-0579-2728

Received: 28 August 2017; Accepted: 30 September 2017; Published: 8 October 2017

Abstract: In a smart grid, several optimization techniques have been developed to schedule load in the residential area. Most of these techniques aim at minimizing the energy consumption cost and the comfort of electricity consumer. Conversely, maintaining a balance between two conflicting objectives: energy consumption cost and user comfort is still a challenging task. Therefore, in this paper, we aim to minimize the electricity cost and user discomfort while taking into account the peak energy consumption. In this regard, we implement and analyse the performance of a traditional dynamic programming (DP) technique and two heuristic optimization techniques: genetic algorithm (GA) and binary particle swarm optimization (BPSO) for residential load management. Based on these techniques, we propose a hybrid scheme named GAPSO for residential load scheduling, so as to optimize the desired objective function. In order to alleviate the complexity of the problem, the multi dimensional knapsack is used to ensure that the load of electricity consumer will not escalate during peak hours. The proposed model is evaluated based on two pricing schemes: day-ahead and critical peak pricing for single and multiple days. Furthermore, feasible regions are calculated and analysed to develop a relationship between power consumption, electricity cost, and user discomfort. The simulation results are compared with GA, BPSO and DP, and validate that the proposed hybrid scheme reflects substantial savings in electricity bills with minimum user discomfort. Moreover, results also show a phenomenal reduction in peak power consumption.

Keywords: demand side management; demand response; home energy management system; meta-heuristic techniques

---

1. Introduction

It has been observed that the residential area is a major cause of energy consumption and greenhouse gas (GHG) emissions. In China, it is considered the second highest sector which is responsible for energy consumption and GHG emission. Around 40% of energy is consumed by the residential sectors in Arabian countries. While, In Palestine 60% of energy is consumed [1]. With the emergence of new types of demand (i.e., electric vehicles, smart appliances etc.) and economic development in the past couple of decades, a drastic increase in the energy consumption by the residential area has been noticed [2]. This shows that the residential sector has a significant role in energy consumption which in turn threatens the reliability and efficiency of the power grid. To fulfil the increasing demand, there is a need to install bulk generation and transmission infrastructure, which is a very cumbersome and expensive process. The electricity market has also increased the electricity prices in response to increased user demand. The concept of smart grid (SG) has emerged which introduces information and communication technology in the traditional grid infrastructure.

A smart home plays a very vital role to overcome the above mentioned challenges. It is equipped with smart appliances, smart meter and an energy management controller (EMC). Basically, the main motives behind the utilization of energy management programs include environmental concerns, capacity limits and reliability of utility infrastructure systems, maintainance and operation, and to meet the financial needs of consumers. The energy management in entire electrical network is classified into two categories: supply side management (SSM) and demand side management (DSM). The SSM is responsible for generating and delivering reliable energy to the consumers. Conversely, DSM utilizes the potential of advance communication and control infrastructure. It is one of the key components of the SG that aims at utilizing the available energy effectively and optimally. DSM designs demand response (DR) programs which entice the consumers to actively participate in load shifting mechanism in response to time varying prices. By shifting laod from on-peak to off-peak hours, the electricity consumer achieves significant reduction in cost but has to tolerate the discomfort in the form of delayed operation of appliances [3].

In the literature, a lot of efforts have been done to tackle the above mentioned challenges. We categorise the literature review into two main threads. In the first thread, we will discuss the concerns related to the minimization of electricity cost, peak load and users' discomfort. Rastegar et al. in [4] proposed an idea of cost minimization along with the value of lost load (VOLL). The idea behind VOLL is to enhance the consumers' priorities and minimize the difference between the actual and the predetermined energy consumption of appliances. Authors in [5,6] considered energy cost as an objective function to be optimized by efficiently utilizing the available energy. The trade-off analysis between privacy and cost is addressed in [7], whereas [8] demonstrated a trade-off between consumers' comfort and operation delay of devices. In [9], Vardakas et al. uses the recursive process for peak load calculation. The authors develop four control scenarios under real-time pricing (RTP) environment. Ref. [10] considered user satisfaction in the proposed approach while restricting the total cost under the predefined budget. However, in this approach, devices with high power ratings are neglected. The results validate that the proposed models have efficiently and optimally reduced the electricity consumption cost of the consumers. In [11,12], the proposed techniques aim to minimize the energy consumption cost while taking into account the users' convenience. Thermal comfort is taken as a metric of users' satisfaction in the proposed work. Muralitharan et al. in [13] aim to minimize the consumption cost while considering the waiting time of consumers. ToU pricing mechanism is used in the proposed scheme. The results validate the trade-off between cost and waiting time of consumers. In [14], the authors developed a novel concept of cost efficiency (CE): the ratio of total energy consumption benefits to the total electricity payments. CE is considered as an indicator for consumers to adjust their energy consumption pattern. Moreover, the effects of DERs and service fee on CE are analyzed. The performance results show that CE is increased with increasing DERs and decreased with increasing service fee. Authors in [15] designed a model to minimize the consumption cost and balance the energy consumption under ToU pricing scheme. Moreover, renewable and storage systems are efficiently addressed, so in this way the surplus energy can be sold back to the grid. Zhang et al. [16] proposed a model to minimize the electricity cost and reduce carbon emission. A cluster of 30 houses is taken under consideration and each house having 12 devices subjected to control. In [17] authors deal with the problem of unanticipated peaks. In [18–22], electricity consumption cost and operational delay of devices are addressed as an optimization problem. Minimization of end users' electricity consumption cost and comfort maximization are the two conflicting objectives to achieve, simultaneously.

While in the second thread, we discuss the scheduling techniques used for managing energy in the smart homes. In DSM, several optimization techniques exist to efficiently manage the energy consumption behavior of consumers. Many researchers focused at both mathematical and heuristic optimization techniques which are capable to optimally schedule the consumers' load. In [10,23,24], the authors applied genetic algorithm (GA) as an optimization technique, in which electricity cost is taken as a primary objective function to be minimized. The MINLP along with dynamic pricing scheme is used in [25] to manage energy in a smart home. Bahrami et al. in [26] approximate the users' optimal scheduling policy by using Markov perfect equilibrium (MPE). The authors have developed an online load scheduling learning (LSL) algorithm which helps to determine the users' MPE policy. Samadi et al. in [27] propose a novel real-time pricing algorithm for the future smart grid by creating an interaction between smart meters and energy providers and exchanging the real-time price and energy consumption information of subscribers'. In [28,29], residential load scheduling problem is solved by using MINLP. In [30], the authors used game theoretic approach for cost minimization problem, whereas in [31], a variant of ant colony optimization (ACO) is used to solve the energy management problem. Yi et al. in [18] proposed an opportunistic based optimal stopping rule (OSR) for scheduling of home appliances. Chakraborty et al. [32] devised a system for energy optimization by the integration of Photovoltaic (PV) and a wind turbine as renewable energy sources (RESs) In order to address uncertainties occurred by RESs integration fuzzy logic is considered. For optimal scheduling and dispatching of energy, an efficient quantum evolutionary algorithm (EA) is implemented while considering the economic and environmental impacts. Moreover, scheduling is performed optimally in order to alleviate the cost of production and carbon emission resource. In [33], the authors have used the optimization techniques: teacher learning based optimization (TLBO) and shuffled frog leap (SFL) to efficiently manage energy in smart home. MILP is applied in [34] and [35] for efficient utilization of available energy. Gupta et al. in [34] proposed a model based on cost minimization problem. MILP is used for problem formulation, whereas, load scheduling is performed via genetic algorithm (GA). Authors in [36] developed a dynamic model for home energy management system (HEMS). The developed model employs the game theoretic approach for efficient scheduling of residential load. Safdarian et al. in [37] categorized DSM infrastructure into two stages. In first stage, decentralized system is considered and the aim is to minimize the electricity cost of consumers. MILP is used to formulate the problem and is solved by using general algebraic modeling system (GAMS). In second stage, the aim of the proposed model is to benefice the utility by modifying the load profile while preserving the constraints of cost and comfort. Mixed integer quadratic programming is used to achieve the objective of modifying load profile curve. In [38], the authors proposed a model based on a large number of residential appliances. PL-Generalized Bender's technique is used for scheduling the residential load and protecting the private information of the consumers. This model efficiently handled the consumption cost of the residential consumers along with the protection of privacy. The interval number optimization technique is proposed in [39] to handle residential load scheduling problem, thermostatically controlled and interruptible loads are considered in this scheme. Moreover, BPSO combined with integer linear programming (ILP) is used for load scheduling.

In the literature, as discussed earlier, zillion of methods are introduced for efficient utilization of available energy by using DSM infrastructure. The entire electrical network can be made well balanced and reliable, by managing the energy consumption, electricity cost, peak to average ration (PAR) and users' satisfaction. The work discussed above addresses the challenges of electricity consumption cost, consumers' convenience and peak demand reduction. However, some of the challenges are yet to be addressed by the research community; both by industry and academia. The real time implementation of the current system still requires a lot of advanced research efforts. Dynamic and adaptive control systems should be developed to predict and monitor the energy consumption behavior of occupants, comfort level of consumers and more importantly, stability of the entire grid. In [40,41], authors proposed a model for the scheduling of large number of devices with an objective of cost minimization and reduction of peak power consumption. Load scheduling strategy is applied in order to achieve an optimal energy consumption pattern. Evolutionary algorithm (EA) is implemented to apportion the consumers' load aptly over the time horizon. The proposed models perform well in terms of cost minimization and peak demand reduction, however, consumers' comfort is not addressed, which is a key component for end users' to participate in DR programs. In [42], minimization of electricity consumption cost and user discomfort are considered as an objective function. Time flexible and power flexible appliances are considered for efficient utilization of energy. The scheduling problem is formulated as convex optimization and electricity price is defined by the utility on day ahead basis. The simulations results show that the proposed technique achieved a desire trade-off between both the parameters of an objective function. However, by increasing the size of problem the computational complexity also increases. In order to address these challenges: cost and discomfort minimization along with peak demand reduction, a hybrid technique is proposed. The contributions of this paper are as follows:

- GAPSO: In this paper, we focus on designing a load shifting technique with day-ahead pricing (DAP) mechanism. We demonstrate the performance of a traditional optimization technique and two heuristic optimization techniques. After analyzing GA and BPSO, it is observed that these techniques show pre mature convergence when dealing with high dimension problems. So, there is a need to develop such an optimization method which can improve search efficiency and precision and adequate to handle multiple constraints. Based on these heuristic techniques, a hybrid technique is proposed with the objectives of cost and discomfort minimization. Extensive simulations are conducted to validate the results. The efficiency of the proposed technique is validated by analyzing the performance metrics, which show high cost savings with minimum user discomfort. Furthermore, our proposed model has less computational complexity and more generality.
- We formulate the binary optimization problem through multiple knapsack problem (MKP). MKP helps in the effort of finding an optimal solution while employing GAPSO and respecting the total capacity of available amount of power.

The rest of the paper is organized as follows. Section 2 elaborates the system model. The problem formulation is briefly discussed in Section 3. Heuristic optimization techniques used in this work is given in Section 4. In Section 5 proposed technique is discussed. Section 6 contains simulations and discussions. Section 7 concludes the work.

2. System Model

In this research work, we consider multiple smart homes in a residential area where the appliances in each smart home have low energy consumption ratings and short length of operation. There are 2604 controllable appliances available in this sector from 14 different types of appliances. All types of appliances have different energy consumption pattern and operation time. As in this area consumers have low priorities regarding the time when the energy has to be utilized, so more savings can be achieved in residential sector. The amount of incentives given to consumers depend on how much discomfort the consumer is willing to undergo. In the proposed model, we considered shiftable appliances. However, devices fulfill their length of operation time without exceeding the maximum allowable delay. Additionally, in the proposed model comfort level of consumer is incorporated, as a result of which the consumption cost is increased. Moreover, half an hour time slot is considered in the proposed model. The power ratings of appliances and their length of operation are given in Table 1.

The system model comprises of energy management controller, smart homes, communication networks and pricing model. The system model is demonstrated in Figure 1.

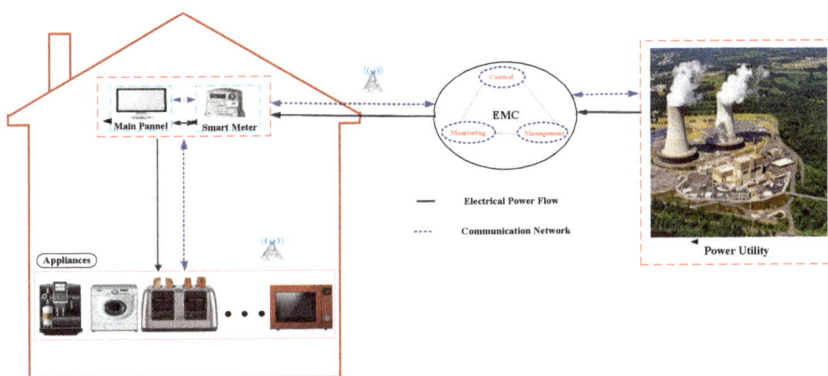

Figure 1. An Overview of Home Energy Management System Model.

Table 1. Appliances parameters.

Appliance's Type	Power Rating (kW)	Length of Operation Time (hours)	Total Devices
Dryer	1.2	4.0	189
Dish Washer	0.7	3.0	288
Washing Machine	2.0	2.5	268
Oven	1.3	3.0	279
Iron	1.0	2.0	340
Vacuum Cleaner	2.0	2.0	158
Fan	0.2	24	288
Kettle	2.0	4.0	406
Toaster	0.9	3.0	048
Rice Cooker	0.85	4.0	059
Hair Dryer	1.5	2.0	058
Blender	0.3	1.5	066
Frying Pan	1.1	1.5	101
Coffee Maker	0.8	1.5	056
Total	-	-	2604

2.1. Energy Management Controller

In this model, DSM focuses on efficient utilization of energy in residential sector. The power utility is directly connected to EMC and exchanges bidirectional information and unidirectional power flow in real time.The central EMC receives the price information from the power utility and performs the appropriate action. At the same time it contains the information from the consumer's end. It acts as a gate way between power utility and several homes. The main functionalities of EMC are monitoring, controlling and managing the residential load. In this case DSM uses load shifting as a basic scheme that can be implemented by using the central EMC. In this way EMC is capable to handle large number of residential appliances in well informed and organized manner. The residential devices send their arrival requests to the EMC and then requests are processed based on the availability of time slot. The scheduling mechanism is performed on day ahead basis.

2.2. Communication Network

The communication network includes wide area networks (WANs), neighbourhood area networks (NANs) and home area networks (HANs). The residential appliances are connected to smart meter via HANs. The residential appliances share their information to the smart meter and then this information is forwarded to the central EMC. The smart meters of different homes are connected to the central EMC via NANs. Through NANs the collective information is reached to the main EMC. The EMC exchanges the received information to the power utility via WANs. Through WANs the demand response and the price information is exchanged between power utility and the main EMC.

2.3. Pricing Schemes

In this work two pricing schemes are used, and based on these schemes we analyzed the performance of our proposed model.

2.3.1. DAP Model

The basic purpose of providing the pricing model on day ahead basis is to facilitate the consumers to take well-informed decisions. In this way consumers can adjust their electricity consumption pattern while taking care of comfort level. This helps consumers to reduce their electricity bills and these pricing models are readily available to consumers via advanced metering infrastructure. In this work, DAP is used similar to [41], and is shown in Figure 2. The pricing signal portrays three main regions: on-peak, off-peak and shoulder-peak hours. The load can be altered by observing the pricing signal offered by the utility.

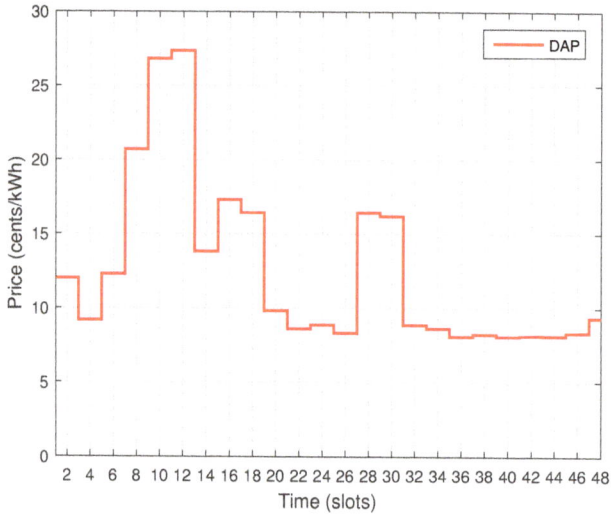

Figure 2. Electricity Price-DAP.

2.3.2. CPP Model

To validate and generalize the performance of the proposed model, we extend our approach by implementing the critical peak pricing (CPP) for load scheduling purpose. In CPP, depending upon the utility policies, electricity prices are double or even higher at critical peak hours. More specifically in this case we have considered a hot summer day, having critical peak hours from 12:00 p.m. to 3:00 p.m. where prices are almost double than usual. Critical events occurred very rarely in entire season or a year due to intense hot or cold weather. Figure 3 portrays the CPP pricing signal.

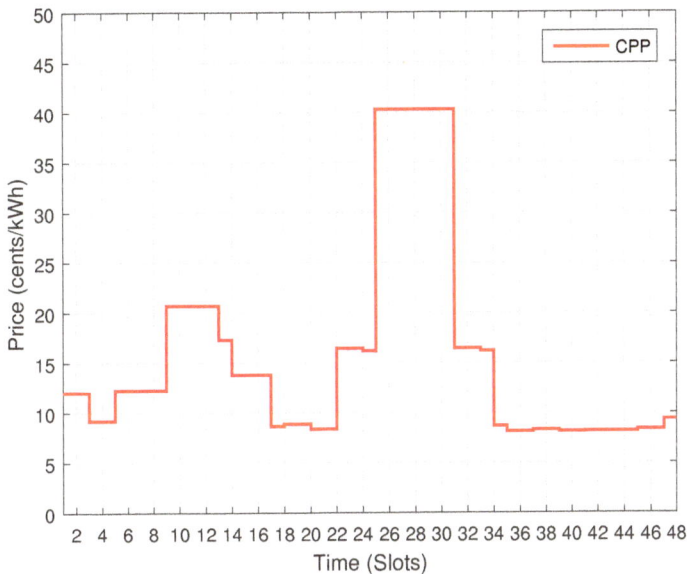

Figure 3. Electricity Price-CPP.

3. Problem Formulation

In this section, energy scheduling problem is formulated for an objective function and constraints. The aim is to minimize the consumption cost and maximize the users' comfort while respecting all the constraints. In objective function we formulate the maximization of user comfort as minimization of user discomfort, so both the terms are used interchangeably.

3.1. Multiple Knapsack Problem

The energy scheduling is one of the core issues in energy management system. In this work, multiple knapsack problem (MKP) is used to address the scheduling problem of a residential load. Knapsack is a combinatorial problem in which a number of objects, each having weight and value, must be packed in a bin of a specific capacity, in such a way that the total profit inside the bin is maximum. MKP is a resource allocation problem, and every resource has a specific capacity constraint. In this way, the system finds an optimal combination of household appliances operation modes while respecting the total capacity of available amount of power [43]. The reasons for using MKP are as follows,

1. It can be referred as a simplest integer linear programming (LP).
2. It can be viewed as subproblems in many complex problems.
3. It may represent the great practical situations.

For the sake of simplicity the abbreviations used in mathematical formulation are given in Table 2.

Table 2. Abbreviations.

Variables	Description
T	Time period of a day
T_u^t	User defined time
T_s^t	Scheduler defined time
EMC	Energy Management Controller
λ_i	Electricity price
T_{OTI}^i	Operation time interval of appliance i
α_i	Start time of an appliance i
β_i	End time of an appliance
T_{LoT}^i	Length of operation time of device i
Cap_T	Maximum allowable energy that can be used for each hour of the day

3.2. MKP in Energy Management System

The relation between the key terms used in MKP and energy management system can be developed as in [44] and is given below,

1. m knapsacks = m time interval.
2. n objects = n appliances.
3. w weight of an object = E_i Energy consumed by an appliance i.
4. value of an object = consumption cost of an appliance at time t.
5. Capacity of knapsack = user demand with respect to the maximum amount of energy that can be drawn from the grid at time t.

The mathematical formulation of an objective function and constraints is performed and the performance metrics of the considered problem are computed.

The electricity consumption cost, consumers' discomfort, total energy consumption and PAR are calculated and based on these equations we modelled the proposed approach and addressed the challenges of residential load.

The energy consumed by a single residential device over 24 h time horizon can be calculated by using the following equation,

$$E = \sum_{t=1}^{m} P_r^i \times \Delta_{i,t} \tag{1}$$

where P_r^i is power rating of a device i and $\Delta_{i,t}$ is status of device i at time slot t which can be given as,

$$\Delta_{i,t} = \begin{cases} 1 & \text{if device of type } i \text{ at time } t \text{ is ON;} \\ 0 & \text{otherwise.} \end{cases} \tag{2}$$

The total residential energy consumed by n number of smart devices can be calculated as follows,

$$E_e^r = \sum_{t=1}^{m} \left(\sum_{i=1}^{n} P_r^i \times \Delta_{i,t} \right) \tag{3}$$

Similarly, the total energy consumption cost of all the devices is given by

$$C_c^r = \sum_{t=1}^{m} \sum_{i=1}^{n} P_r^i (\Delta_{i,t} \times \lambda_t) \tag{4}$$

As maximizing the user comfort and minimizing the user discomfort can be used alternatively. So, for simplicity of our objective function i.e., Equation (7), we used "minimization" for both the parameters collectively.

The consumers' discomfort is represented as Γ and calculated by using Equation (5) similar to [42], similarly Equation (5) also demonstrates that ρ and k are the real numbers that represent the operational characteristics of devices, and Ts_i^t and Tu_i^t are the operational time of appliances set by the scheduler and consumer respectively.

$$\Gamma = \sum_{i=1}^{n} \rho(Ts_i^t - Tu_i^t)^k \tag{5}$$

where,

$$0 < \rho < 1 \,,\, k \geq 1$$

The value of ρ lies between this interval because we aim to minimize the discomfort caused by delaying the operation of appliances. In case of violation of this limit, the electricity consumer will be suffered with more discomfort in the form of more delay in the operation of appliances. Where, the real number k represents the behaviour of appliance.

PAR can be calculated as follows

$$PAR = \frac{\max_{t \in T} \left(\sum_{i=1}^{n} (P_r^i \times \Delta_{i,t}) \right)}{\frac{1}{m} \left(\sum_{t=1}^{m} \sum_{i=1}^{n} (P_r^i \times \Delta_{i,t}) \right)} \; \forall \, T = \{1, 2, ..., m\} \tag{6}$$

We have used the linear weighted sum method (scalarization approach) in which both parameters; cost and discomfort get normalized values, range between 0 and 1. So, both cost and discomfort are comparable. Furthermore, we have assigned equal weights (i.e., 0.5) to both cost and discomfort and their sum is equal to 1, $w_1 + w_2 = 1$.

The mathematical formulation of an objective function and constraints can be given as,

$$\text{Minimize} \quad \left(w_1 \times C_t^r + w_2 \times \Gamma \right) \tag{7}$$

Subject to

$$\sum_{i=1}^{n} (P_r^i \times \Delta_{i,t}) \leq Cap_T \; \forall \, t = \{1, 2, ..., m\} \tag{7a}$$

$$PAR^{\text{ with EMC}} \leq PAR^{\text{ without EMC}} \tag{7b}$$

$$\sum_{t=1}^{m} \Delta_{i,t} = T_{LoT}^i \; \forall \, i = \{1, 2, ..., n\} \tag{7c}$$

$$\alpha_i \leq T_{OTI}^i \leq \beta_i \tag{7d}$$

$$C_c^{r \text{ with EMC}} \leq C_c^{r \text{ without EMC}} \tag{7e}$$

$$E_e^{r \text{ with EMC}} = E_e^{r \text{ without EMC}} \tag{7f}$$

Equation (7) shows an objective function to be minimized and comprises of electricity cost and discomfort of consumers. Equations (7a)–(7f) define the constraint functions. In Equation (7), the cost and user discomfort are assigned equal weights w_1 and w_2 respectively. However, the values of weights can be varied in the range between [0, 1] or $w_1 + w_2 = 1$. In Equation (7a), limit shows the maximum allowable capacity that can be utilized at any hour of the day. The boundary limit ensures the stability of a grid by restricting the consumers to a limited amount of energy consumption. In Equation (7b), PAR is addressed to avoid the peaks creation at any hour of the day so that stability of a grid remains un-jeopardized.

Equation (7c) shows that length of the operation time of each device must be completed to avoid the users' frustration. Equation (7d) depicts that device must fulfill operation time after its start time and before end time in order to mitigate the user discomfort. In Equation (7e), it is depicted that the total consumption cost with EMC must be less than the total consumption cost without EMC. Equation (7f) illustrates that the total energy consumption with and without EMC must be the same.

4. Optimization Techniques

The optimization techniques used in this work are briefly discussed.

4.1. Heuristic Optimization Techniques

We have considered two heuristic optimization techniques. On the basis of these techniques we have proposed our own technique.

4.1.1. Brief Description of GA

GA has been successfully used in problems such as scheduling job shops and travelling salesman. It is a robust adaptive optimization technique which is based on biological paradigm. It performs efficient search on-poorly defined spaces which motivates the application of this technique to solve the binary optimization problem. It aims at finding the best candidates from the entire population. The fittest candidates are ranked higher in the population, whereas the least fit candidates are ranked lower in the population. In the end, one fittest candidate is selected which is called global best. The entire chronological process is followed as, random generation of population, fitness evaluation, elitism, selection, crossover and mutation. The population is updated by using the aforementioned parameters, the fittest chromosomes are survived and least fit candidates are weeded out in the next population. More detailed knowledge about GA can be found in [45,46].

4.1.2. Brief Description of BPSO

BPSO is an established version of PSO, it is a heuristic optimization technique inspired by the social behavior of bird flocking and fish schooling. This technique has been extensively used for a variety of binary optimization problems which motivated us to solve our binary optimization problem employing this technique. BPSO aims at finding the best possible solution to a problem from entire search space. The velocity and position of particles are randomly initialized, then updated by using their respective equations. The particles traverse through the entire space so that an optimal solution can be found. The evaluation of all the particles are performed and the global best and personal best positions are updated if required. At the end of stipulated iterations one global best is opted which is considered as a solution to the problem [47]. Each particle is associated with its position and velocity. The position of particle at any point in search space can be determined as follows:

$$\vec{X}_k(t) = \vec{X}_k(t-1) + \vec{V}_k(t) \tag{8}$$

Each particle is associated with the velocity vector, containing the information of local and global best positions achieved so far. The updated velocity of a particle can be given as,

$$\vec{V}_k(t) = \varphi \vec{V}_k(t-1) + \Omega_1.rand1.(\vec{P}_k - \vec{X}_k(t-1))$$
$$+ \Omega_2.rand2.(\vec{P}_g - \vec{X}_k(t-1)) \tag{9}$$

where φ is the inertia constant or weight of the particles, $k \in 1, 2, ..., M$ is the number of particles, Ω_1 and Ω_2 are constant numbers and $\Omega_1 + \Omega_2 = 4$. $\vec{P}_k$ and $\vec{P}_g$ are local and global best solutions achieved so far. $\vec{X}_k(t-1)$ and $\vec{X}_k(t)$ are previous and current positions of particle in the search space. The velocity update expression composed of three main components.

- The first component is often known as "inertia" or "momentum", it tends to move a particle in the same direction as it was travelling in. The inertia component can be scaled with a constant factor known as inertia constant. The inertia constant controls the velocity of a particle so that the particle cannot move beyond or below the scope of optimal search space. Mathematically inertia constant can be given as,

$$\varphi = \varphi_f + (\varphi_f - \varphi_i) \times \left(\frac{k^{th} iteration}{maximum\ iterations} \right) \tag{10}$$

- The second component represents the local best solution found for the first time in search space. It tends to converge the solution toward local optima.
- The third component can be referred as the linear attraction towards the global best solution from the entire search space. It tends to fetch the optimum solution by using group knowledge of all the particles.

If the value of velocity exceeds the maximum or minimum limits, then it can be written as follows:

$$\vec{V}_k(t) = \begin{cases} \vec{V}_{max} & \text{if } \vec{V}_k(t) > \vec{V}_{max}; \\ \vec{V}_{min} & \text{if } \vec{V}_k(t) < \vec{V}_{min}. \end{cases} \tag{11}$$

The position of each member of particle is updated by using the following equation,

$$\vec{X}_k(t) = \begin{cases} 1 & \text{if } sig(\vec{V}_k(t) > rand); \\ 0 & \text{otherwise}. \end{cases} \tag{12}$$

where, $sig(\vec{V}_k(t)) = \frac{1}{(1+exp(\vec{V}_k(t)))}$.

Sigmoid function converts the value of velocity to a binary format by comparing it with randomly generated number in range between $[0, 1]$. The maximum and minimum extremes of velocity are $[\vec{V}_{max}$ and $\vec{V}_{min}]$.

4.2. Deterministic Optimization Technique

The description of deterministic technique is given as follows:

Brief Description of DP

The dynamic programming is used in order to solve the knapsack problem. DP was basically introduced by Bellman [48] to solve the knapsack problem. It has the ability to divide a problem into sub-problems and memorizing. DP allows the knapsack problem to be divided into *n* sub problems. The solution of each problem is maintained in a table. In this work, small items having largest values i.e., electricity cost are selected for Off-peak hours, where large item with small values are used to represent On-peak hours.

5. Proposed Technique

Residential sector has large number of appliances of different types, and all the appliances have different power ratings and consumption patterns. DSM needs such a technique that can efficiently handle these complexities. In literature, mathematical techniques such as linear programming (LP) and DP are used for this purpose, these techniques require more computational time and additionally, inadequate to handle multiple constraints [41,49–51]. Evolutionary heuristic techniques have shown capabilities to cope with such complex scenarios.

GAPSO

In the beginning, two heuristic optimization techniques: GA and BPSO are implemented and their performance is analyzed. Both the heuristic optimization techniques are briefly discussed in Section 4. After analyzing the performance of these algorithms, it is observed that the two above mentioned techniques both show pre mature convergence when dealing with high dimension problems. As a result of pre mature convergence, it is difficult and even impractical to solely rely on such optimization criteria. So, there is a need to develop such an optimization method which can improve search efficiency and precision and adequate to handle multiple constraints. In this work, therefore, GAPSO is proposed so as to obviate the problem of pre mature convergence. The proposed technique is intended to solve the residential load scheduling problem in more accurate and economic way. Moreover, in comparison to traditional optimization methods, the proposed model has lower computational complexity.

The positive traits of each technique are merged together to overcome the problem of convergence at local optima. The proposed technique is then capable of searching feasible space more effectively. The steps involve in the proposed hybrid model are given as; in the beginning a random population is generated, and then evaluation of the fitness function is performed by using 7. Tournament base selection criteria is used for selecting the parents from the population. Binary crossover and two bits mutation is used in this work. The crossover and mutation is done on selected parents to modify the population. Crossover, mutation and fitness evaluation are performed similar to [52]. Elitism is the process of remembering the good solution achieved so far. At this stage, position and velocity of particles are further updated by using Equations (8) and (9) respectively. While updating the velocity, sigmoid function which is discussed in Section 4 is used to convert the values in binary format. The evolution of population by using the innate traits of BPSO: position and velocity, further explores the search space. This results in mitigating the problem of pre mature convergence. The entire process is repeated until the termination criteria is reached. The termination criteria depends on the stipulated number of iterations or when the variations in fitness are not more than a predefined limit (i.e., 10^{-10}) for numerous (i.e., 50) successive generations. Algorithm 1 shows the working of the proposed technique.

In this way, the proposed scheme has significantly affected the desired performance parameters. The user comfort in term of waiting time is also taken into consideration, since it is of great importance. User comfort along with reimbursement is the only factor which enticed the consumers to actively participate in DR program. So, the proposed technique is considered to manage electricity cost and user comfort along with peak consumption. The parameters used in the proposed technique are given in Table 3.

Table 3. Variables used in proposed technique.

Variables	Values
Probability of crossover	0.9
Probability of mutation	0.1
Insite	1.0
Vmax	4.0
Vmin	−4
Ω_1	2.0
Ω_2	2.0
Population size	200
Maximum Iterations	600
ρ	0.001
k	3.0

Algorithm 1: GAPSO

Input- Initialize population size, length of chromosome, selection criteria, crossover and
mutation rates (p_c, p_m), maximum and minimum velocities, maximum number of iterations,
local and global pulls, inertia constant

Initialization - Generate initial population

while *stopping criteria is not met* **do**

end

Evaluate the fitness of population

Perform elitism to save the best chromosome

Apply tournament base selection criteria to select two parents from the population

if *$p_c \leq 0.9$* **then**

| Select crossover point of both the selected parents

| Reproduce the offsprings by applying crossover operation

end

if *$p_m \leq 0.1$* **then**

| Select a chromosome after crossover operation

| Randomly invert a bit of selected individual

end

Calculate velocity of particles as,

$\vec{V}_k(t) = \varphi \vec{V}_k(t-1) + \Omega_1.rand1.(\vec{P}_k - \vec{X}_k(t-1)) + \Omega_2.rand2.(\vec{P}_g - \vec{X}_k(t-1))$

Sigmoid Function:

$sig(\vec{V}_k(t)) = \frac{1}{(1+exp(\vec{V}_k(t)))}$

Update position of particles as,

$\vec{X}_k(t) = \vec{X}_k(t-1) + \vec{V}_k(t)$

Evaluate Fitness using Equation (7)

if *current fitness value is better than previous* **then**

| Set current value as local best ($\vec{P}_k$)

end

Choose the particle with best fitness value of all the particles as global best ($\vec{P}_g$)

Return Best

---

6. Simulations and Discussions

In this section, the performance of GA, BPSO, and GAPSO is discussed in detail.
While implementing the heuristic optimization techniques for scheduling of residential load,
various factors are observed regarding cost minimization, efficient power consumption, peak reduction
and user comfort.

6.1. Performance Parameters Definitions

The cost minimization can be referred to as the amount of reduction in electricity bills of consumers.
The consumers pay this amount to the utility on hourly consumption basis at the completion of a
predefined period. The efficient power consumption can be defined as intelligent utilization of
available power in such a way that the total demand never exceeds the generation capacity. Due to
synchronization among consumers' energy utilization pattern, peaks are formed which may damage
the stability of a grid. The user comfort of consumers can be defined as the minimum electricity cost
and minimum interruption of devices in daily routine life.

6.2. Peak Power Consumption

Figures 4 and 5 show the power consumption behavior under four optimization techniques
with DAP mechanism on daily and monthly basis respectively. The energy consumption depends

on power rating and length of the operation time of devices. The performance of GA in term of peak power consumption is analyzed. It is demonstrated and validated that GA is less efficient when dealing with peak power consumption. This is due to the global exploration mode of GA which always focuses on minimum electricity price offered by the utility. This resulted in peaks formation at off-peak hours while user satisfaction is not taken under consideration. In this way GA scheduled most of the residential devices at hours where electricity prices are low regardless of taking peak power consumption into account. Whereas, BPSO performed well in term of reducing peak power consumption, because BPSO scheduled less number of devices at off-peak hours as compared to that of GA, this results in significant reduction in peak power consumption. In GAPSO, the peak power consumption is analyzed and it is observed that peak consumption is reduced to a significant amount. Additionally, the results of DP are also analyzed, and it is observed that DP performed better in term of peak demand reduction. It is observed that GA, BPSO, GAPSO and DP have hourly peak consumption of 1572.3 kW, 1232.3 kW, 1085.3 kW and 1108.8 kW respectively. Results validated that the proposed technique has efficient response for time varying price signal.

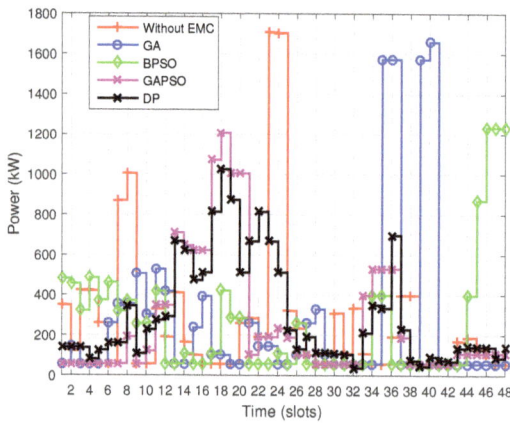

Figure 4. Daily Power Consumption-DAP.

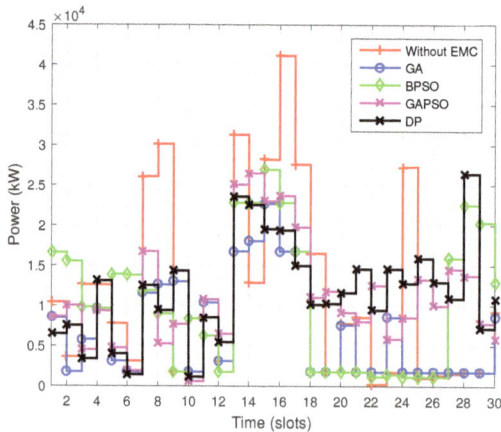

Figure 5. Monthly Power Consumption-DAP.

Figures 6 and 7 show that the proposed scenario is implemented for CPP. For CPP, it is observed that during a hot or cold day, most of residents consume energy during critical peak hours as a result of which more peaks are created during this time. The overall residential energy consumption behavior is demonstrated in Tables 4 and 6.

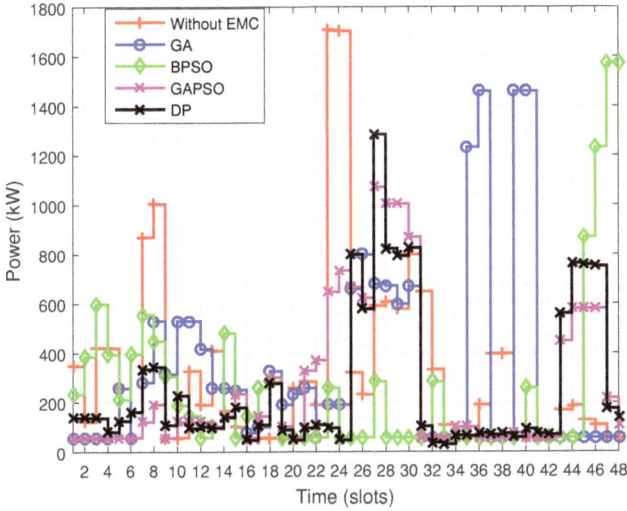

Figure 6. Daily Power Consumption-CPP.

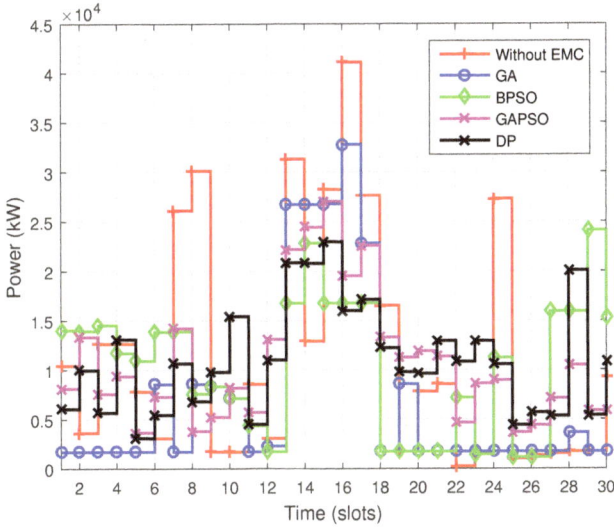

Figure 7. Monthly Power Consumption-CPP.

Table 4. Daily Energy Consumption Cost and Peak Load.

Technique	Parameters	Without EMC	With EMC	Reduction (%)
GA	Cost ($)	1581.9	1480.7	29.9702
	Peak-Load (kW)	1706.3	1572.3	7.8532
BPSO	Cost ($)	1581.9	1591.2	24.0470
	Peak-Load (kW)	1706.3	1232.3	27.7794
GAPSO	Cost ($)	1581.9	1181.8	25.2923
	Peak-Load (kW)	1706.3	1085.3	36.39
DP	Cost ($)	1581.9	1297.2	25.6467
	Peak-Load (kW)	1706.3	1108.8	35.0172

6.3. Electricity Cost

Electricity consumption cost under different techniques is demonstrated in Figures 8 and 9 for DAP mechanism. It is observed from the figures that the performance of GA shows substantial savings in electricity bills. The results validate that GA achieved 29.9702% reduction in electricity consumption cost. Whereas, BPSO achieved the reduction of 24.0470% in electricity consumption cost. Because both the techniques shifted the residential load from on-peak hours to off-peak hours where prices are minimum regardless of waiting time, and hence results in reduction in electricity cost. Through out the ample simulations it is shown that GAPSO successfully managed to reduce the consumption cost up to 25.2923% with minimum waiting time. Although the proposed technique is less efficient than GA in term of cost reduction, however, with optimized consumers' satisfaction. The reason associated with this fact is the inverse relationship between electricity bills and user satisfaction. The performance of the proposed model is also compared with DP. The results demonstrate that the proposed model has comparable performance with DP, however, with less computational complexity and storage space.

Since GA finds an optimal or near optimal solution from the entire search space and schedules the residential devices where consumers pay minimum electricity expenses. It is an inherent trait of GA that it can deal with complexities and non-linearities. It is capable of fulfilling the length of operation time of all the devices. Due to all these characteristics GA efficiently manages to reduce the electricity consumption cost. The performance of BPSO in term of cost minimization is analyzed and in this work it is shown that BPSO achieved less savings in electricity bill as compared to that of GA. It is attributed to the fact that BPSO uniformly scheduled the residential load over the time period to avoid the peaks creation. Although BPSO shifted the load at off-peak hours, however, the shifted load is comparatively less than that of GA. It is worth mentioning that by delaying an operation of devices, more reduction in electricity cost can be achieved at end consumers'. While analyzing the performance of the proposed model in terms of cost minimization, it is observed that GAPSO has optimally achieved the objective of cost minimization. The results show that GAPSO achieved 4.6779% less reduction in electricity consumption cost as compared to that of GA. Moreover, it is also observed that GAPSO achieved 1.2453% more reduction in electricity cost than BPSO, because in proposed technique both the parameters: consumption cost and user discomfort are taken into consideration. It results in fewer savings in electricity bills with improved consumers' lifestyle. To substantiate the performance of the proposed work, results are compared with DP. It is observed that both the techniques performed efficiently while reducing the energy consumption cost. DP achieved a bit higher savings because it converges to the optimal results, however, at the expense of time and storage space.

Figures 10 and 11 show the energy consumption cost for CPP signal on daily and monthly basis. It is noted that during critical hours consumers are charged with high electricity prices. It is also observed that in case of CPP energy consumption cost is significantly increased as compared to that of DAP, because utility offers maximum electricity prices during critical hours (i.e., 12 p.m.–3 p.m.). Energy consumption cost is analyzed for both the scenarios in Tables 4 and 6.

Energies **2017**, *10*, 1546

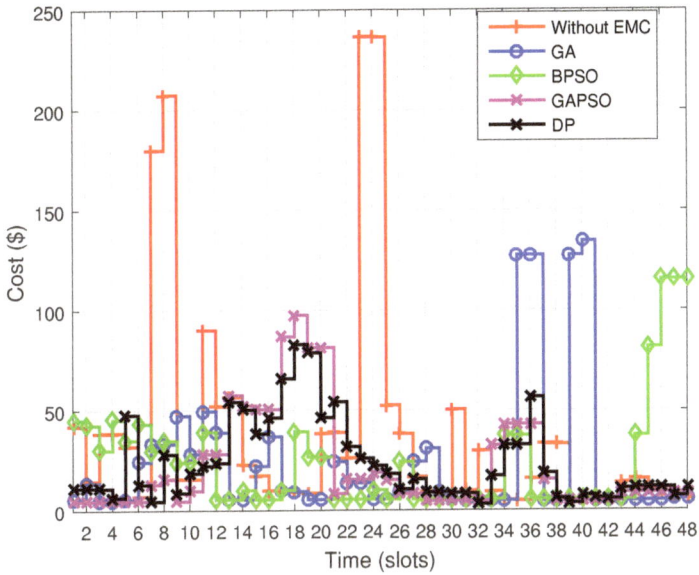

Figure 8. Daily Electricity Cost-DAP.

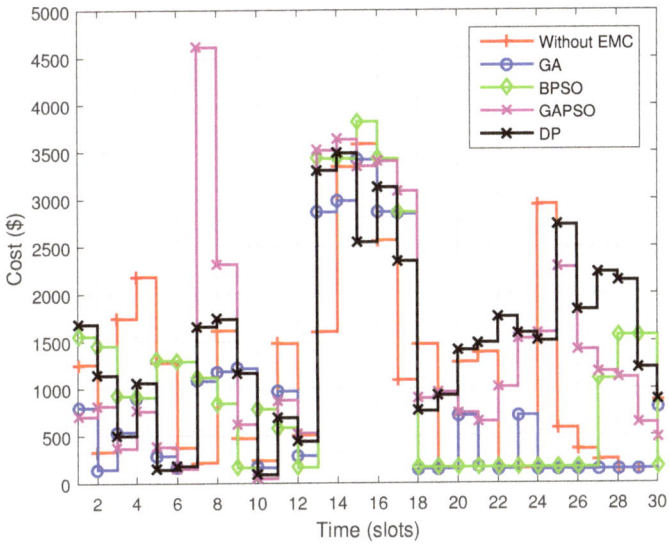

Figure 9. Monthly Electricity Cost-DAP.

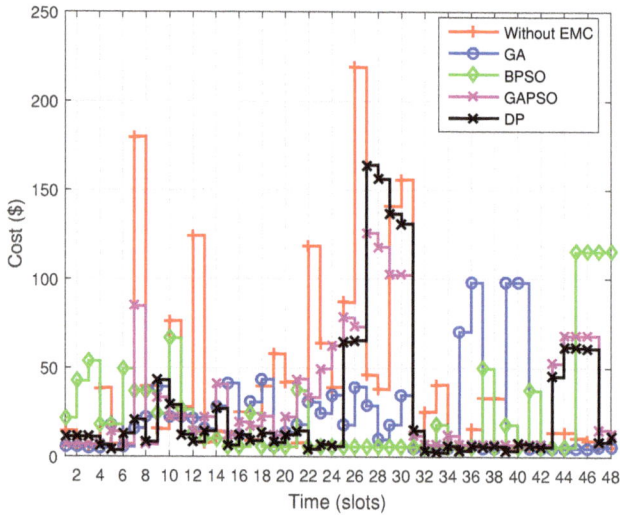

Figure 10. Daily Electricity Cost-CPP.

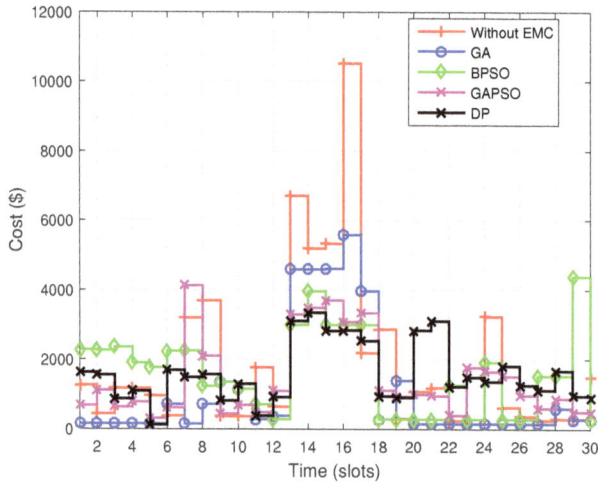

Figure 11. Monthly Electricity Cost-CPP.

6.4. PAR

The stability and reliability of a grid can be ensured by analyzing the PAR. Figures 12 and 13 depict PAR on daily and monthly basis when considering DAP as a pricing scheme. Figures infer that GA and BPSO achieved 7.8532% and 27.7794% reduction in peak power consumption respectively. Both heuristic techniques scheduled residential load from on-peak hours to off-peak hours. It is validated from the results that these heuristic techniques scheduled the load where electricity price is minimum. Whereas, GAPSO reduced 36.39% peak power consumption. This is due to the fact that GAPSO managed to distribute the entire residential load over 24 h time horizon. The load is distributed in such a manner that no peaks are created while respecting the waiting time of devices. Moreover, the performance of DP is also analyzed and compared with the proposed approach, it is observed that DP performed better in terms of peak demand reduction.

For CPP, Figures 14 and 15 show the PAR for a day and a month respectively. It is deduced that for a single day, the BPSO performed well as compared to other techniques, however, GAPSO outperformed rest of the techniques when compared with rest of the techniques for a month. It is due to the fact that, GAPSO scheduled the most prior load at critical hours. So as to maintain the stability of entire electrical network during critical peak hours, while taking into account the users' discomfort.

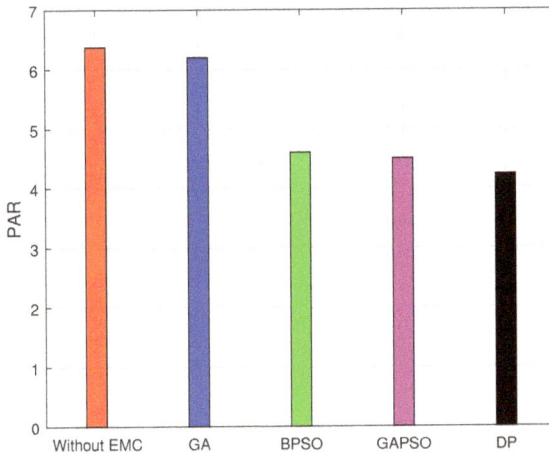

Figure 12. Daily PAR-DAP.

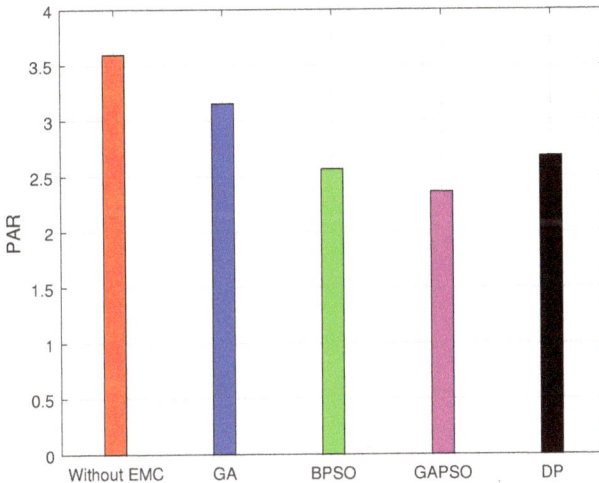

Figure 13. Monthly PAR-DAP.

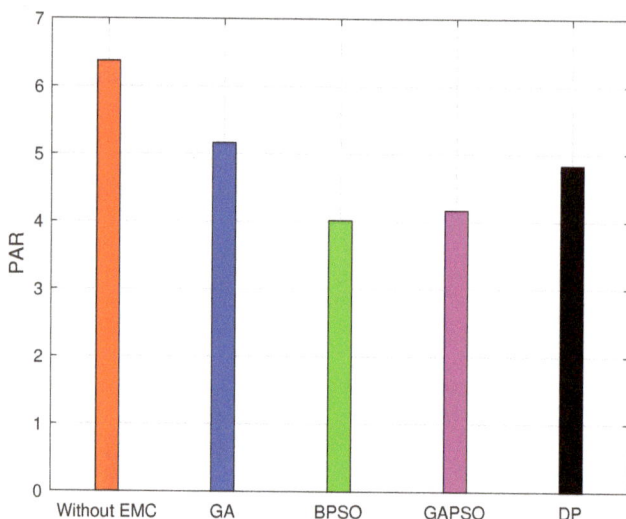

Figure 14. Daily PAR-CPP.

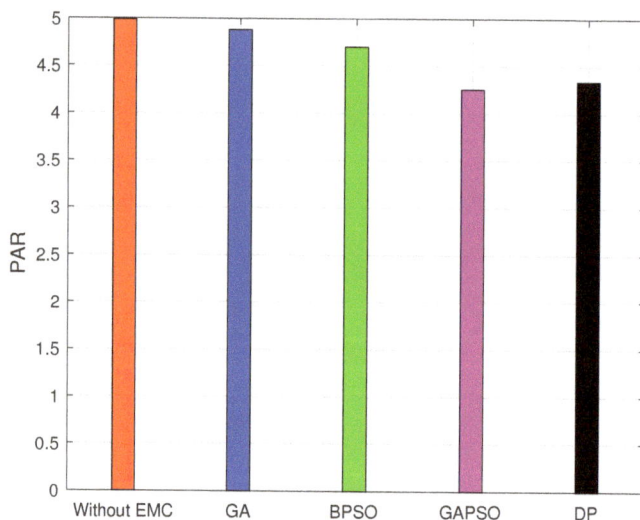

Figure 15. Monthly PAR-CPP.

6.5. User Comfort

The user comfort is associated with minimum consumption cost, minimum waiting time for the operation of devices, maintaining desired indoor temperature level, illuminance level, air quality and humidity etc. In this work, waiting time is considered as user comfort and thus to be optimized.

While implementing the GA for the residential load scheduling problem, user comfort is not taken into consideration. It results in maximum load scheduled at end hours and reduced maximum consumption cost. Similarly, in BPSO user comfort is not taken into consideration, and operation time of most of the devices are shifted to later hours. User comfort in terms of user discomfort and waiting time can be given as follows:

1. Since in this model, the maximization of user comfort is considered equivalent to the minimization of user discomfort, so both the terms can be used interchangeably. Figure 16 portrays the user discomfort of all the residential devices over the 24 h time horizon. Through performing extensive simulations it has been noticed that by minimizing the user discomfort, electricity cost is increased. The waiting time associated with discomfort is also analyzed and discussed.
2. Figure 17 demonstrates the waiting time of all the appliances. The average waiting time of 5 h is considered in the proposed scheme. Moreover, in this work, the length of operation time of fan is 24 h and it is demonstrated that the associated waiting time is zero for this device. Generally, by delaying the appliance's operation time more monetary benefits are achieved at consumers' end. It is also observed in the proposed technique, that with the incorporation of user comfort, comparatively less savings are achieved. In the proposed scenario half an hour is considered as an operational time slot of appliances (i.e., 1 slot = 30 min).
3. Figure 16 shows the discomfort faced by each corresponding residential device. Whereas, Figure 17 shows that average waiting time for each device. No comparison is being made in these figures, as the purpose of these figures is to demonstrate the user discomfort and average waiting for each corresponding residential device.

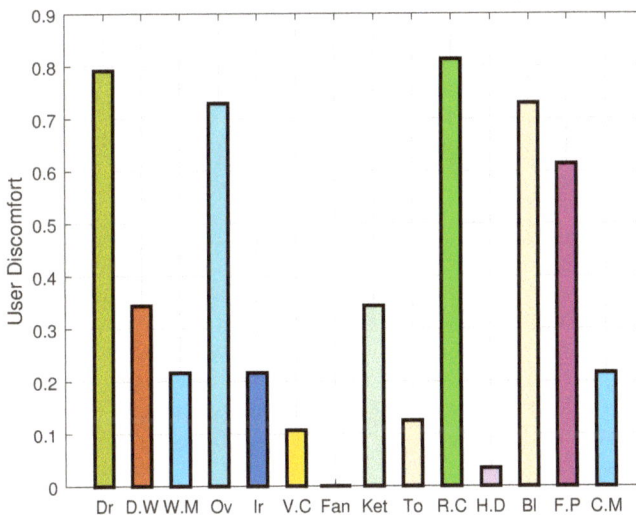

Figure 16. User Discomfort (Dr: Dryer, D.W: Dish Washer, W.M: Washing Machine, Ov: Oven, Ir: Iron, V.C: Vacuum, Ket: Kettle, To: Toaster, R.C: Rice Cooker, H.D: Hair Dryer, Bl: Blender, F.P: Frying Pan, C.M: Coffee Maker).

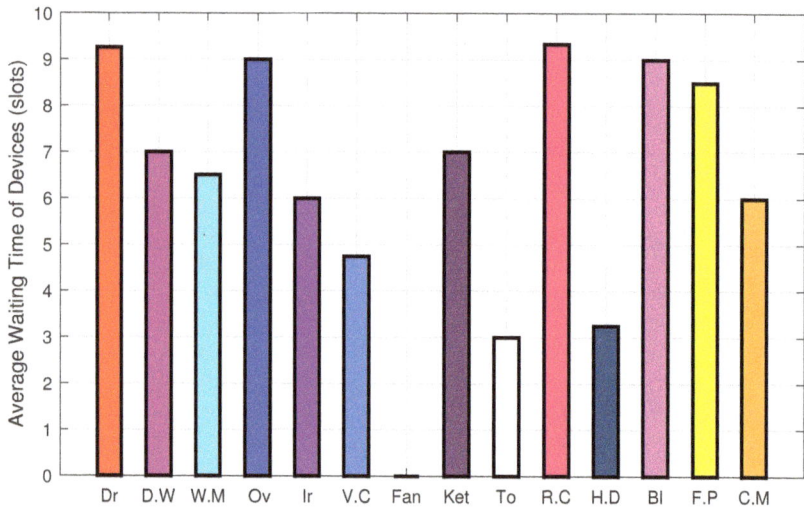

Figure 17. Average Waiting Time (Dr: Dryer, D.W: Dish Washer, W.M: Washing Machine, Ov: Oven, Ir: Iron, V.C: Vacuum, Ket: Kettle, To: Toaster, R.C: Rice Cooker, H.D: Hair Dryer, Bl: Blender, F.P: Frying Pan, C.M: Coffee Maker).

6.6. Feasible Region

A region comprises a set of points having a possible solution for a problem is known as a feasible region. Generally, feasible region is associated with the concept of optimization. In this work, feasible region is considered as an area containing all the possible solutions for an optimization problem. The evaluated performance parameters are analyzed graphically with the help of feasible region.

6.6.1. Feasible Region for Consumption Cost and Power

Electricity cost and power consumption are two directly linked parameters, varying consumption behavior and electricity price affect the electricity cost. A region bounded by a set of four points: P1(57.6, 4.6656), P2(57.6, 15.7536), P3(1706.3, 138.2103) and P4(1706.3, 466.67) represents a feasible region for electricity consumption cost and is shown in Figure 18. Point P1(57.6, 4.6656) denotes a minimum power consumption at minimum electricity cost over the entire day. Whereas, P2(57.6, 15.7536) shows minimum power consumption at maximum electricity cost offered by the utility. In P3(1706.3, 138.2103), it is demonstrated that the maximum consumption at minimum electricity cost. Whereas, P4(17063, 466.67) depicts an extreme point in a feasible region where both electricity cost and power consumption are maximum. However, P5(1706.3, 207.6448) shows maximum power consumption and electricity cost for our proposed model. Feasible region infers that by tailoring the consumption behavior consumers can minimize the consumption cost.

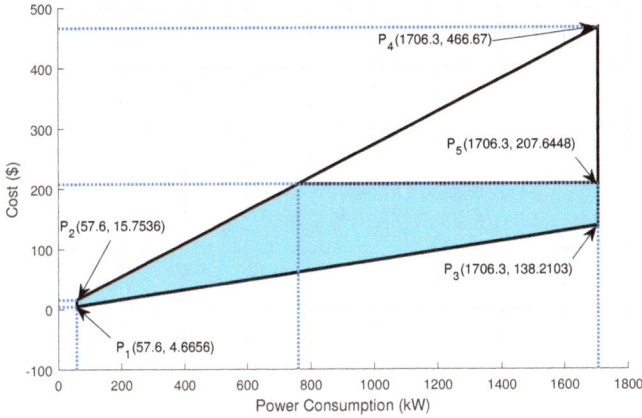

Figure 18. Feasible Region: Cost and Power Consumption.

6.6.2. Feasible Region for Cost and Waiting Time

In our proposed scenario, the user discomfort is discussed in term of waiting time of devices. The maximum allowable waiting time for residential devices is 10 slots (i.e., 5 h). Figure 19 portrays the trade-off between the consumption cost and waiting time. User discomfort and electricity cost are inversely proportion to each other, by decreasing user discomfort electricity cost increases and vice versa. P1(0, 4.6656) and P2(0, 207.6448) show minimum and maximum consumption cost at zero waiting time. Consumers achieve maximum comfort at zero delay for the operational time of their devices. Whereas, P3(10, 4.6656) and P4(10, 97.23) denote minimum and maximum consumption at maximum waiting time.

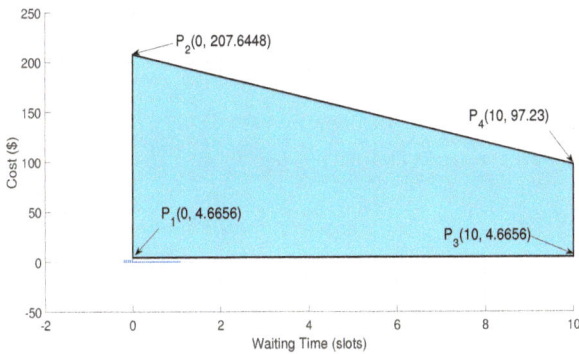

Figure 19. Feasible Region: Cost and Waiting Time.

6.7. Performance Trade-Off

It is deduced from the results that with the incorporation of user comfort in term of waiting time, the performance parameters are also affected. It can be viewed vividly from the same figure (i.e., GA and BPSO) that the user has achieved maximum monetary benefits, however, compromised on consumers' convenience. Similarly, it is shown that GAPSO achieved comparatively less savings in electricity bills with maximum comfort level. In this way, electricity cost and user comfort both are efficiently addressed in the proposed model. The savings in electricity bills are decreased by

4.6779%, this decrement in savings is due to the fact that electricity cost and user comfort are inversely proportional to each other. By increasing the user comfort, savings in electricity bills are decreased and vice versa. The tradeoff between user comfort and cost is obvious since without sacrificing the convenience consumers are incapable of achieving the reduction in consumption cost.

In this work, we have considered uncontrolled parameters (without EMC) as bench mark; however, the results of the proposed technique are also compared with DP. The performance of the proposed approach is analyzed and is demonstrated in Table 5. Table shows the upper and lower ranges of energy consumption cost, user discomfort and peak demand reduction. By analyzing the deviations between upper and lower values, it is deduced that the proposed model achieved the desired objective with 95% confidence interval. Moreover, optimality of the proposed model is also analyzed, as the DP provides optimal results. The difference between the performance parameters of proposed technique and that of DP provides the optimality gap. Table 6 provides the monthly energy consumption cost and the peak load. It is clear from the figures provided in this table that as compared to GA and BPSO, GAPSO benefits the consumers by reducing their cost significantly. We have noticed that there is no significant difference between cost reduction by GAPSO and DP; however, we still prefer GAPSO over DP due to its computational efficiency which is clear from Table 7. The computational time of the proposed technique for 112 residential devices is also analyzed and compared with other considered techniques. Moreover, Table 7 portrays the time analysis of the proposed heuristic technique with DP, thus depicting the efficiency of the proposed technique with the deterministic approach in terms of computational time. The results clearly elucidate that our proposed technique; GAPSO solves the formulated problem with least amount of time.

Table 5. The Comparison of Performance Metrics for a Day.

Technique	Parameters	Lower Value	Upper Value
GA	Cost ($)	1106.7	1116.0
	Discomfort	0.1240	0.8941
	PAR	5.8204	6.1599
BPSO	Cost ($)	1201.4	1205.2
	Discomfort	0.2310	0.8421
	PAR	4.5706	4.7336
GAPSO	Cost ($)	1179.6	1182.8
	Discomfort	0.1102	0.8100
	PAR	4.4858	4.5283
DP	Cost ($)	1175.6	1175.6
	Discomfort	0.1102	0.8100
	PAR	4.2350	4.2315

Table 6. Monthly Energy Consumption Cost and Peak Load.

Technique	Parameters	Without EMC	With EMC	Reduction (%)
GA	Cost ($)	57,584	45,771	20.5143
	Peak-Load (kW)	41,088	32,136	21.7873
BPSO	Cost ($)	57,584	48,550.5	15.6883
	Peak-Load (kW)	41,088	26,928	34.4626
GAPSO	Cost ($)	57,584	43,765	23.9979
	Peak-Load (kW)	41,088	27,476	33.1288
DP	Cost ($)	57,584	43,840	23.8677
	Peak-Load (kW)	41,088	27,400	33.331

Table 7. Computational Time of Employed Heuristic Techniques.

Techniques	Computation Time (Seconds)
GA	0.68
BPSO	0.59
GAPSO	0.55
DP	0.7

7. Conclusions

In this paper, we have modelled a residential energy management system proposing a hybrid technique for residential load scheduling. The scheduling problem is formulated through MKP mainly focusing on achieving the objectives of minimizing the electricity cost and consumers' discomfort. We analysed the performance of our proposed model under four different parameters: power consumption, electricity cost, PAR and user discomfort. Furthermore, the performance of the proposed technique is analysed and compared with GA, BPSO and DP. Results demonstrate that the performance of the proposed model is comparable to that of DP. However, the proposed model is efficient as it requires less computational time and storage. The proportional relation between performance parameters is calculated and shown with the help of feasible regions. Simulation results show that the proposed hybrid scheme, GAPSO, performed better in terms of cost and occupants' discomfort minimization along with reduction of peak power consumption compared to its counterpart schemes GA and BPSO.

Acknowledgments: This project was full financially supported by the King Saud University, through the Vice Deanship of Research Chairs.

Author Contributions: Nadeem Javaid, Fahim Ahmed, and Ibrar Ullah proposed, implemented, and wrote the optimization schemes. Samia Abid, Wadood Abdul, Ahmad Almogren, and Atif Alamri wrote technical sections of the manuscript. All authors refined the manuscript and responded to the queries of the respected reviewers.

Conflicts of Interest: The authors declare no conflict of interest.

References

1. Geng, Y.; Chen, W.; Liu, Z.; Chiu, A.S.; Han, W.; Liu, Z.; Zhong, S.; Qian, Y.; You, W.; Cui, X. A bibliometric review: Energy consumption and greenhouse gas emissions in the residential sector. *J. Clean. Prod.* **2017**, *159*, 301–316.

2. Samadi, P.; Bahrami, S.; Wong, V.W.; Schober, R. Power dispatch and load control with generation uncertainty. In Proceedings of the 2015 IEEE Global Conference on Signal and Information Processing (GlobalSIP), Orlando, FL, USA, 14–16 December 2015; pp. 1126–1130.

3. Gelazanskas, L.; Gamage, K.A. Demand side management in smart grid: A review and proposals for future direction. *Sustain. Cities Soc.* **2014**, *11*, 22–30.

4. Rastegar, M.; Fotuhi-Firuzabad, M.; Zareipour, H. Home energy management incorporating operational priority of appliances. *Int. J. Electr. Power Energy Syst.* **2016**, *74*, 286–292.

5. Yalcintas, M.; Hagen, W.T.; Kaya, A. An analysis of load reduction and load shifting techniques in commercial and industrial buildings under dynamic electricity pricing schedules. *Energy Build.* **2015**, *88*, 15–24.

6. Zazo, J.; Zazo, S.; Macua, S.V. Robust Worst-Case Analysis of Demand-Side Management in Smart Grids. *IEEE Trans. Smart Grid* **2017**, *8*, 662–673.

7. Tan, O.; Gómez-Vilardebó, J.; Gündüz, D. Privacy-cost trade-offs in demand-side management with storage. *IEEE Trans. Inf. Forensics Secur.* **2017**, *12*, 1458–1469.

8. Ahmed, N.; Levorato, M.; Li, G.P. Residential Consumer-Centric Demand Side Management. *IEEE Trans. Smart Grid* **2017**, doi:10.1109/TSG.2017.2661991.

9. Vardakas, J.S.; Zorba, N.; Verikoukis, C.V. Power demand control scenarios for smart grid applications with finite number of appliances. *Appl. Energy* **2016**, *162*, 83–98.

10. Ogunjuyigbe, A.S.O.; Ayodele, T.R.; Akinola, O.A. User satisfaction-induced demand side load management in residential buildings with user budget constraint. *Appl. Energy* **2017**, *187*, 352–366.

11. Shirazi, E.; Jadid, S. Optimal residential appliance scheduling under dynamic pricing scheme via HEMDAS. *Energy Build.* **2015**, *93*, 40–49.

12. Althaher, S.; Mancarella, P.; Mutale, J. Automated demand response from home energy management system under dynamic pricing and power and comfort constraints. *IEEE Trans. Smart Grid* **2015**, *6*, 1874–1883.

13. Muralitharan, K.; Sakthivel, R.; Shi, Y. Multiobjective optimization technique for demand side management with load balancing approach in smart grid. *Neurocomputing* **2016**, *177*, 110–119.

14. Ma, J.; Chen, H.H.; Song, L.; Li, Y. Residential load scheduling in smart grid: A cost efficiency perspective. *IEEE Trans. Smart Grid* **2016**, *7*, 771–784.

15. Kusakana, K. Energy management of a grid-connected hydrokinetic system under Time of Use tariff. *Renew. Energy* **2017**, *101*, 1325–1333.

16. Zhang, D.; Shah, N.; Papageorgiou, L.G. Efficient energy consumption and operation management in a smart building with microgrid. *Energy Convers. Manag.* **2013**, *74*, 209–222.

17. Muratori, M.; Rizzoni, G. Residential demand response: Dynamic energy management and time-varying electricity pricing. *IEEE Trans. Power Syst.* **2016**, *31*, 1108–1117.

18. Yi, P.; Dong, X.; Iwayemi, A.; Zhou, C.; Li, S. Real-time opportunistic scheduling for residential demand response. *IEEE Trans. Smart Grid* **2013**, *4*, 227–234.

19. Yaagoubi, N.; Mouftah, H.T. User-aware game theoretic approach for demand management. *IEEE Trans. Smart Grid* **2015**, *6*, 716–725.

20. Liu, Y.; Yuen, C.; Yu, R.; Zhang, Y.; Xie, S. Queuing-based energy consumption management for heterogeneous residential demands in smart grid. *IEEE Trans. Smart Grid* **2016**, *7*, 1650–1659.

21. Liu, Y.; Yuen, C.; Huang, S.; Hassan, N.U.; Wang, X.; Xie, S. Peak-to-average ratio constrained demand-side management with consumer's preference in residential smart grid. *IEEE J. Sel. Top. Signal Process.* **2014**, *8*, 1084–1097.

22. Fakhrazari, A.; Vakilzadian, H.; Choobineh, F.F. Optimal energy scheduling for a smart entity. *IEEE Trans. Smart Grid* **2014**, *5*, 2919–2928.

23. Miao, H.; Huang, X.; Chen, G. A genetic evolutionary task scheduling method for energy efficiency in smart homes. *Int. Rev. Electr. Eng. (IREE)* **2012**, *7*, 5897–5904.

24. Zhao, Z.; Lee, W.C.; Shin, Y.; Song, K.B. An optimal power scheduling method for demand response in home energy management system. *IEEE Trans. Smart Grid* **2013**, *4*, 1391–1400.

25. Anvari-Moghaddam, A.; Monsef, H.; Rahimi-Kian, A. Optimal smart home energy management considering energy saving and a comfortable lifestyle. *IEEE Trans. Smart Grid* **2015**, *6*, 324–332.

26. Bahrami, S.; Wong, V.W.; Huang, J. An Online Learning Algorithm for Demand Response in Smart Grid. *IEEE Trans. Smart Grid* **2017**, doi:10.1109/TSG.2017.2667599.

27. Samadi, P.; Mohsenian-Rad, A.H.; Schober, R.; Wong, V.W.; Jatskevich, J. Optimal real-time pricing algorithm based on utility maximization for smart grid. In Proceedings of the 2010 First IEEE International Conference on Smart Grid Communications (SmartGridComm), Gaithersburg, MD, USA, 4–6 October 2010; pp. 415–420.

28. Erdinc, O.; Paterakis, N.G.; Mendes, T.D.; Bakirtzis, A.G.; Catalão, J.P. Smart household operation considering bi-directional EV and ESS utilization by real-time pricing-based DR. *IEEE Trans. Smart Grid* **2015**, *6*, 1281–1291.

29. Agnetis, A.; de Pascale, G.; Detti, P.; Vicino, A. Load scheduling for household energy consumption optimization. *IEEE Trans. Smart Grid* **2013**, *4*, 2364–2373.

30. Belhaiza, S.; Baroudi, U. A game theoretic model for smart grids demand management. *IEEE Trans. Smart Grid* **2015**, *6*, 1386–1393.

31. Marzband, M.; Yousefnejad, E.; Sumper, A.; Domínguez-García, J.L. Real time experimental implementation of optimum energy management system in standalone microgrid by using multi-layer ant colony optimization. *Int. J. Electr. Power Energy Syst.* **2016**, *75*, 265–274.

32. Chakraborty, S.; Ito, T.; Senjyu, T.; Saber, A.Y. Intelligent economic operation of smart-grid facilitating fuzzy advanced quantum evolutionary method. *IEEE Trans. Sustain. Energy* **2013**, *4*, 905–916.

33. Derakhshan, G.; Shayanfar, H.A.; Kazemi, A. The optimization of demand response programs in smart grids. *Energy Policy* **2016**, *94*, 295–306.

34. Gupta, A.; Singh, B.P.; Kumar, R. Optimal provision for enhanced consumer satisfaction and energy savings by an intelligent household energy management system. In Proceedings of the 2016 IEEE 6th International Conference on Power Systems (ICPS), New Delhi, India, 4–6 March 2016 .

35. Zhang, D.; Evangelisti, S.; Lettieri, P.; Papageorgiou, L.G. Economic and environmental scheduling of smart homes with microgrid: DER operation and electrical tasks. *Energy Convers. Manag.* **2016**, *110*, 113–124.

36. Reka, S.S.; Ramesh, V. A demand response modeling for residential consumers in smart grid environment using game theory based energy scheduling algorithm. *Ain Shams Eng. J.* **2016**, *7*, 835–845.

37. Safdarian, A.; Fotuhi-Firuzabad, M.; Lehtonen, M. Optimal residential load management in smart grids: A decentralized framework. *IEEE Trans. Smart Grid* **2016**, *7*, 1836–1845.

38. Moon, S.; Lee, J.W. Multi-Residential Demand Response Scheduling with Multi-Class Appliances in Smart Grid. *IEEE Trans. Smart Grid* **2016**, doi:10.1109/TSG.2016.2614546.

39. Wang, J.; Li, Y.; Zhou, Y. Interval number optimization for household load scheduling with uncertainty. *Energy Build.* **2016**, *130*, 613–624.

40. Bharathi, C.; Rekha, D.; Vijayakumar, V. Genetic Algorithm Based Demand Side Management for Smart Grid. *Wirel. Pers. Commun.* **2017**, *93*, 481–502.

41. Logenthiran, T.; Srinivasan, D.; Shun, T.Z. Demand side management in smart grid using heuristic optimization. *IEEE Trans. Smart Grid* **2012**, *3*, 1244–1252.

42. Ma, K.; Yao, T.; Yang, J.; Guan, X. Residential power scheduling for demand response in smart grid. *Int. J. Electr. Power Energy Syst.* **2016**, *78*, 320–325.

43. Naoyuki, M. Energy-on-Demand System Based on Combinatorial Optimization of Appliance Power Consumptions. *J. Inf. Process.* **2017**, *25*, 268–276.

44. Kumaraguruparan, N.; Sivaramakrishnan, H.; Sapatnekar, S.S. Residential task scheduling under dynamic pricing using the multiple knapsack method. In Proceedings of the 2012 IEEE PES Innovative Smart Grid Technologies (ISGT), Washington, DC, USA, 16–20 January 2012.

45. Kim, D.H.; Abraham, A.; Cho, J.H. A hybrid genetic algorithm and bacterial foraging approach for global optimization. *Inf. Sci.* **2007**, *177*, 3918–3937.

46. Arabali, A.; Ghofrani, M.; Etezadi-Amoli, M.; Fadali, M.S.; Baghzouz, Y. Genetic-algorithm-based optimization approach for energy management. *IEEE Trans. Power Deliv.* **2013**, *28*, 162–170.

47. Del Valle, Y.; Venayagamoorthy, G.K.; Mohagheghi, S.; Hernandez, J.C.; Harley, R.G. Particle swarm optimization: basic concepts, variants and applications in power systems. *IEEE Trans. Evolut. Comput.* **2008**, *12*, 171–195.

48. Bellman, R. *Dynamic Programming*; Princeton University Press: Princeton, NJ, USA, 1957.

49. Ng, K.H.; Sheble, G.B. Direct load control-A profit-based load management using linear programming. *IEEE Trans. Power Syst.* **1998**, *13*, 688–694.

50. Kurucz, C.N.; Brandt, D.; Sim, S. A linear programming model for reducing system peak through customer load control programs. *IEEE Trans. Power Syst.* **1996**, *11*, 1817–1824.

51. Hsu, Y.Y.; Su, C.C. Dispatch of direct load control using dynamic programming. *IEEE Trans. Power Syst.* **1991**, *6*, 1056–1061.

52. Azadeh, A.; Ghaderi, S.F.; Tarverdian, S.; Saberi, M. Integration of artificial neural networks and genetic algorithm to predict electrical energy consumption. *Appl. Math. Comput.* **2007**, *186*, 1731–1741.

![energies logo] *energies*

MDPI

Article

An Innovative Hybrid Model Based on Data Pre-Processing and Modified Optimization Algorithm and Its Application in Wind Speed Forecasting

Ping Jiang [1], Zeng Wang [1,*], Kequan Zhang [2] and Wendong Yang [1]

[1] School of Statistics, Dongbei University of Finance and Economics, Dalian 116025, China; pjiang@dufe.edu.cn (P.J.); hshwendong@hotmail.com (W.Y.)
[2] Key Laboratory of Arid Climatic Change and Reducing Disaster of Gansu Province, College of Atmospheric Sciences, Lanzhou University, Lanzhou 730000, China; zhangkq@lzu.edu.cn
* Correspondence: wzsdutdufe@163.com; Tel.: +86-183-4222-2536

Academic Editor: Wei-Chiang Hong
Received: 22 June 2017; Accepted: 4 July 2017; Published: 9 July 2017

Abstract: Wind speed forecasting has an unsuperseded function in the high-efficiency operation of wind farms, and is significant in wind-related engineering studies. Back-propagation (BP) algorithms have been comprehensively employed to forecast time series that are nonlinear, irregular, and unstable. However, the single model usually overlooks the importance of data pre-processing and parameter optimization of the model, which results in weak forecasting performance. In this paper, a more precise and robust model that combines data pre-processing, BP neural network, and a modified artificial intelligence optimization algorithm was proposed, which succeeded in avoiding the limitations of the individual algorithm. The novel model not only improves the forecasting accuracy but also retains the advantages of the firefly algorithm (FA) and overcomes the disadvantage of the FA while optimizing in the later stage. To verify the forecasting performance of the presented hybrid model, 10-min wind speed data from Penglai city, Shandong province, China, were analyzed in this study. The simulations revealed that the proposed hybrid model significantly outperforms other single metaheuristics.

Keywords: back propagation (BP); forecasting accuracy; modified firefly algorithm; wind speed; singular spectrum analysis

1. Introduction

Wind power is one of the most significant recycled energy resources presently being applied [1]. Recently, due to the pollution of the global environment, recyclable energy [2] and non-polluting sources such as wind energy have been gaining extensive attention [3]. Wind energy, which is one of the most promising and active recyclable sources, is providing an increasingly strong supplement to traditional energy sources [4]. When it comes to the accurate forecasting of wind speed and its wide use in wind power, we encounter great challenges, because the wind is a periodical phenomenon [5] with a nonlinear, anomalistic, and stochastic nature. Wind speed forecasting is applied in several domains, for instance, target tracking, shipping, weather forecasting, agricultural production, and electric load forecasting. To dispatch wind energy before wind power grid integration, it is very important for a wind farm operator to accurately determine the wind speed. This is because the local wind speed is always the foremost factor affecting wind power generation, and can be used for wind turbine selection and for wind farm layout [6]. In addition, wind speed can enhance the power system's schedule and strengthen resource configuration, promoting the reliability of the power grid. Predictions made with higher precision can allow power system operators to dispatch power efficiently in order to properly meet the demands of consumers [7].

Given a more precise wind speed value, the power operator is able to forecast power delivery. This is extremely helpful for power systems in terms of optimizing storage capacity, making sensible and proper programs, and dispatching electric energy well. Because of the wind's irregularity and complex fluctuations, variations in wind speed forecasting may result in quick changes in the prediction results of wind power. This feature indicates that accurate wind speed forecasting is highly important. The wind speed forecast plays a vital role in utilizing wind power appropriately and efficiently. Various methods have been proposed to promote the accuracy of wind speed prediction. Three of the most extensively used methods are the physical forecasting method, the conventional statistical method, and the artificial intelligence method. Given a series of meteorological parameters, the physical forecasting method uses physical variables to derive a time series forecast. Therefore, higher prediction accuracy can be obtained using this method [8]. However, their extremely intricate computations always lead to it being largely a waste of time. Numerical weather forecasting is one of the most widely used physical forecasting methods, consisting of a computer program that aims to solve questions through meteorological data processing and describe how the atmosphere changes as time goes on [9]. In addition, the traditional approaches include the regression analysis method, the auto-regressive integrated moving average (ARIMA) [10] model, the non-parametric estimation method, exponential smoothing [11], the state-space model [12], Box-Jenkins models [13], the spatial correlation model, and the difference method. Furthermore, support vector machines (SVMs) [14] such as non-neural networks are also frequently applied in wind speed forecasting.

Among the above methods, artificial neural networks (ANNs) have been frequently and widely applied. By imitating the human brain in handling information with a sequence of neurons, ANNs obtain a distinguished capacity for mapping, and their complex and highly nonlinear input and output modes with making nothing of the type of real model can establish some simple models and compose different networks depending on different connections. Therefore, ANNs demonstrate the following advantages: high adaptability, excellent ability to learn using cases, and ability to summarize. It is well known that the multi-layered perceptron (MLP) is one of the most broadly used ANN methods. The vast majority of available methods that can be used to train ANNs pay close attention only to the alteration of connection weights in a certain topology, which usually leads to defective results.

MLPs are prosperously applied in many fields, such as pattern classification [15], digit recognition [16], image processing, coal price prediction [17], function approximation, measurement of object shape [18], and adaptation control. The back-propagation (BP) algorithm [19] performs most effectively of all training algorithms for MLP methods. The selection of a suitable structure for the forecasting question and the alteration of connection weights of the network constitute the two parts of training MLPs for the problem. Several studies have been successfully used to solve these issues.

A great deal of research has been conducted to precisely forecast the wind energy and the local value of wind speed. Wind power and speed forecast is a fundamental problem for wind farm operation, best power flow between the electric system and wind power plant, market price, electric power system dispatching, and wind power resource reserve, and storage programming and dispatching. Over the last few decades, the ANN [20,21] has been the superior model, and has frequently been applied to forecast time series.

The ANN is a pragmatic calculation method, similar to the human biological neuron. Various improvable neural networks exist, of which the following two are the most frequently employed: feed forward neural networks and feedback neural networks. Feed forward neural networks have no feedback. On the contrary, feedback neural networks possess a feedback. Back-propagation (BP) neural networks, perceptrons, and radial basis function (RBF) networks play an important role in feed forward networks. Recurrent Neural Networks (RNNs) [22] and pulsed neural networks are two important models of feedback networks. The feedback networks mainly consist of RNN and spiking neural networks [23]. In this paper, we pay more attention to feed forward neural networks [24].

Energies **2017**, *10*, 954

The BP algorithm has various significant advantages; for example, it can help to roughly estimate a great many functions, it is relatively simple to implement, and it can be used as a reference method. In addition, its most effective characteristic is that the momentum parameter and the learning rate factor can be altered, thereby enhancing the innovation speed of the traditional BP algorithm.

To gain good forecasting accuracy and low deviation, many studies [25–32] have been conducted to determine the optimal weights of neural networks. However, an original hybrid model system—a traditional hybrid method based on the rapid searching theory developed by Xiao et al. [33]—has been put into use. An extensive study was conducted by Xiao et al. [33] using four test functions to evaluate the optimization algorithm's capacity for development, searching, avoiding partial optima, and convergence velocity, and the results of this experiment demonstrated that the modified method is more sufficient and excellent than the original algorithm. In recent years, a number of developmental optimization algorithms have been applied to help confirm the threshold values of a prediction method. Particle swarm optimization (PSO) was applied by Liu et al. [25] to optimize the parameters of the prediction technique for short-term electric load prediction in micro-grids. Wang et al. [26] employed a modified PSO to optimize the weight distribution of their proposed combined model developed for electric load prediction. The cuckoo search (CS) algorithm [27–29] was applied to determine the parameters of the proposed model for electric load forecasting. Wang et al. [30] modified the CS method to optimize the parameters of multi-step-ahead wind speed forecasting models. Xiao et al. [31] applied the genetic algorithm (GA) to optimize the parameters of the proposed model. In the present paper, a highly valid optimization method, the Broyden-Fletcher-Goldfarb-Shanno-Firefly Algorithm (BFGS-FA), is used to determine the parameters of the proposed hybrid model.

Recently, numerous continuous and novel improvements have been made to promote the effectiveness of the FA for optimizing neural networks, including the binary, Gaussian, firefly, high-dimensional firefly, Lévy flight, simultaneous firefly, and chaos-based FA [34,35]. Though most of these improvements to the FA enhance its performance successfully, few of them have been introduced to optimize the parameters of hybrid models. This paper intends not only to enhance the research and development abilities of the FA, but also to minimize the drawback of the partial optima seeking capacity, which appeared in the CS algorithm. On the basis of the BFGS quasi-Newton method, an original improvement of the FA was proposed to enhance the diversity of species of fireflies. Obviously, increasing the convergence standard may result in individual fireflies likely being caught in partial optima; however, it decreased when this optimized algorithm was used. Of course, the decomposition of the original wind sequence is a significant process for data filtering. This can always effectively promote the prediction accuracy of the model to obtain better forecast results [36]. Important techniques, such as empirical mode decomposition (EMD) [37], wavelet decomposition (WD) [38], and singular spectrum analysis (SSA) are often applied to remove the noise series. However, the wavelet de-noising algorithm is sensitive to the determination of the threshold, and the EMD may lead to mode confusion [39], which may result in a badly decomposed performance. In addition, SSA has many advantages, and overcomes the disadvantages of EMD and WD in terms of decomposition. Moreover, we analyzed some articles in the literature [40–44] that deal with wind forecasting by applying neural networks, and that are in line with the theme of the present paper. From these studies, we found that some data preprocessing or optimization algorithms are insufficient, and the details are listed in Table 1. Therefore, based on the discussed limitations, this manuscript proposes a characteristic hybrid model that unites the BP algorithm, SSA theory, and BFGS-FA. Ten-minute wind speed values collected from Penglai city, Shandong province, China, were applied to verify the unique hybrid model. The results of tests and practices in this study indicate that the hybrid model considerably outperformed the other three models. This demonstrates that the hybrid method could be applied to calculate wind speeds, which would be beneficial for enabling wind power system to make optimal decisions, such as providing better sites of wind power, taking early measures to reduce losses that can be caused due to bad weather, reducing production costs, and minimizing energy consumption (coal, etc.). This model is also useful for helping wind power companies to make correct decisions in real life. Thus, the hybrid forecasting method with high accuracy represents a model that will have potential application in the near future. Furthermore, the practical hybrid model can also be applied to other forecasting domains, such as target tracking, stock index forecasting, environment forecasting, shipping, weather forecasting, agricultural production, and electric load forecasting.

The primary contributions and novelties of this manuscript are listed as follows: (a) The BFGS-FA method, back propagation neural network (BPNN), and the concept of the de-noising algorithm were combined to form two new models: singular spectrum analysis-back propagation (SSA-BP), and singular spectrum analysis-Broyden-Fletcher-Goldfarb-Shanno-Firefly Algorithm-back propagation (SSA-BFGS-FA-BP). (b) This paper evaluates the developed models on the basis of two aspects: forecasting accuracy and stability. The results indicate that BFGS-FA-BP is a better model when considering accuracy only, but the hybrid SSA-BFGS-FA-BP is a better model overall: even with the low cost of calculation, the accuracy remained high. (c) The novel combined BFGS-FA algorithm successfully avoids the shortcomings of FA while optimizing, during the later period, the low velocity and the poor convergence performance. (d) The proposed hybrid approach integrates the advantages of other individual models. (e) A time sequence pre-processing method was applied to de-noise the raw data successfully.

The remainder of this paper is designed as follows. Section 2 presents the single prediction method developed according to the BPNN and the hybrid forecasting method theory. This section also describes the optimization algorithms BFGS, FA, and their combination BFGS-FA, which are applied to confirm the parameters of the hybrid forecasting model. SSA theory and the Diebold-Mariano (DM) test, which can help to determine the forecasting effectiveness of the developed hybrid method, are introduced at the end of Section 2. In Section 3, the wind speed time sequences collected from three separate sites are used to test the proposed hybrid model. Subsequently, the wind resources and the evaluation criteria of the forecasting model are described. In Section 4, we give a discussion about this study. In the end, Section 5 concludes this paper.

Table 1. Summary of intelligent and hybrid forecasting methods and theoretical comparison of existing forecasting models. RBF: radial basis function; ENN: Elman neural network; WNN: Wavelet neural network; GA: genetic algorithm; BP: back propagation; FA: firefly algorithm; PSO: particle swarm optimization.

Models	Variables	Date Set	Results	Advantages	Disadvantages	Ref.
RBF	Wind speed	March 2012 in Jiangsu, Ningxia and Yunnan, China.	The proposed methods can provide higher quality prediction intervals than the conventional method.	Has a good local approximation and computer implementation is easy.	Has a low forecasting performance.	[40]
ENN	Wind speed	January 2010 to March 2011 in Gansu, China.	The proposed approach is an effective way to improve prediction accuracy.	Has a good fitting effect and can deal with nonlinear data.	Has a narrow forecasting scale.	[41]
WNN	Wind speed	720 samplings in Inner Mongolia, China	The proposed model can be a robust method for wind speed forecasting.	Has a good forecasting performance and higher computational efficiency.	The model is unstable and highly dependent on time series.	[42]
Hybrid GA and BP	Wind speed	January 2011 in Inner Mongolia, China.	The method can improve both forecasting accuracy and computational efficiency.	The simulation method is easy to understand and combines with other methods.	It is easy to get into local optimum.	[43]
Hybrid FA/PSO Elman	Wind speed	January 2011 to November 2011 in Shandong, China.	The proposed model outperforms other comparative models in forecasting.	Has good adaptability, and can take advantage of other models.	The process of building models is relatively complex.	[44]

2. Methodology

Since McCulloch and Pitts [45] proposed the neural network mathematical model in 1943, ANNs have been applied in numerous fields, including signal processing, market analysis and forecasting, pattern recognition, and automatic control. In this part, separate theories of this innovative hybrid model will be introduced in detail.

2.1. BP Algorithm

In mathematically simulating the human brain system, the BP algorithm benefits from its underlying processes, fuzzy information processing, and chaotic performance. On account of the error BP algorithm and the multilayer neural network, the BP neural network performs excellently in training ANNs. An input layer, one or more hidden layers, and an output layer constitute a representative BP network. The BP algorithm is always applied to adjust the thresholds, in which the errors from the output are propagated back into the network, transforming the thresholds as it goes, in order to keep the error [46] from emerging again. Its topology and flow structure are as follows:

The main procedures of the BP algorithm can be generalized as follows:

Step 1. We obtain the wind speed time sequence and corresponding parameter values from the wind power plant. The inputs have exhaustive information on historical values. The input value is often affected by the site, surrounding temperature, air pressure, time, and even the collectors. Our primary task is to make full use of four different parameters collected from the wind power plant.

Step 2. We transform the original value into the requested form (0 to 1). The normalization method is summarized as follows:

$$(Value)_{normalized} = [(Value)_{actual} - (Value)_{min}] / [(Value)_{max} - (Value)_{min}]$$

Step 3. We build the BP algorithm and set its parameters, which include the number of neurons in the input layer, hidden layer, and output layer; the learning rate; the maximum training times; and training requirement accuracy. The training can be summarized as learning from the historical values to discover the implied information among the previous time series data, which can be applied to forecast the future wind speed.

Step 4. We use the testing set to assess the effectiveness of the trained BP network.

Step 5. In the end, the future wind speed value (output) is forecast by the neural network.

The key parameters that emerged in this study are not sensitive in small intervals; therefore, the key parameters of these algorithms are determined by repeated trails. The corresponding experimental parameters of the method are summarized in Tables 2 and 3.

Table 2. The experimental set points of BP.

Experimental Set Point	Default Value
The number of units in the input layer	6
The number of units in the hidden layer	7
The number of units in the output layer	1
The learning rate	0.1
The maximum training time	1000
Required accuracy of training	0.00001

2.2. Broyden-Fletcher-Goldfarb-Shanno

The BFGS [47] algorithm is an excellent method, and one of the most useful nonlinear quasi-Newton procedures.

Definition 1. *Let $\mathbf{x}^t$ be the consequence at the representative iteration t and $\mathbf{x}^{t+1} = \mathbf{x}^t + \lambda_t \mathbf{d}_t$ be a recursive function in which λ_t is the step size. The hunting path is $\mathbf{d}_t = -\mathbf{D}_t \nabla f(\mathbf{x}^t)$, in which $\mathbf{D}_t$ is an $n \times n$ positive certain symmetric matrix as a proximity of the inverse matrix of the real Hessian matrix at $\mathbf{x}^t$.*

Definition 2. *The new path of BFGS can be designed as follows:*

$$\mathbf{D}_{t+1} = \left(\mathbf{I} - \rho_t \mathbf{p}_t \mathbf{q}_t^T\right) \mathbf{D}_t \left(\mathbf{I} - \rho_t \mathbf{q}_t \mathbf{p}_t^T\right) + \rho_t \mathbf{p}_t \mathbf{p}_t^T ; \rho_t = 1/\mathbf{p}_t^T \mathbf{q}_t \tag{1}$$

where

$$\mathbf{p}_t = \lambda_t \mathbf{d}_t = \mathbf{x}^{t+1} - \mathbf{x}^t \tag{2}$$

$$\mathbf{q}_t = \nabla f\left(\mathbf{x}^{t+1}\right) - \nabla f(\mathbf{x}^t) \tag{3}$$

In addition, the primary BFGS algorithm is generalized in Appendix A.

2.3. Firefly Algorithm

The FA was first proposed by Xin-She Yang in 2008 [48]. The FA was inspired by the flashing nature of fireflies [49,50]. The firefly will be shining while flying, which can be regarded as a signal to attract other companions. The method has three regulations:

(1) All fireflies are unisexual; in addition, any firefly can be attracted by others.
(2) Attraction is directly proportional to their brightness; that is, for any two fireflies, the less bright one will be attracted by the brighter one, and will move towards it; the brightness will decrease as the distance between them increases.
(3) If there are no brighter fireflies around a known firefly, it will fly at random. The brightness of the firefly must be tightly related to the objective function.

The experimental set points of the FA are described in Table 3.

Table 3. The experimental set points of FA.

Experimental Set Point	FA
Maximum iteration	1000
Population size	20
Alpha	0.25
Beta	0.2
Gamma	1
Convergence tolerance	0.00001

The FA is a developed computational method that is also used to optimize controller parameters. Each firefly in the FA indicates a solution to the problem, which is defined on the basis of position. In a d-dimensional vector space, the present location of the ith firefly is acquired by $x_i = (x_{i1}, \ldots, x_{in}, \ldots, x_{id})$. The random positions of m fireflies are initialized within the specified range. The position updating equation for the ith firefly, which is attracted to move to a brighter firefly j, is given as follows:

$$x_i(t+1) = x_i(t) + \beta_0 \exp(-\gamma r_{ij}^2)(x_j - x_i) + \alpha(\mathbf{rand} - 0.5) \tag{4}$$

In addition, the position updating equation for the brightest firefly is given as follows:

$$xbest_i(t+1) = xbest_i(t) + \alpha(\mathbf{rand} - 0.5) \tag{5}$$

where the first terms $x_i(t)$ and $xbest_i(t)$ of Equations (4) and (5) are the current positions of a less bright firefly and the brightest firefly, respectively. The second term in Equation (4) is the firefly's attraction to light intensity. β_0 is the original attraction at $r = 0$, γ is the absorption parameter in the range [0, 1], and r_{ij} is the distance between any two fireflies i and j, at position x_i and x_j, respectively, and can be formulated as a Cartesian or Euclidean distance as follows:

$$r_{ij} = \sqrt{\sum_{n=1}^{d} (x_{in} - x_{jn})^2} \tag{6}$$

where x_i and x_j are the position vectors for fireflies i and j, respectively, with x_{in} representing the position value for the dimension, and the third term in Equation (4) and the second term in Equation (5) are used to reduce the randomness; that is, the movement of the fireflies is gradually reduced according to $\alpha = \alpha_0 \delta^t$, where α_0 is in the range [0, 1]. δ is the random reduction parameter where $0 < \delta < 1$, and t is the iteration number. Every new position must be evaluated by a fitness function, which is assumed to be integral square error. The flow chart of the FA is presented in Figure 1, and the original FA algorithm is summarized in Appendix B.

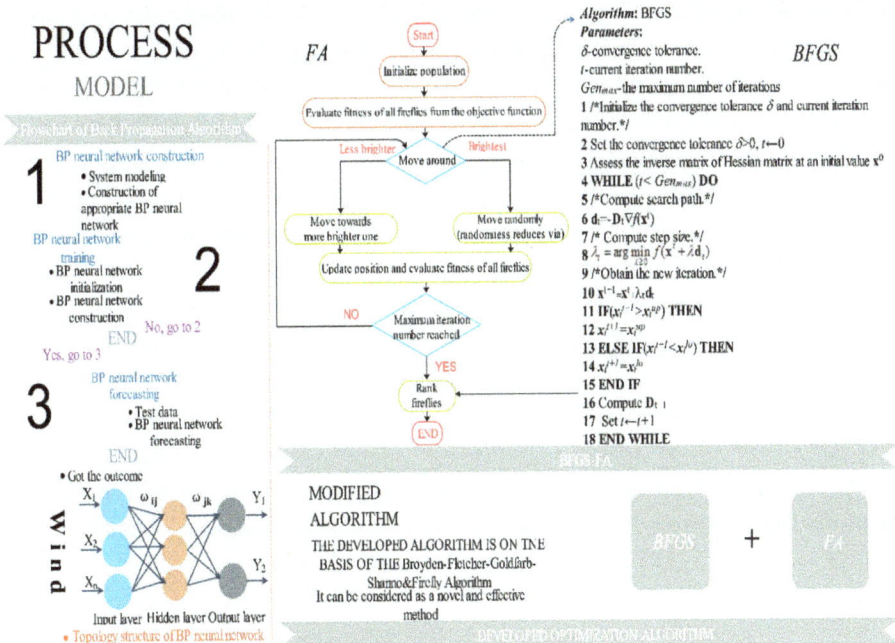

Figure 1. Topology structure and flow chart of BP neural network and the flow chart and structure of the combined BFGS-FA Algorithm.

2.4. BFGS-FA

FA possesses good global optimization and development capacities; however, it will usually be manifest a low velocity and poor convergence performance while optimizing in the later stage. Therefore, as shown in Figure 1, the BFGS is applied while FA renews the answers after a generation to search for a sub-optimization solution, which can be used to promote the partial optimization capacity and the rate of partial convergence of the total method. The primary method of BFGS-FA is generalized in Appendix C.

2.5. Singular Spectrum Analysis (SSA)

In America and England, SSA has been exploited separately based on singular spectrum analysis, whereas in Russia, it was proposed under the name Caterpillar-SSA [51]. SSA possesses the superiority of statistics and probability theory; meanwhile, it assimilates the knowledge of power systems and signal processing ideas.

Suppose that $\mathbf{y} = [y_1, y_2, \ldots, y_T]$ is a time sequence with T elements. The SSA method contains two parts: decomposition and reconstruction [52,53].

2.5.1. Decomposition

In decomposition, an observed unidimensional time series data $\mathbf{y} = [y_1, y_2, \ldots, y_T]$ is converted into its trajectory matrix. Subsequently, $\mathbf{XX}^T$ and its corresponding singular value decomposition (SVD) are computed. This can be divided into two steps: embedding and SVD.

Step 1. The primary aim of this step is to propose the concept of the trajectory matrix or deferred edition of the initial time sequence $\mathbf{y}$. The main purpose of this step is to propose the concept of a trajectory matrix or a hysteretic version of the initial time sequence $\mathbf{y}$. The resulting matrix has a window width W ($W \leq T/2$), which is usually determined by the operator. Suppose that $P = T - W + 1$, the trajectory matrix is denoted as follows:

$$\mathbf{X} = [\mathbf{X}_1, \ldots, \mathbf{X}_P] = \begin{pmatrix} y_1 & y_2 & y_3 & \cdots & y_P \\ y_2 & y_3 & y_4 & \cdots & y_{P+1} \\ y_3 & y_4 & \cdots & \cdots & \cdots \\ \cdots & \cdots & \cdots & \cdots & \cdots \\ y_W & y_{W+1} & y_{W+2} & \cdots & y_T \end{pmatrix} \tag{7}$$

In fact, this trajectory matrix is a Hankel matrix; that is, all the elements of the diagonal $i + j = $ const are equal [54].

Step 2. We obtain the covariance matrix $\mathbf{XX}^t$ from $\mathbf{X}$. $\mathbf{XX}^t$ processed by the SVD will result in a group of L eigenvalues $\lambda_1 \geq \lambda_2 \geq \ldots \geq \lambda_L \geq 0$ and their corresponding eigenvectors $\mathbf{U}_1, \mathbf{U}_2, \ldots, \mathbf{U}_L$, which are often defined by empirical orthogonal functions. Therefore, the SVD of the trajectory matrix could be denoted as $\mathbf{X} = \mathbf{E}_1 + \mathbf{E}_2 + \ldots + \mathbf{E}_d$, where $\mathbf{E}_i = \sqrt{\lambda_i} \mathbf{U}_i \mathbf{V}_i^t$, d is the rank of $\mathbf{XX}^t$ (the total amount of non-zero characteristic values) and $\mathbf{V}_1, \mathbf{V}_2, \ldots, \mathbf{V}_d$ are the corresponding principal components, which are denoted by $\mathbf{V}_i = \mathbf{X}^t \mathbf{U}_i / \sqrt{\lambda_i}$. The set $(\sqrt{\lambda_i}, \mathbf{U}_i, \mathbf{V}_i)$ is the ith eigenvalue of the matrix $\mathbf{X}$. Suppose that $T = \sum_{i=1}^{d} \lambda_i$, then λ_i / T—the ratio of the variance of $\mathbf{X}$—which is defined by $\mathbf{E}_i : \mathbf{E}_1$, has the highest contribution [55], and E_d has the minimum contribution. The SVD will consume more elapsed time if the length of the time sequence is long enough (i.e., $T > 1000$).

2.5.2. Reconstruction

We compute $\mathbf{XX}^T$ and its SVD to obtain its L eigenvalues: $\lambda_1 \geq \lambda_2 \geq \ldots \geq \lambda_L \geq 0$ and its corresponding eigenvectors. Each signal, as represented by the eigenvalue, is analyzed and assembled to reconstruct the new time series. This section can be resolved into two steps: grouping and averaging.

Step 1. Here, the designer chooses r out of d eigenvalues. Define $I = \{i1, i2, \ldots, ir\}$ to be a set of r chosen eigenvalues and $\mathbf{X}_I = \mathbf{X}_{i1} + \mathbf{X}_{i2} + \ldots + \mathbf{X}_{ir}$, in which $\mathbf{X}_I$ is connected to the "information" of $\mathbf{y}$; nevertheless, the remaining $(d–r)$ eigenvalues, which are not selected, represent the error term ε.

Step 2. The set of r elements chosen from the foregoing section is then applied to regroup the definitive elements of the time sequence. The fundamental concept is to convert each of the terms $\mathbf{X}_{i1}, \mathbf{X}_{i2}, \ldots, \mathbf{X}_{ir}$ into the reconstructed data time series $y_{i1}, y_{i2}, \ldots, y_{ir}$ by using the Hankelization process $\mathbf{H}(\mathbf{Z})$ or diagonal averaging: assume Z_{ij} is an element of the ordinary matrix $\mathbf{Z}$, then the kth term of the rebuilt time sequence could be acquired by averaging Z_{ij}, on the precondition of $i + j = k + 1$. Obviously, $\mathbf{H}(\mathbf{Z})$ is a time sequence with T elements rebuilt by matrix $\mathbf{Z}$.

After averaging, we can obtain the approximation of $\mathbf{y}$, which is the regrouped time series, and is given as follows:

$$\mathbf{y} = \mathbf{H}(\mathbf{X}_{i1}) + \mathbf{H}(\mathbf{X}_{i2}) + \ldots + \mathbf{H}(\mathbf{X}_{ir}) + \varepsilon \tag{8}$$

From the whole time series, a singular eigenvalue will be reconstructed as suggested by Alexandrov and Golyandina. This indicates that SSA is not an awkward algorithm, and is therefore strong to abnormal values.

In addition, as shown in Figure 2, the original wind speed preprocessed by SSA is forecast by the BP algorithm, and its parameters are optimized by BFGS-FA.

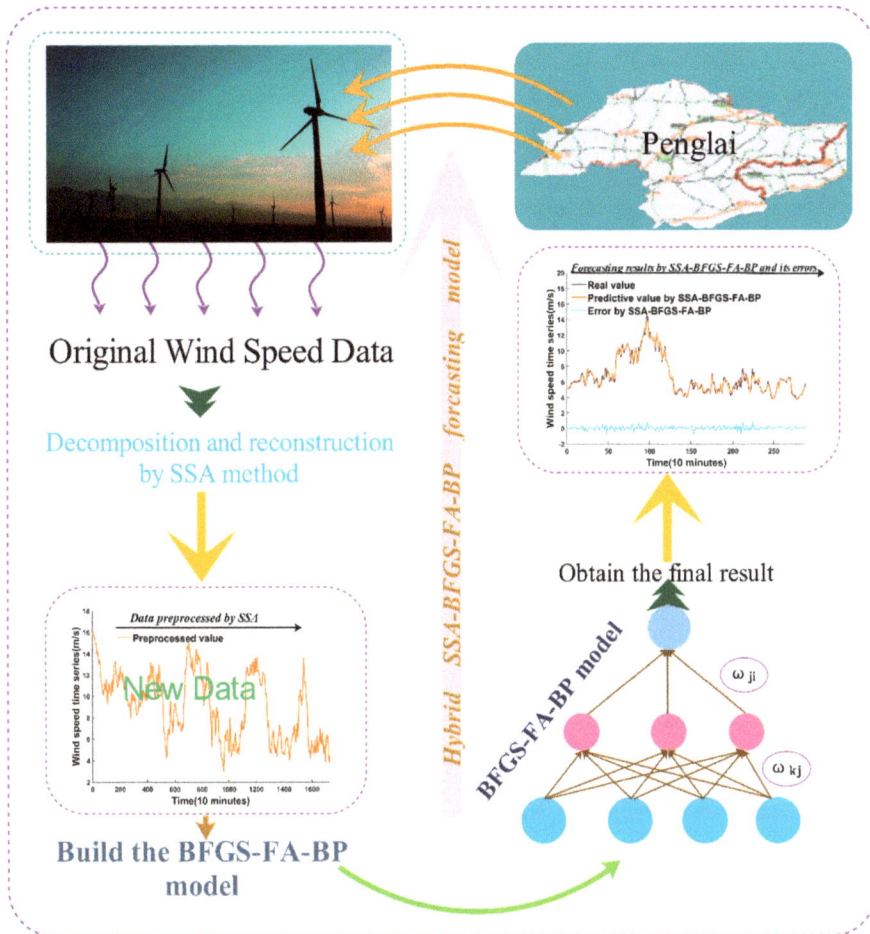

Figure 2. The flow chart of the hybrid SSA-BFGS-FA-BP model.

2.6. Proposed Hybrid Model

The BP algorithm is selected as the forecasting method to forecast the wind speed time series in this paper. However, because of its unstable structure, we could not obtain more accurate forecasting results with minor error; therefore, it is important to determine the optimal parameters and threshold values of the BP network to promote the predictive effectiveness. BFGF-FA is proposed to determine the weight and threshold. In addition, large amounts of noise present in the original wind sequence will lead to a poor forecasting performance. Therefore, we choose the SSA to remove the noise from the raw time sequence. The corresponding basic procedures are presented as follows, and are depicted in Figure 2.

Step 1. SSA is used to remove the noise from the raw data. It also aims to remove the high frequency of the original sequence after decomposing, and then reconstructs them into new experimental data.

Step 2. BFGS-FA is used to determine the weight and threshold of the BP neural network. Thus, the ability of the global optimization of the BP algorithm is greatly promoted.

Step 3. The optimized BP neural network is applied to predict the wind speed time sequence.

Step 4. The proposed hybrid model indeed outperforms the single models in forecasting time sequences based on historical values. Multi-step forecasting also proves that the proposed hybrid method has a higher effectiveness, and their forms can be described as follows:

(1). *One-step prediction*: The predictive value $\hat{p}(t+1)$ is calculated on the basis of the past time sequence $\{p(1), p(2), \ldots, p(t-1), p(t)\}$, where t is the sample size of the wind speed time sequence.

(2). *Two-step prediction*: The predictive value $\hat{p}(t+2)$ is calculated on the basis of the past time sequence $\{p(1), p(2), \ldots, p(t-1), p(t)\}$ and the former predictive value $\hat{p}(t+1)$.

(3). *Three-step prediction*: The predictive value $\hat{p}(t+3)$ is calculated on the basis of the past time sequence $\{p(1), p(2), \ldots, p(t-1), p(t)\}$ and the former predictive value $\hat{p}(t+1)$ and $\hat{p}(t+2)$.

(4). Higher-step forecasting value will be obtained on the basis of the above form.

2.7. Testing Method

In this paper, we also employed a testing method called the Diebold-Mariano (DM) test to estimate the proposed model.

The Diebold-Mariano (DM) test [56], which is focused on predictive accuracy, compares and evaluates the predictive effectiveness of the proposed hybrid method with other simple models. In practical applications, there will be two or more time sequence models available for predicting a specific variable of interest.

Real values:

$$\{c_n; m = 1, \cdots, t+l\} \tag{9}$$

Two predictions:

$$\left\{\hat{c}_m^{(1)}; m = 1, \cdots, t+l\right\}; \quad \left\{\hat{c}_m^{(2)}; m = 1, \cdots, t+l\right\} \tag{10}$$

The prediction errors according to the two models can be described as follows:

$$\varphi_{t+g}^{(1)} = c_{t+g} - \hat{c}_{t+g}^{(1)}, \quad g = 1, 2, \cdots, l. \tag{11}$$

and:

$$\varphi_{t+g}^{(2)} = c_{t+g} - \hat{c}_{t+g}^{(2)}, \quad g = 1, 2, \cdots, l. \tag{12}$$

The precision of each forecasting model is evaluated by an appropriate loss function, $L\left(\varphi_{t+g}^{(i)}\right); i = 1, 2.$

The most widespread and available loss function is square error loss, and its formulation is as follows:

Square error loss:

$$L\left(\varphi_{t+g}^{(i)}\right) = \left(\varphi_{t+g}^{(i)}\right)^2 \tag{13}$$

The DM test statistic assesses the prediction according to the random loss function $L(p)$:

$$DM = \frac{\sum\limits_{g=1}^{l} k_g}{l\sqrt{S^2/l}} s^2 \tag{14}$$

where S^2 is the estimated value of the variance of $k_g = L\left(]\varphi_{t+g}^{(1)}\right) - L\left(\varphi_{t+g}^{(2)}\right)$, and the null hypothesis is:

$$H_0 : E(k_g) = 0\,\forall m \tag{15}$$

in contrast, the alternative hypothesis is:

$$H_1 : E(k_g) \neq 0 \tag{16}$$

Under the null hypothesis, the two predictions possess uniform precision. In contrast, the alternative hypothesis has different standards, namely, the two predictions differ in accuracy. If the null hypothesis is right, the Diebold-Mariano statistic will be an asymptotically standard normal distribution $N(0,1)$. The null hypothesis should not be refused if the calculation of DM statistic falls inside the interval $[-Z_{\alpha/2}, Z_{\alpha/2}]$, otherwise we must reject it; that is, the reject region is $(-\infty, -Z_{\alpha/2})\&(Z_{\alpha/2}, +\infty)$, which is defined as follows:

$$|DM| > Z_{\alpha/2} \tag{17}$$

where $Z_{\alpha/2}$ is the positive Z-value from the standard normal table according to half of the confidence level α of the experiment.

3. Experimental Design and Results

In this section, the wind speed data gathered from three sites are forecast by the developed hybrid method. The data location and effectiveness of the prediction estimation standard are also presented. All the experiments in this paper were conducted in MATLAB R2014b on Windows 7 with 3.30 GHz Intel (R) Core (TM) i5 4590 CPU, 64 bit and 8 GB RAM.

3.1. Data Sets

The hybrid SSA-BFGS-FA-BP method was tested using data from experiments of wind speed prediction time sequences at three sites. A data set gathered at 10-min intervals from Penglai city, Shandong province, China, was used. Figure 3a displays the geographical position of Penglai city in China.

In this study, wind speeds are taken from three different sites, and we chose 1728 of them as observation values. Of these, 1440 values were used to train the network, and the remaining 288 values were selected as the testing set for each station.

The original data from the three sites are shown in Figure 3b, which illustrates the inordinance, wave, and mutability of the original time series.

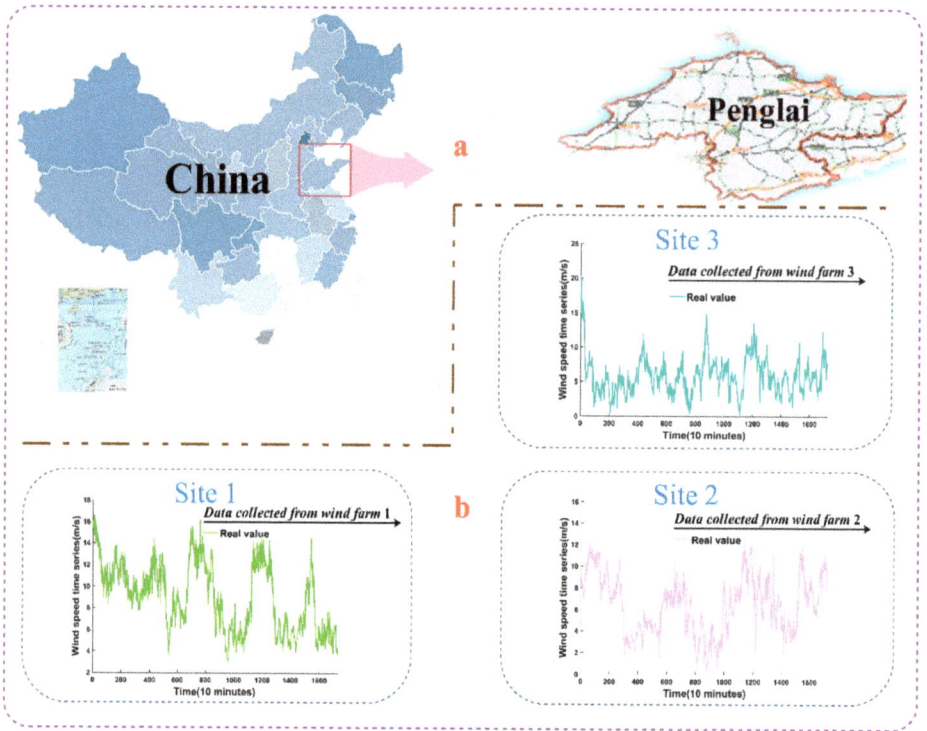

a Geographical location of this study area
b Real values of three sites

Figure 3. Geographical position of the survey regions and actual values of three stations.

3.2. Forecast Error Metrics

Forecasting errors are applied to assess the ability of the applied forecasting approaches and to evaluate the effectiveness of the proposed method on account of on-site/true measures.

The metric equations in Table 4 show us the universal error index applied to most forecasting models for renewables. The mean absolute error (MAE), the root mean square error (RMSE) [57], and the mean absolute percentage error (MAPE) are used to estimate the forecasting effectiveness of the proposed method. They are denoted as follows:

p_t and $\hat{p}_t$ are the true value and the predicted value, respectively. T is the total number of elements in this data array. The MAE depends on p_t and $\hat{p}_t$, the RMSE depends on p_t and $\hat{p}_t$, and furthermore, the MAPE gives the relative error between $|p_t - \hat{p}_t|$ and p_t. Quantified by these three frequently used indices, we can clearly and concisely perceive the difference between the predicted and exact wind speed values. A smaller difference value indicates that the forecasting method has a better performance. Nevertheless, MAPE, a unit-free estimator, has better sensitivity for small-scale variation, does not reveal some weak characteristics of data, such as asymmetry, and has lower abnormal value protection. Therefore, a better MAPE will be chosen as the standard in this paper.

Table 4. The metric equations.

Error Index	Definition	Formula
MAE	The mean absolute error of T times predictive results	$\frac{1}{T}\sum\limits_{t=1}^{T}\left\|p_t - \hat{p}_t\right\|$
RMSE	The root mean square error	$\sqrt{\frac{1}{T}\sum\limits_{t=1}^{T}(p_t - \hat{p}_t)^2}$
MAPE	The mean absolute percentage error	$\frac{1}{T}\sum\limits_{t=1}^{T}\left\|\frac{p_t - \hat{p}_t}{p_t}\right\| \times 100\%$

3.3. Comparison Method and Its Corresponding Results

Our main contribution is not only to provide an optimization algorithm, but also to propose a novel hybrid wind speed forecasting model. Experimental results prove that the proposed method can be perfectly used for short-term wind speed forecasting, and that it has considerable practical value and strong operability in wind farms and grid management. In addition, we performed a comparative experiment to compare the proposed model with other forecasting approaches, and the corresponding results are presented in Table 5, revealing that the proposed hybrid method achieves higher forecasting accuracy than the other methods.

Table 5. The results of the hybrid model, ARIMA and SVM model at three sites.

Model	Site 1			Site 2			Site 3		
	MAE	**RMSE**	**MAPE**	**MAE**	**RMSE**	**MAPE**	**MAE**	**RMSE**	**MAPE**
ARIMA	0.5210	0.6832	7.9275	0.4496	0.5995	6.8094	0.5404	0.7606	10.4179
SVM	0.5069	0.6789	7.5806	0.4526	0.6118	6.8365	0.5379	0.7573	10.3284
Hybrid model	0.1849	0.2461	4.8496	0.2337	0.2465	4.5512	0.2407	0.2851	6.7548

3.4. Case Studies

This study consists of three classic experiments. Each of them is grouped into two sections, one of which uses the primary wind speed data and the other applies data preprocessed by the SSA approach. Original data are forecast by the single BP algorithm, and the BP algorithm is optimized using combined BFGS-FA (BFGS-FA-BP); decomposed data are also predicted by the single BP and BFGS-FA (BFGS-FA-BP).

The two model aim to compare the single BP with the optimized BFGS-FA-BP to determine the performance of the hybrid method. The parameters of SSA are presented in Table 6.

Table 6. The experimental parameters of singular spectrum analysis (SSA).

Experimental Parameters	SSA
Embedding dimension	50
Components	10
Method of calculating the covariance matrix	Unbiased (N-K weighted) Biased (N-weighted or Yule-Walker) BK(Broomhead/King type estimate)

3.4.1. Case Study 1

In this section, all results will be clearly demonstrated in the figures and tables to reveal the effectiveness of each model. First, the predicted values of wind speed in the three locations are presented in Tables 7–9. Considering the random disturbances of the forecasts, it is necessary to repeat each experiment many times to ensure the reliability of results. Therefore, in our study, we performed each experiment 20 times and then used the average values as the final results, to make sure that the results are dependable.

Figure 4a shows the data of the estimated performance with and without using the SSA approach at site 1. The effectiveness of the experiment that used real values is displayed at the top left side of the chart, and was forecast by models without using the SSA approach. The effectiveness of the experiment that used processed data is described at the bottom left side of the Figure, and was forecast by models using the SSA approach. We can conclude from the Figure that the SSA-BP model predicts values close to the true values, especially SSA-BFGS-FA-BP. In other words, the experiment that used processed data showed better performance than the other.

Figure 4b shows the difference between the forecast values and the exact values collected from site 1, and their corresponding errors. We can clearly see that the error of the SSA-BFGS-FA-BP model is much lower than that of the BP, BFGS-FA-BP, and SSA-BP models, which implies that the SSA-BFGS-FA-BP method performs much better than other models.

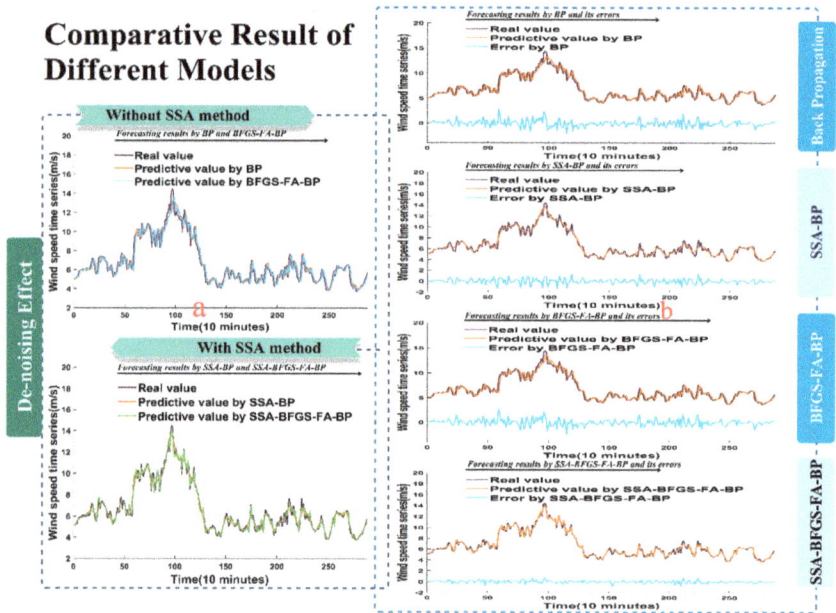

Figure 4. The forecast results for wind speed collected from site 1 at 10-min intervals. (a) The forecast effect without SSA algorithm and with SSA algorithm; (b) The comparison between the forecast values and raw data and their corresponding errors.

3.4.2. Results of Analysis

In this section, another two samples are described in Figures 5a and 6a, whose predictive performance with and without using the SSA method are compared. Furthermore, Figures 5b and 6b illustrate the difference between the forecast values and the real values collected from the other two sites and their corresponding errors.

Similar to what was described above, and as shown in Figures 5a and 6a, the BP and BFGS-FA-BP models closely approached the actual values, but the SSA-BP model and especially the SSA-BFGS-FA-BP model performed much better in forecasting. Therefore, we can conclude that the experiment using the SSA approach outperforms the other. As revealed in Figures 5b and 6b, the deviations of the SSA-BFGS-FA-BP model are much smaller than those of the BP, BFGS-FA-BP, and SSA-BP models. In particular, it is very clear that the SSA-BFGS-FA-BP model gets extremely close to the exact wind speed, and has higher performance than the other three models.

To test the accuracy of the experiment and guarantee the practicability and feasibility of the developed method, we performed another three experiments. As shown in Table 10, data were taken during four seasons (spring, summer, autumn, and winter) from a fixed location to verify the stability of the model. The results presented in this indicate that (1) the variance of the proposed hybrid SSA-BFGS-FA-BP method is minimal; and (2) the predicted value of the hybrid SSA-BFGS-FA-BP model is closer to the true value, having higher stability than the other three models.

In another experiment, the new predicted value based on historical data was used as the new real value to test. Using this, we performed three-step iterative, six-step iterative, and twelve-step iterative experiments. The final experimental results are shown in Table 10. The results indicate that the variance of the hybrid SSA-BFGS-FA-BP model is smaller, and that the predicted value within six steps of hybrid SSA-BFGS-FA-BP model is closer to the true value, having higher stability than the other three models. However, in the 12-step iterative experiment, the proposed SSA-BFGS-FA-BP model did not show better accuracy than the other three models. This indicates that the optimal results will be worse with an increase in the number of iterations beyond a certain extent, because of the increase in randomness. This also verifies that, with an increase in the number of iterations, the accuracy of prediction is low and the deviation is high.

Finally, we collected data at different time intervals (10, 30 and 60-min intervals) from a fixed location to conduct an experiment, and the results of the experiment are shown in Table 11. We can conclude from the table that the effects of the optimization method will run into a bottleneck when the time interval of the data becomes too great. The hybrid SSA-BFGS-FA-BP model did not show a better performance. The error of forecasting reached a high value when the time interval was so long.

In conclusion, the SSA-BFGS-FA-BP has much higher effectiveness than the single BP, BFGS-FA-BP, and SSA-BP models. We can confirm that the hybrid SSA-BFGS-FA-BP model can make a more accurate prediction on account of the original time sequence.

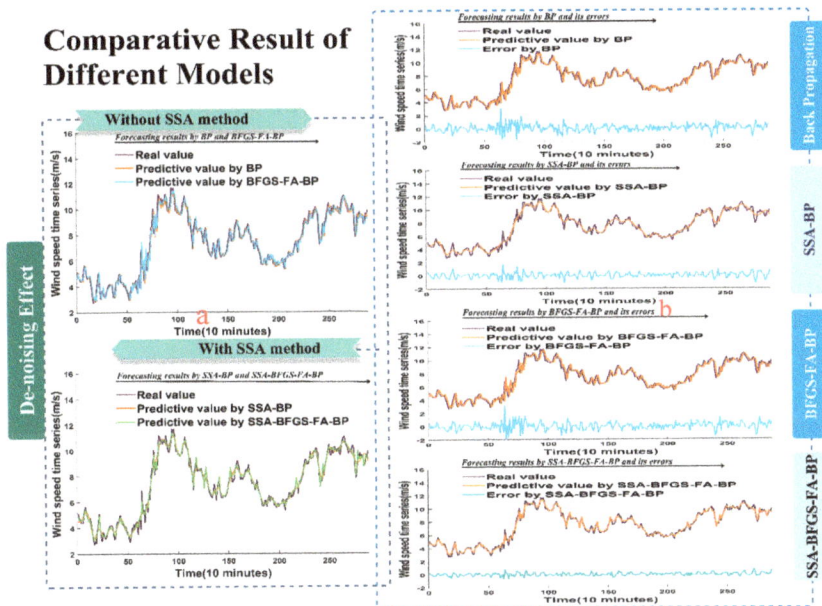

Figure 5. The forecast results for wind speed collected from site 2 at 10-min intervals. (**a**) The forecast effect without SSA algorithm and with SSA algorithm; (**b**) The comparison between the forecast values and the raw data and their corresponding errors.

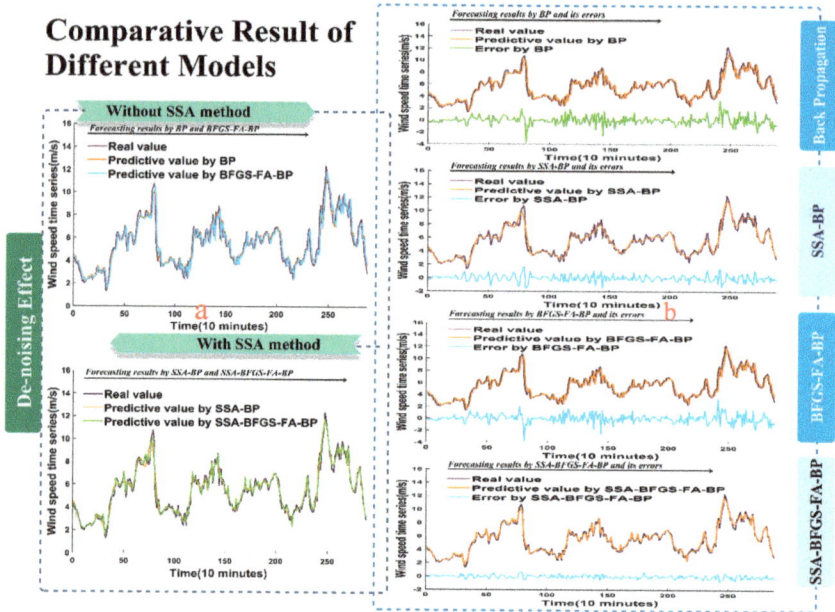

Figure 6. The forecast results for wind speed collected from site 3 at 10-min intervals. (**a**) The forecast effect without SSA algorithm and with SSA algorithm; (**b**) The comparison between the forecast values and the raw data and their corresponding errors.

3.5. The Results of the DM Test

The DM test was employed to verify the levels of accuracy forecasted by the proposed hybrid method and the other three single models. Table 12 shows that the values of the DM statistics between the proposed hybrid model and the BP, SSA-BP and BFGS-FA-BP models are 8.1064, 8.0468 and 8.1696, respectively. Under a 1% confidence level, the upper limit value is much smaller than these DM statistics; therefore, we cannot accept the null hypothesis and we have to admit the alternative hypothesis. Thus, we can conclude that the hybrid method outperforms the other single methods.

Remark. *We could learn from the results in terms of estimations on the basis of the DM test that the novel hybrid method achieves a more precise and stable prediction capacity than the other three models, and that the forecasting effectiveness of the hybrid method differs from that of the BP, SSA-BP, and BFGS-FA-BP models.*

Table 7. MAE, RMSE, and MAPE of site 1 for each forecasting model.

Numbers of Test	BP			BFGS-FA-BP			SSA-BP			SSA-BFGS-FA-BP		
	MAE	RMSE	MAPE	MAE	RMSE	MAPE	MAE	RMSE	MAPE	MAE	RMSE	MAPE
1	0.5017	0.6789	7.6543	0.5113	0.6892	7.5143	0.3482	0.4643	5.2833	0.1835	0.2453	4.9052
2	0.5217	0.6871	7.7668	0.5145	0.6868	7.6611	0.3479	0.4632	5.2902	0.1853	0.2454	4.8152
3	0.5038	0.6769	8.2622	0.508	0.6758	7.5722	0.3487	0.4645	5.3063	0.1831	0.2461	4.7851
4	0.5064	0.6767	8.0176	0.5098	0.678	7.6388	0.3472	0.4633	5.2917	0.1833	0.246	4.0152
5	0.5364	0.6951	8.041	0.5225	0.6869	7.6698	0.3475	0.4638	5.292	0.1883	0.2462	4.8752
6	0.5118	0.6848	7.7139	0.5122	0.684	7.6718	0.3483	0.4636	5.2923	0.1813	0.2465	4.6995
7	0.5051	0.6848	7.9229	0.5234	0.692	7.5886	0.3478	0.4636	5.302	0.1873	0.2462	4.8905
8	0.5133	0.6827	7.9593	0.5121	0.6826	7.5832	0.3473	0.4637	5.303	0.1835	0.2461	4.7775
9	0.5243	0.6855	7.7893	0.5155	0.6805	7.6507	0.3474	0.4632	5.2966	0.1839	0.2462	4.9824
10	0.5161	0.6838	8.2	0.5068	0.6807	7.6527	0.3466	0.4631	5.2993	0.1983	0.2461	4.9867
11	0.5333	0.6941	7.6998	0.5109	0.6829	7.5984	0.348	0.4636	5.2924	0.1884	0.2461	4.2015
12	0.5091	0.6785	7.708	0.5121	0.6835	7.6684	0.3475	0.4637	5.2859	0.1823	0.246	4.2152
13	0.5277	0.6877	7.9615	0.5063	0.6757	7.5189	0.3467	0.4631	5.2933	0.1835	0.2459	4.5952
14	0.5106	0.6864	7.74	0.5123	0.6824	7.6302	0.3475	0.4634	5.2884	0.1833	0.2465	4.8951
15	0.5218	0.6858	8.1374	0.5042	0.6834	7.6242	0.3471	0.4633	5.2926	0.1803	0.2458	4.3895
16	0.5111	0.6874	7.7658	0.5027	0.6812	7.6604	0.3471	0.4631	5.3006	0.1893	0.2464	4.8679
17	0.5086	0.6832	7.6895	0.5117	0.6853	7.6	0.3481	0.4636	5.2913	0.1836	0.2462	4.2509
18	0.5078	0.6773	7.7802	0.5075	0.6767	7.6615	0.3469	0.4633	5.2944	0.1834	0.2463	4.5995
19	0.5199	0.6811	7.8523	0.508	0.6802	7.5538	0.3471	0.4633	5.3011	0.1813	0.2466	4.7655
20	0.515	0.6859	7.9596	0.501	0.6662	7.663	0.3474	0.4634	5.2851	0.1843	0.2457	4.4792
Average value	0.5153	0.6842	7.8811	0.5106	0.6817	7.6191	0.3475	0.4635	5.2941	0.1849	0.2461	4.8496

Table 8. MAE, RMSE, and MAPE of site 2 for each forecasting model.

Numbers of Test	BP			BFGS-FA-BP			SSA-BP			SSA-BFGS-FA-BP		
	MAE	RMSE	MAPE	MAE	RMSE	MAPE	MAE	RMSE	MAPE	MAE	RMSE	MAPE
1	0.473	0.6309	7.0188	0.444	0.5972	6.6955	0.3272	0.426	4.9888	0.2327	0.2427	4.3499
2	0.4704	0.6253	7.0698	0.4447	0.608	6.6478	0.3268	0.4254	4.9841	0.2033	0.2343	4.45
3	0.4785	0.6279	7.0767	0.444	0.594	6.6344	0.3272	0.4261	4.9866	0.2433	0.2243	4.65
4	0.4688	0.6207	7.0969	0.4423	0.5955	6.6792	0.3272	0.4258	4.9942	0.2433	0.2543	4.7499
5	0.4563	0.6109	6.8203	0.4471	0.6016	6.6969	0.3286	0.428	5.0096	0.2533	0.3278	4.4995
6	0.468	0.6236	6.9768	0.445	0.5972	6.7103	0.3276	0.4268	4.994	0.2733	0.2043	4.5105
7	0.4767	0.6394	7.1899	0.4424	0.5932	6.6757	0.3282	0.4269	5.0038	0.2133	0.2427	4.992
8	0.4682	0.6276	6.9898	0.4519	0.6107	6.7899	0.3277	0.4266	4.998	0.2432	0.2333	4.3489
9	0.4848	0.6525	7.3656	0.4419	0.5975	6.6656	0.3272	0.4262	4.9893	0.2328	0.2554	4.4655
10	0.4608	0.6133	6.8227	0.4482	0.6016	6.7711	0.3275	0.4262	4.9952	0.2328	0.2412	4.651
11	0.462	0.6164	6.8987	0.4455	0.6022	6.7745	0.3272	0.4265	4.9918	0.2631	0.2943	4.8215
12	0.4547	0.6126	6.9756	0.4455	0.6025	6.721	0.3274	0.4265	4.996	0.2328	0.2342	4.3499
13	0.4592	0.612	6.8545	0.4439	0.598	6.6876	0.3276	0.4268	4.994	0.3026	0.2444	4.585
14	0.509	0.6582	7.4376	0.4444	0.6028	6.6422	0.3268	0.4258	4.9876	0.2533	0.2453	4.251
15	0.5148	0.6677	7.3453	0.4438	0.5961	6.7134	0.3269	0.4256	4.9842	0.2132	0.2433	4.815
16	0.4474	0.5996	7.0924	0.4467	0.5991	6.7905	0.328	0.4274	4.9991	0.2031	0.2121	4.2501
17	0.4674	0.6244	6.9551	0.4426	0.5946	6.679	0.3276	0.427	4.9964	0.2234	0.2543	4.495
18	0.4688	0.6245	7.0109	0.441	0.5963	6.6643	0.3274	0.4261	4.9922	0.2132	0.2314	4.625
19	0.4704	0.6391	7.0898	0.4427	0.5936	6.6678	0.3265	0.4252	4.9808	0.1935	0.2333	4.4635
20	0.4685	0.615	7.1212	0.444	0.5965	6.6865	0.3277	0.4272	5.0004	0.2032	0.2768	4.7011
Average value	0.4714	0.6271	7.0604	0.4446	0.5989	6.6997	0.3274	0.4264	4.9933	0.2337	0.2465	4.5512

Table 9. MAE, RMSE, and MAPE of site 3 for each forecasting model.

Numbers of Test	BP			BFGS-FA-BP			SSA-BP			SSA-BFGS-FA-BP		
	MAE	RMSE	MAPE	MAE	RMSE	MAPE	MAE	RMSE	MAPE	MAE	RMSE	MAPE
1	0.5404	0.7577	10.7941	0.5488	0.7665	10.4888	0.3887	0.5128	7.5209	0.2389	0.2851	6.752
2	0.5612	0.7752	10.9005	0.5623	0.7899	10.4192	0.3891	0.5143	7.5248	0.2389	0.2905	6.7538
3	0.5503	0.7679	10.784	0.5426	0.7681	10.5803	0.3905	0.5148	7.5595	0.2439	0.3051	7.0753
4	0.5519	0.7718	11.1111	0.5472	0.7707	10.6574	0.3882	0.5124	7.5094	0.2139	0.2885	6.6716
5	0.5509	0.7757	10.7029	0.5475	0.7694	10.6845	0.3902	0.5152	7.5699	0.2563	0.2768	6.9975
6	0.5518	0.7843	10.7331	0.5485	0.7601	10.6767	0.3888	0.5132	7.5312	0.2609	0.2905	7.1275
7	0.552	0.7628	10.8619	0.5645	0.7787	10.4793	0.3893	0.5142	7.5439	0.2389	0.279	6.8768
8	0.5561	0.7801	10.7375	0.5531	0.772	10.6667	0.3893	0.5135	7.5346	0.2204	0.2676	6.6675
9	0.553	0.7762	11.4947	0.534	0.7523	10.6886	0.3906	0.5154	7.5668	0.2539	0.2595	6.9075
10	0.5407	0.7588	10.6916	0.5515	0.7753	10.4897	0.3913	0.5158	7.6121	0.2339	0.3151	6.1558
11	0.5583	0.7745	10.9143	0.5571	0.7755	10.5834	0.3888	0.5135	7.5325	0.239	0.3005	6.8776
12	0.5615	0.7698	10.8	0.5462	0.7687	10.307	0.3897	0.5127	7.5142	0.2432	0.2982	6.6275
13	0.5447	0.7655	10.9335	0.5554	0.7754	10.5456	0.3897	0.514	7.526	0.2733	0.2775	6.8098
14	0.5491	0.7628	10.7102	0.5522	0.7697	10.5567	0.3915	0.5157	7.586	0.214	0.2591	6.4725
15	0.5466	0.77	10.8156	0.5456	0.7632	10.5918	0.3884	0.5134	7.537	0.2399	0.2835	6.7357
16	0.5415	0.7591	10.7132	0.549	0.7696	10.5576	0.3896	0.5142	7.5376	0.2238	0.2499	7.1752
17	0.5526	0.7702	10.9634	0.5483	0.7623	10.4912	0.3902	0.5151	7.5778	0.2546	0.29	6.3975
18	0.5608	0.7855	10.9059	0.5423	0.7615	10.4807	0.389	0.5124	7.5297	0.2434	0.2945	6.7595
19	0.5486	0.7667	11.2402	0.5378	0.7585	10.6654	0.3885	0.5126	7.5293	0.2386	0.2795	6.9808
20	0.5426	0.7708	10.9645	0.5437	0.7586	10.5158	0.3874	0.5122	7.4898	0.2444	0.3125	6.2753
Average value	0.5507	0.7703	10.8886	0.5489	0.7683	10.5563	0.3894	0.5139	7.5417	0.2407	0.2851	6.7548

Table 10. MAE, RMSE, and MAPE of four seasons and various iterations in one site for each forecasting model.

Model	Error Index	Season				Iteration			
		Spring	Summer	Autumn	Winter	One	Three	Six	Twelve
BP	MAE	0.5153	0.5975	0.4804	0.6708	0.5153	0.7955	1.0541	1.2740
	RMSE	0.6842	0.8103	0.6054	1.0172	0.6842	1.0229	1.3421	1.6878
	MAPE	7.8811	9.5102	13.7179	7.4235	7.8811	12.4853	16.7727	20.3831
BFGS-FA-BP	MAE	0.5106	0.6049	0.4629	0.5272	0.5106	0.7745	1.0051	1.3240
	RMSE	0.6817	0.7996	0.5854	0.6651	0.6817	1.0064	1.2991	1.7399
	MAPE	7.6191	9.2133	13.1582	6.1367	7.6191	12.0691	16.1433	21.2589
SSA-BP	MAE	0.3475	0.4486	0.3368	0.3625	0.3475	0.3651	0.6742	1.3673
	RMSE	0.4635	0.5659	0.4243	0.4544	0.4635	0.4824	0.8456	1.7684
	MAPE	5.2941	7.1654	9.9706	4.2700	5.2941	5.4943	10.6671	21.2998
SSA-BFGS-FA-BP	MAE	0.1849	0.3875	0.2863	0.3159	0.1849	0.3354	0.6370	1.4349
	RMSE	0.2461	0.4961	0.3941	0.4018	0.2461	0.4389	0.8154	1.8355
	MAPE	4.8496	5.2463	7.5297	3.4269	4.8496	4.8671	9.9499	22.5317

Table 11. MAE, RMSE, and MAPE of various intervals in one site for each forecasting model.

Time (minutes)	Model Error	BP	BFGS-FA-BP	SSA-BP	SSA-BFGS-FA-BP
10	MAE	0.5153	0.5106	0.3475	0.1849
	RMSE	0.6842	0.6817	0.4635	0.2461
	MAPE	7.8811	7.6191	5.2941	4.8496
30	MAE	0.9769	0.9444	0.6193	0.5854
	RMSE	1.3655	1.2808	0.8087	0.7426
	MAPE	9.9414	9.4053	6.6134	5.4780
60	MAE	0.9253	0.8890	0.6244	0.6043
	RMSE	1.1464	1.1116	0.7881	0.7392
	MAPE	28.3766	26.0956	19.1245	17.4441

Table 12. Results of the DM test and operational time (s).

Performance Metric	Compared Model	Average Value
DM-test	BP	8.1064 *
	SSA-BP	8.0468 *
	BFGS-FA-BP	8.1696 *
Operational time	BP	0.3
	SSA-BP	0.6
	BFGS-FA-BP	490.3
	Hybrid model	157.1

Note: * Indicates the 1% significance level.

4. Discussion

In this section, we initially discuss the application of SSA in preprocessing the original data, which influences the forecasting performance. We also examine the MAPEs of decreased relative percentage (DRP) between the proposed model and other forecasting approaches. Furthermore, we present and discuss variations of the data selection.

4.1. Data Pre-Processing

In general, plenty of noise and high-frequency time series lie in the raw wind speed time sequence. Therefore, the decomposition of the original data sequence is a significant process in data filtering. This can always effectively enhance the prediction accuracy of the model to obtain better forecast results. Through the comparison between BP and SSA-BP, we can assess the effectiveness of the data pre-processing using a new metric called DRP (%), and its corresponding defining equation is summarized as follows:

$$\text{DRP} = \frac{\text{MAPE}_i - \text{MAPE}_j}{\text{MAPE}_i} \times 100 \tag{17}$$

The experimental results show that the method significantly enhances the forecasting effectiveness: it decreases the MAPE by 32.8%, 29.3% and 30.7% for site 1, site 2 and site 3, respectively.

4.2. Neural Networks

In the field of practical engineering, the quality of a model depends on its effectiveness, rather than its complexity. However, the question of how to seek an effective forecasting method to enhance performance is not only a problem that is in urgent need of a solution, but also a critical problem in the field of forecasting. The relevant study [58] showed that there was no one unified model for forecasting time series, and model effectiveness under different circumstances should be analyzed and understood, with incremental improvements being made on the basis of the knowledge gained; therefore, it is impossible to find one model to solve all forecasting problems.

Thus, our attention should be more focused on the DRP of the error forecast by different approaches using different data sets, in order to find a relatively good model for forecasting wind speed time series. Through analyzing the difference between the proposed model and the other comparative models, we can find that the proposed model improves effectiveness by 38.4654%, 35.5391 and 37.9645% for site 1, site 2 and site 3, respectively. The detailed results are presented in Table 13. From the Table, we can see that the developed method has a very good performance in decreasing wind speed forecasting error.

Table 13. The DRP of MAPE of the proposed model and comparison models.

Cases	DRP of MAPE (%)			
	ENN	WNN	PSO/FA Elman	Proposed Model
case 1	15.6250	10.2506	13.9158	38.4654
case 2	24.0000	32.1063	12.7855	35.5391
case 3	47.8261	-	15.4897	37.9645
case 4	35.2941	-	-	-

4.3. Data Selection

According to the forecast results, the 10-min interval data sequence achieves the best forecasting effectiveness for all three observation sites, with an MAPE of approximately 6%; therefore, the proposed hybrid model shows excellent performance in forecasting the wind speed time sequence at 10-min intervals. The 10-min interval time series at each observation site decreases the forecasting error by 12.59%, 10.14% and 11.41%, respectively.

For the time series data with a 60-min interval, the forecast results are good for all three observation sites, while the forecasting performance is worse than for the data with a 10-min interval. Therefore, the SSA-BFGS-FA-BP is more applicable to forecasting the wind speed time sequence with a 10-min interval, and the data selection will have a serious effect on forecasting effectiveness. However, regardless of the time interval, the forecasting effectiveness is in an acceptable range. Many works apply the wind speed series with time resolutions including 10, 30 and 60 min for the purpose of the forecast, which is representative for studying wind speed forecasting. The detailed comparison results are presented in Table 14.

Table 14. Comparison results of three observation sites with different time intervals.

Observation Sites	10-min	30-min	60-min
Site 1	4.8496	5.4780	17.4441
Site 2	4.5512	9.8730	14.6964
Site 3	6.7548	12.9558	18.1605

5. Conclusions and Future Work

As a kind of non-polluting and renewable energy source, wind energy has been increasingly applied in the development of industry and agriculture, and its forecasting is becoming increasingly important for wind farms. Recently, academia and wind farm projects have been gradually paying more attention to wind speed forecasting. Perfect prediction can not only reduce costs and enhance personal safety, but also help wind farm management develop more effective programs. The accuracy of a model is as important as its stability in forecasting. It is of great interest to propose an outstanding method for wind speed prediction with high accuracy and long-term stability. Nevertheless, wind speed prediction has been generally considered a challenging task in terms of the effects of various intangible factors, such as temperature, location, tides, atmospheric pressure, and other factors. In this paper, to overcome these difficulties, a hybrid model that combines the SSA approach, BFGS-FA algorithm, and BP method is presented.

The results based on evaluation criteria such as the MAE, RMSE, MAPE and a statistical test are shown in a sequence of charts, in which the superior qualities of the developed hybrid method are revealed most vividly. From the data in the tables and figures, we can draw the conclusion that the proposed hybrid method achieves the best forecasting effectiveness and a higher stability and reliability.

SSA is a practical decomposition approach, which can remove the noise from the raw data, leaving the principal component for forecasting. The BP model, based on feed forward neural networks,

Energies **2017**, *10*, 954

has increasingly turned into a fairly distinguished tool. It is shown that the BP model can get its final predictive results in a remarkably short time.

In brief, the hybrid model always has the lowest MAPE value compared with other single forecasting methods, which implies that the hybrid method has the best performance and higher reliability. Improvements in forecasting accuracy and stability can not only help to save large amounts of energy and money, but also help to reduce the time the system requires. The experiments performed in the present study show that the developed hybrid method is a potential algorithm with high accuracy. In addition, the hybrid method could be applied to other fields of practical engineering, such as electric load forecasting, stock price prediction, and solar resource forecasting.

Acknowledgments: This work was funded by the National Natural Science Foundation of China (Grant No. 71573034).

Author Contributions: Zeng Wang wrote the whole manuscript, and carried out the data analysis; Ping Jiang gave overall guidance; Kequan Zhang and Wendong Yang conducted the programming and part of the data processing.

Conflicts of Interest: The authors declare no conflict of interest.

Abbreviations

ARIMA	auto-regressive integrated moving average	Gen_{max}	the maximum number of iterations
BFGS	Broyden-Fletcher-Goldfarb-Shanno	**MAPE**	mean absolute percentage error
MLP	multi-layer perceptron	**MAE**	mean absolute error
BP	back propagation	**RMSE**	root mean square error
RBF	radial basis function	**betamin**	minimum value of beta
RNN	Recurrent Neural Networks	**gamma**	absorption coefficient
BPA	back propagation algorithm	**DM**	Diebold-Mariano
PSO	particle swarm optimization	**SVM**	Support Vector Machine
ANN	artificial neural network	T	the length of time series
FA	firefly algorithm	$Value_{min}$	the minimum real value
BPNN	back propagation neural network	K	the length of the trajectory matrix
CS	cuckoo search	**SVD**	singular value decomposition
MCS	modified cuckoo search	**GA**	genetic algorithm
$Value_{actual}$	real value	H_0	the null hypothesis
SSA	singular spectrum analysis	H	continuous function
EMD	empirical mode decomposition	**WD**	wavelet decomposition
$Value_{normalized}$	data after linear transformation	H_1	the alternative hypothesis
$Value_{max}$	the maximum real value	$Z_{\alpha/2}$	upper (or positive) Z-value
Z_{ij}	an element of a generic matrix Z	**DRP**	decreased relative percentage
ENN	Elman neural network	**WNN**	Wavelet neural network

Appendix A

Algorithm A1. BFGS.

Parameters:
δ–the tolerance of convergence. t –present iterative times.
Gen_{max}–the max iterative times.

1: /*Initialize the convergence tolerance δ and present iterative times.*/
2: Set the convergence tolerance $\delta > 0$, $t \leftarrow 0$
3: Assess the inverse matrix of Hessian matrix at an initial value x°
4: **WHILE** ($t < Gen_{max}$) **RUN**
5: /* Compute search path.*/
6: $\mathbf{d}_t = -\mathbf{D}_t \nabla f(\mathbf{x}^t)$

Algorithm A1. *Cont.*

7:	/* Compute step size.*/
8:	$\lambda_t = \underset{\lambda \geq 0}{\arg\min} f(\mathbf{x}^t + \lambda \mathbf{d}_t)$
9:	/*Obtain the new iteration.*/
10:	$\mathbf{x}^{t+1} = \mathbf{x}^t + \lambda_t \mathbf{d}_t$
11:	**IF**$(x_p^{t+1} > x_p^{up})$ **WELL**
12:	$X_p^{t+1} = x_p^{up}$
13:	**ELSE IF**$(x_p^{t+1} < x_p^{lo})$ **WELL**
14:	$X_p^{t+1} = x_p^{lo}$
15:	**END**
16:	Compute $\mathbf{D}_{t+1}$
17:	Set $t \leftarrow t+1$
18:	**END**

Appendix B

Algorithm A2. FA.

Input:

$x_h^0 = (x^0(1), x^0(2), \ldots, x^0(k))$–a series of data for training.

$x_m^0 = (x^0(k+1), x^0(k+2), \ldots, x^0(k+d))$–a series of data for verifying.

Output:

x_b–the corresponding value of x when it acquires the optimal fitness among all fireflies.

Parameters:

Gen_{max}–the max iterative times. n–the total number of fireflies.

F_p–the fitness function according to firefly p. x_p–nest p. g–the present number of iterations.

L_p–the brightness of firefly p. d–the dimension of the parameter.

1::	/* Define all the parameters related to FA.*/
2:	/* Initialize the species of fireflies $x_p (p = 1, 2, \ldots, n)$ at random.*/
3:	**FOR** $p = 1{:}n$ **RUN**
4:	Assess the relevant fitness function F_p
5:	**END**
6:	/* Confirm light intensity. */
7:	**FOR** $p = 1{:}n$ **RUN**
8:	Confirm the brightness L_p through $F(x_p)$
9:	**END**
10:	**WHILE** $(g < Gen_{max})$ **RUN**
11:	**FOR** $p = 1{:}n$ **RUN**
12:	**FOR** $q = 1{:}n$ **RUN**
13:	/* Adjust the firefly from p to q in any direction.*/
14:	**IF** $(L_q > L_p)$ **WELL**
15:	$r_{pq} = \|x_p - x_q\| = \sqrt{\sum_{t=1}^{d} (x_{p,t} - x_{q,t})^2}$
16:	$x_p = x_p + \beta_0 e^{-\gamma r^2}(x_q - x_p) + \alpha(\mathbf{rand} - 0.5)$
17:	**END**
18:	Attraction changes with the distance r via e^{-r^2}
19:	**END**
20:	**END**
21:	/*Renew the best nest x_m of the d generation*/
22:	**FOR** $p = 1{:}n$ **RUN**
23:	**IF** $(F_m < F_b)$ **WELL**
24:	$x_b \leftarrow x_m$;
25:	**END**
26:	**END**
27:	**END**
28:	**RETURN** x_b

Appendix C

Algorithm A3. BFGS-FA.

Input:

$x_h^{(0)} = (x^{(0)}(1), x^{(0)}(2), \ldots, x^{(0)}(k))$ –a series of data for training.

$x_m^{(0)} = (x^{(0)}(k+1), x^{(0)}(k+2), \ldots, x^{(0)}(k+d))$ –a series of data for verifying

Output:

x_b–the corresponding value of x when it acquires the optimal fitness among all fireflies.

Parameters:

Gen_{max}–the max iterative times. n–the total number of fireflies.

F_p–the fitness function according to firefly p. x_p–nest p. g–the present number of iterations.

L_p–the brightness of firefly p. d–the dimension of the parameter.

1:    /*Define all the parameters related to FA and BFGS.*/

2:    /* Initialize population of n fireflies $x_p (p = 1, 2, \ldots, n)$ at random.*/

3:    **FOR**$p = 1$:n**RUN**

4:        Assess the relevant fitness function F_p

5:    **END**

6:    /* Confirm light intensity*/

7:    **FOR** $p = 1$:n **RUN**

8:        Confirm brightness L_p through $F(x_p)$

9:    **END**

10:   **WHILE** ($g < Gen_{max}$) **RUN**

11:       **FOR** $p = 1$:n **RUN**

12:          **FOR** $q = 1$:n **RUN**

13:            /* Adjust the firefly from p to q in any direction */

14:            **IF**($L_q > L_p$) **WELL**

15:                $r_{pq} = |x_p - x_q| = \sqrt{\sum_{t=1}^{d} (x_{p,t} - x_{q,t})^2}$

16:                $x_p = x_p + \beta_0 e^{-\gamma r^2}(x_q - x_p) + \alpha(\textbf{\textit{rand}} - 0.5)$

17:            **END**

18:            Attraction changes with the distance r via e^{-r^2}

19:            Apply **BFGS** to help to renew the new site of fireflies $x_p (p = 1, 2, \ldots, n)$ quickly.

20:            /*Assess the new position and renew the new light intensity L_p.*/

21:            **FOR** $p = 1$:n **RUN**

22:               Assess the relevant fitness function F_p

23:            **END**

24:            /* Update the brightness*/

25:            **FOR** $p = 1$:n **RUN**

26:               Confirm the brightness L_p through $F(x_p)$

27:            **END**

28:          **END**

29:       **END**

30:       /*Update best nest x_m of the d generation.*/

31:       **FOR** $p = 1$:n**RUN**

32:          **IF**($F_m < F_b$)**WELL**

33:            $x_b \leftarrow x_m$

34:          **END**

35:       **END**

36:   **END**

37:   **RETURN** x_b

References

1.   Manzano-Agugliaro, F.; Zapata-Sierra, A.; Herna, Q. The wind power of Mexico. *Renew. Sustain. Energy Rev.* **2010**, *14*, 2830–2840.

2. Manzano-Agugliaro, F.; Sanchez-Muros, M.J.; Barroso, F.G.; Martínez-Sánchez, A.; Rojo, S. Insects for biodiesel production. *Renew. Sustain. Energy Rev.* **2012**, *16*, 3744–3753. [CrossRef]

3. Hernández-Escobedo, Q.; Saldaña-Flores, R.; Rodríguez-García, E.R. Wind energy resource in Northern Mexico. *Renew. Sustain. Energy Rev.* **2014**, *32*, 890–914. [CrossRef]

4. World Wind Energy Association (2009). World Wind Energy Report. Available online: http://www.wwindea. org/home/images/stories/worldwindenergyreport2009_s.pdf (accessed in 10 March 2017).

5. Hernandez-Escobedo, Q.; Manzano-Agugliaro, F.; Gazquez-Parra, J.A.; Zapata-Sierra, A. Is the wind a periodical phenomenon? The case of Mexico. *Renew. Sustain. Energy Rev.* **2011**, *15*, 721–728. [CrossRef]

6. Montoya, F.G.; Manzano-Agugliaro, F.; López-Márquez, S.; Hernández-Escobedo, Q.; Gil, C. Wind turbine selection for wind farm layout using multi-objective evolutionary algorithms. *Expert Syst. Appl.* **2014**, *41*, 6585–6595. [CrossRef]

7. Ernst, B.; Oakleaf, B.; Ahlstrom, M.L.; Lange, M.; Moehrlen, C.; Lange, B.; Focken, U.; Rohrig, K. Predicting the wind. *IEEE Power Energy Mag.* **2007**, *5*, 78–89. [CrossRef]

8. Liu, H.; Tian, H.Q.; Liang, X.F.; Li, Y.F. Wind speed forecasting approach using secondary decomposition algorithm and Elman neural networks. *Appl. Energy* **2015**, *157*, 183–194. [CrossRef]

9. Al-Yahyai, S.; Charabi, Y.; Gastli, A. Review of the use of numerical weather prediction (NWP) models for wind energy assessment. *Renew. Sustain. Energy Rev.* **2010**, *14*, 3192–3198. [CrossRef]

10. Amral, N.; Ozveren, C.S.; King, D. Short term load forecasting using Multiple Linear Regression. In Proceedings of the 42th International Universities Power Engineering Conference, Brighton, UK, 4–6 September 2007; pp. 1192–1198.

11. Li, W.; Zhang, Z.G. Based on time sequence of ARIMA model in the application of short-term electricity load forecasting. In Proceedings of the International Conference on Research Challenges in Computer Science, Shanghai, China, 28–29 December 2009; pp. 11–14.

12. Christiaanse, W.R. Short-Term Load Forecasting Using General Exponential Smoothing. *IEEE Trans. Power Appar. Syst.* **1971**, *PAS-90*, 900–911. [CrossRef]

13. Irisarri, G.D.; Widergren, S.E.; Yehsakul, P.D. On-line load forecasting for energy control center application. *IEEE Trans. Power Appar. Syst.* **1982**, *PAS-101*, 71–78. [CrossRef]

14. Du, P.; Jin, Y.; Zhang, K. A Hybrid Multi-Step Rolling Forecasting Model Based on SSA and Simulated Annealing—Adaptive Particle Swarm Optimization for Wind Speed. *Sustainability* **2016**, *8*, 754. [CrossRef]

15. Hu, Y.C. Pattern classification by multi-layer perceptron using fuzzy integral-based activation function. *Appl. Soft Comput.* **2010**, *10*, 813–819. [CrossRef]

16. Peeling, S.M.; Moore, R.K. Isolated digit recognition experiments using the multi-layer perceptron. *Speech Commun.* **1988**, *7*, 403–409. [CrossRef]

17. Fan, X.; Wang, L.; Li, S. Predicting chaotic coal prices using a multi-layer perceptron network model. *Resour. Policy* **2016**, *50*, 86–92. [CrossRef]

18. Ganotra, D.; Joseph, J.; Singh, K. Profilometry for the measurement of three-dimensional object shape using radial basis function, and multi-layer perceptron neural networks. *Opt. Commun.* **2002**, *209*, 291–301. [CrossRef]

19. Beigy, H.; Meybodi, M.R. Backpropagation algorithm adaptation parameters using learning automata. *Int. J. Neural Syst.* **2001**, *11*, 219–228. [CrossRef] [PubMed]

20. Jain, A.; Srinivas, E.; Rauta, R. Short Term Load Forecasting using Fuzzy Adaptive Inference and Similarity. In Proceedings of the World Congress on Nature & Biologically Inspired Computing, Coimbatore, India, 9–11 December 2009.

21. Kumar, R.; Aggarwal, R.K.; Sharma, J.D. Comparison of regression and artificial neural network models for estimation of global solar radiations. *Renew. Sustain. Energy Rev.* **2015**, *52*, 1294–1299. [CrossRef]

22. Rather, A.M.; Agarwal, A.; Sastry, V.N. Recurrent neural network and a hybrid model for prediction of stock returns. *Expert Syst. Appl.* **2015**, *42*, 3234–3241. [CrossRef]

23. Cardoso, M.C.; Silva, M.; Vellasco, M.M.B.R.; Cataldo, E. Quantum-inspired features and parameter optimization of Spiking Neural Networks for a case study from atmospheric. *Procedia Comput. Sci.* **2015**, *53*, 74–81. [CrossRef]

24. Sheela, K.G.; Deepa, S.N. An Intelligent Computing Model for Wind Speed Prediction in Renewable Energy Systems. *Procedia Eng.* **2012**, *30*, 380–385.

25. Liu, N.; Tang, Q.; Zhang, J.; Fan, W.; Liu, J. A hybrid forecasting model with parameter optimization for short-term load forecasting of micro-grids. *Appl. Energy* **2014**, *129*, 336–345. [CrossRef]

26. Wang, J.; Zhu, S.; Zhang, W.; Lu, H. Combined modeling for electric load forecasting with adaptive particle swarm optimization. *Energy* **2010**, *35*, 1671–1678. [CrossRef]

27. Xiao, L.; Wang, J.; Hou, R.; Wu, J. A combined model based on data pre-analysis and weight coefficients optimization for electrical load forecasting. *Energy* **2015**, *82*, 524–549. [CrossRef]

28. Zhao, J.; Guo, Z.H.; Su, Z.Y.; Zhao, Z.Y.; Xiao, X.; Liu, F. An improved multi-step forecasting model based on WRF ensembles and creative fuzzy systems for wind speed. *Appl. Energy* **2016**, *162*, 808–826. [CrossRef]

29. Qin, S.; Liu, F.; Wang, J.; Song, Y. Interval forecasts of a novelty hybrid model for wind speeds. *Energy Reports* **2015**, *1*, 8–16. [CrossRef]

30. Wang, J.; Song, Y.; Liu, F.; Hou, R. Analysis and application of forecasting models in wind power integration: A review of multi-step-ahead wind speed forecasting models. *Renew. Sustain. Energy Rev.* **2016**, *60*, 960–981. [CrossRef]

31. Xiao, L.; Wang, J.; Yang, X.; Xiao, L. A hybrid model based on data preprocessing for electrical power forecasting. *Int. J. Electr. Power Energy Syst.* **2015**, *64*, 311–327. [CrossRef]

32. Xu, Y.; Yang, W.; Wang, J. Air quality early-warning system for cities in China. *Atmos. Environ.* **2017**, *148*, 239–257. [CrossRef]

33. Xiao, L.; Shao, W.; Liang, T.; Wang, C. A combined model based on multiple seasonal patterns and modified firefly algorithm for electrical load forecasting. *Appl. Energy* **2016**, *167*, 135–153. [CrossRef]

34. Fister, I.; Yang, X.S.; Brest, J. A comprehensive review of firefly algorithms. *Swarm Evolut. Comput.* **2013**, *13*, 34–46. [CrossRef]

35. Azad, S.K. Optimum Design of Structures Using an Improved Firefly Algorithm. *Iran Univ. Sci.Technol.* **2011**, *1*, 327–340.

36. Wang, J.; Qin, S.; Zhou, Q.; Jiang, H. Medium-term wind speeds forecasting utilizing hybrid models for three different sites in Xinjiang, China. *Renew. Energy* **2015**, *76*, 91–101. [CrossRef]

37. Guo, Z.; Zhao, W.; Lu, H.; Wang, J. Multi-step forecasting for wind speed using a modified EMD-based artificial neural network model. *Renew. Energy* **2012**, *37*, 241–249. [CrossRef]

38. Chau, K.W.; Wu, C.L. A hybrid model coupled with singular spectrum analysis for daily rainfall prediction. *J. Hydroinform.* **2010**, *12*, 458–473. [CrossRef]

39. Hu, J.; Wang, J.; Zeng, G. A hybrid forecasting approach applied to wind speed time series. *Renew. Energy* **2013**, *60*, 185–194. [CrossRef]

40. Zhang, C.; Wei, H.; Xie, L.; Shen, Y.; Zhang, K. Direct interval forecasting of wind speed using radial basis function neural networks in a multi-objective optimization framework. *Neurocomputing* **2016**, *205*, 53–63. [CrossRef]

41. Wang, J.; Zhang, W.; Li, Y.; Wang, J.; Dang, Z. Forecasting wind speed using empirical mode decomposition and Elman neural network. *Appl. Soft Comput.* **2014**, *23*, 452–459. [CrossRef]

42. Yao, C.; Gao, X.; Yu, Y. Wind speed forecasting by wavelet neural networks: A comparative study. *Math. Probl. Eng.* **2013**, *9*, 681–703. [CrossRef]

43. Wang, S.; Zhang, N.; Wu, L.; Wang, Y. Wind speed forecasting based on the hybrid ensemble empirical mode decomposition and GA-BP neural network method. *Renew. Energy* **2016**, *94*, 629–636. [CrossRef]

44. Yang, Z.; Wang, J. Multi-step wind speed forecasting using a novel model hybridizing singular spectrum analysis, modified intelligent optimization and rolling Elman neural network. *Math. Probl. Eng.* **2016**, *2016*, 1–30.

45. McCulloch, W. S.; Pitts, W.H. A logical calculus of ideas imminent in nervous activity. *BullMath. Biophy.* **1943**, *5*, 115–133.

46. Yan, P.; Zhang, C. *Artificial Neural Network and Simulating-Evolution Computation*; Tsinghua University Press: Beijing, China, 2000.

47. Bazaraa, M.S.; Sherali, H.D.; Shetty, C.M. *Nonlinear Programming: Theory and Algorithms*; Wiley: Hoboken, NJ, USA, 1993; Volume 3.

48. Yang, X.S. Firefly algorithms for multimodal optimization. In *Lecture Notes in Computer Science (including subseries Lecture Notes in Artificial Intelligence and Lecture Notes in Bioinformatics)*; Springer: Berlin/Heidelberg, Germany, 2009; Volume 5792, pp. 169–178.

49. Bendjeghaba, O.; Boushaki, S.I.; Zemmour, N. Firefly algorithm for optimal tuning of PID controller parameters. In Proceedings of the 2013 fourth International Conference on Power Engineering, Energy and Electrical Drives (POWERENG), Istanbul, Turkey, 13–17 May 2013; pp. 1293–1296.

50. Lohrer, M. A Comparison between the Firefly Algorithm and Particle Swarm Optimization. Ph.D. Thesis, University of Oakland, Rochester, MI, USA, 2013.

51. Beneki, C.; Eeckels, B.; Leon, C. Signal extraction and forecasting of the UK tourism income time series: A singular spectrum analysis approach. *J. Forecast.* **2012**, *31*, 391–400. [CrossRef]

52. Elsner, J.B.; Tsonis, A.A. *Singular Spectrum Analysis: A New Tool in Time Series Analysis*; Springer: Berlin, Germany, 1996; Volume 1283, pp. 932–942.

53. Golyandina, N.; Nekrutkin, V.; Zhigljavsky, A. *Analysis of Time Series Structure: SSA and Related Techniques*; Chapman Hall/CRC: Boca Raton, FL, USA, 2001.

54. Hassani, H. Singular Spectrum Analysis: Methodology and Comparison. *J. Data Sci.* **2007**, *5*, 239–257.

55. Hassani, H.; Heravi, S.; Zhigljavsky, A. Forecasting European industrial production with singular spectrum analysis. *Int. J. Forecast.* **2009**, *25*, 103–118. [CrossRef]

56. Diebold, F.X.; Mariano, R.S. Comparing Predictive Accuracy. *J. Bus. Econ. Stat.* **1995**, *13*, 253–265. [CrossRef]

57. Inman, R.H.; Pedro, H.T.C.; Coimbra, C.F.M. Solar Forecasting Methods for Renewable Energy Integration. *Prog. Energy Combust. Sci.* **2013**, *39*, 535–576. [CrossRef]

58. Moghram, I.; Rahman, S. Analysis and evaluation of five short-term load forecasting techniques. *IEEE Trans. Power Syst.* **1989**, *4*, 1484–1491.

energies

MDPI

Article

Wind Speed Forecasting Based on EMD and GRNN Optimized by FOA

Dongxiao Niu [1], Yi Liang [1,*] and Wei-Chiang Hong [2]

1   School of Economics and Management, North China Electric Power University, Beijing 102206, China; ndx@ncepu.edu.cn
2   Department of Information Management, Oriental Institute of Technology, New Taipei 220, Taiwan; samuelsonhong@gmail.com
*   Correspondence: louisliang@ncepu.edu.cn; Tel.: +86-010-61773079

Received: 13 November 2017; Accepted: 28 November 2017; Published: 1 December 2017

Abstract: As a kind of clean and renewable energy, wind power is winning more and more attention across the world. Regarding wind power utilization, safety is a core concern and such concern has led to many studies on predicting wind speed. To obtain a more accurate prediction of the wind speed, this paper adopts a new hybrid forecasting model, combing empirical mode decomposition (EMD) and the general regression neural network (GRNN) optimized by the fruit fly optimization algorithm (FOA). In this new model, the original wind speed series are first decomposed into a collection of intrinsic mode functions (IMFs) and a residue. Next, the inherent relationship (partial correlation) of the datasets is analyzed, and the results are then used to select the input for the forecasting model. Finally, the GRNN with the FOA to optimize the smoothing factor is used to predict each sub-series. The mean absolute percentage error of the forecasting results in two cases are respectively 8.95% and 9.87%, suggesting that the hybrid approach outperforms the compared models, which provides guidance for future wind speed forecasting.

Keywords: wind speed forecasting; empirical mode decomposition; general regression neural network; fruit fly optimization algorithm

1. Introduction

Wind power, as a type of sustainable and clean energy, is one of the most widely used, technologically mature, and commercially produced renewable sources [1,2]. According to the Global Wind Energy Council (GWEC), the cumulative wind generating installed capacity has reached 486,790 MW at the end of 2016 with the share of 34.7% donated by China [3]. The goal that grid-connected wind power installed capacity should reach 200 GW by 2020 [4] indicates that, during the "13th Five-Year Plan" period, China needs to put into operation more than 20 GW of wind power annually. This means that the targets and tasks of wind power development are basically clear, and the wind power industry will maintain a rapid growth for a long period of time. Concerning the benefits of wind power, a prediction system installed in grid-connected wind farms becomes important to effectively reduce the volatility of the voltage and frequency caused by a sudden cut of wind turbines, and to improve the security, reliability, and controllability of an electric power system to realize the economic dispatch. In this sense, an accurate forecast of wind speed is an essential prerequisite, which helps guarantee the construction and operation of the wind power prediction system.

The commonly used methods in terms of wind speed prediction are mainly divided into two categories: statistical analysis models, such as autoregressive moving average (ARMA) models [5] and autoregressive integrated moving average (ARIMA) models [6–8]; and machine learning methods, including artificial neural networks (ANNs) [9–12] and the support vector machine (SVM) [13,14]. Weron [15] explained the complexity of the available solutions, their strengths and weaknesses, and the opportunities and threats that the forecasting tools offer or that may be encountered. Cincotti, et al. [16] proposed and compared three different methods to model prices time series. Amjady and Keynia [17] applied an improved neural network to day-ahead electricity price forecasting. Here, the back propagation neural network (BPNN) is a typical instance of an ANN. Guo [18] introduced a new strategy based on seasonal exponential adjustment and a BPNN to forecast wind speed, where the BPNN was established to predict the wind speed. Liu [19] put forward a BPNN model with empirical mode decomposition (EMD) to forecast hourly wind speed. The experiment was repeated 30 times and took the mean value as the final results to avoid randomness, which indicated that the model performed well. However, the BPNN has a problem with many parameters to set, and it is easy to fall into over-fitting or a local optimum. Compared with a BPNN, the radial basis function neural network (RBFNN) shows a stronger approximation and anti-interference capability with a simple structure. Zhang [20] exploited a novel method based on the wavelet transform (WT) and an RBFNN with the consideration of seasonal factors. The general regression neural network (GRNN) has strong non-linear mapping capabilities and a flexible network structure as well as a high degree of fault tolerance and robustness, which is suitable for solving nonlinear problems. Moreover, it has more advantages than the RBFNN in approach ability and learning speed. Liu [21] proposed a GRNN model on the basis of an integration of a WT and spectral clustering (SC), which presented a high operation efficiency and prediction accuracy. Thus, a GRNN is considered as the forecasting model in this paper.

The selection of the smoothing factor in the GRNN model has an influence on its performance. Intelligent optimization algorithms, such as the genetic algorithm (GA) [22–24] and particle swarm optimization (PSO) [25–28], are usually taken to select the parameters for forecasting models. PSO is designed by simulating the feeding behavior of birds. Assuming that there is only one piece of food in the area (that is, the optimal solution in question), the task of the flock is to find the food source. During the entire search process, members of the flock pass on their own messages to each other so that other birds know their place. Through such collaborations, they can determine whether they are finding the optimal solution or not and at the same time pass the information of the optimal solution to the entire flock. Eventually, the whole flock can gather around the food source, which means that the optimal solution is found. Ren [29] developed an improved PSO-BPNN model with input parameter selection for wind speed prediction. The study showed that the model optimized by PSO had better results than a single BPNN and an ARIMA model. The PSO effectively improved the forecasting accuracy but also showed the malpractice that, under the condition of convergence, since all the particles fly towards the direction of the optimal solution, the particles tend to be the same, which makes the convergence speed of the latter part slow down significantly. Meanwhile, PSO converges to a certain precision, and cannot be further optimized, thus the accuracy is not high. In order to overcome these drawbacks, the fruit fly optimization algorithm (FOA) based on the behaviors of food finding was proposed by Pan in 2011 [30]. This method needs to set less parameters, performs at a relatively high speed for searching for the optimum, and has a wide application [31–33]. Here, the FOA is utilized to adjust the appropriate smoothing factor in the GRNN model.

The strong randomness and volatility of wind speed add difficulties to its accurate prediction, therefore its inherent characteristics must be taken into account. The original wind speed series can be regarded as a combination of sub-series with different frequency which show more regularities. EMD [34,35] decomposes the signal according to the time-scale characteristics of the data itself without any pre-setting basis function, which is essentially different from the Fourier decomposition and wavelet decomposition methods that are based on the priori harmonic basis functions and wavelet basis functions. Precisely because of this characteristic, EMD can theoretically be applied to any type of signal decomposition and has very obvious advantages for processing nonstationary and nonlinear data. In reference [36], an ANN model integrated with EMD was proposed, where EMD was utilized to decompose the original wind speed series to eliminate its irregular fluctuations. Wang [37] hybridized an Elman Neural Network (ENN) method with EMD. The results showed that it indicated a higher prediction accuracy than the single ENN model. Therefore, EMD is applied to decompose the original datasets in this study.

According to the above research, a GRNN model integrated with EMD and FOA is proposed. It is the first time that these three models have been combined in wind speed forecasting, and several comparing methods are utilized to validate the effectiveness of the proposed hybrid model. The paper is organized as follows. Section 2 introduces the implementation process of EMD and the GRNN optimized using the FOA. Section 3 presents the evaluation criteria of the results. Section 4 provides a case to validate the proposed model. Section 5 analyzes another case in a different place at another time to prove the generalization of the forecasting method. Section 6 obtains the conclusion in this paper.

2. Methodology

2.1. EMD

EMD is an adaptive time series decomposition technique proposed by Norden E. Huang [38]. The principle of this signal processing method is to decompose the original time series with various fluctuations into a stationary one with different characteristics. Each series that is obtained after decomposition is treated as an intrinsic mode function (IMF), which satisfies the following two conditions: (1) in the whole time range, the number of local extremal points and over zero must be equal, or the maximum difference is one; and (2) the mean value of the two envelopes formed by the local maxima and local minima, respectively, is zero at any point.

For the original time series $s(t)$, the procedures of EMD are shown as follows:

(1) Apply cubic spline interpolation to connect all the local maxima and minima after identification in the time series $s(t)$ so that the upper envelope $x_{max}(t)$ and lower envelope $x_{max}(t)$ are accordingly formed. Calculate the mean value $n(t)$ of the two envelopes and the difference between $n(t)$ and the original signal $s(t)$:

$$n(t) = \frac{x_{max}(t) + x_{min}(t)}{2} \tag{1}$$

$$p(t) = s(t) - n(t) \tag{2}$$

(2) Identify whether $p(t)$ satisfies the two conditions of IMFs. If it conforms, $p(t)$ can be considered as the first IMF; then, calculate the difference between the original signal $s(t)$ and $c_1(t)$:

$$c_1(t) = p(t) \tag{3}$$

$$r_1(t) = s(t) - c_1(t) \tag{4}$$

If not, repeat the above procedure until it meets the two conditions.

(3) The sifting process above will be repeated n times until r_n is a monotone function. The original signal $s(t)$ can be reconstructed as follows:

$$s(t) = \sum_{i=1}^{n} c_i + r_n \tag{5}$$

where c_i represents the IMFs, and r_n is the final residue.

2.2. GRNN

The GRNN model was proposed by the American scholar Donald F.Specht in 1991 [39]. The GRNN model has strong nonlinear mapping capabilities and flexible network structure as well as a high degree of fault tolerance and robustness, which is suitable for solving nonlinear problems. Moreover, it has more advantages than an RBFNN in approach ability and learning speed. The GRNN model is structurally similar to an RBFNN. It consists of four layers, as shown in Figure 1, which are the input layer, the pattern layer, the summation layer, and the output layer. Corresponding to the network input is $X = [X_1, X_2, \cdots, X_n]^T$, and its output is $Y = [Y_1, Y_2, \cdots, Y_k]^T$.

Input Layer Pattern Layer Summation Layer Output Layer

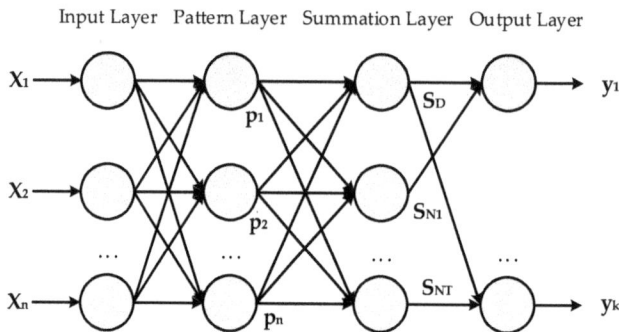

Figure 1. The structure of the general regression neural network (GRNN).

(1) input layer. The number of neurons is equal to the dimension of the input vector in the learning sample. Each neuron is a simple distribution unit that passes the input variable directly to the pattern layer.

(2) pattern layer. The number of neurons is equal to the number of learning samples. Each neuron corresponds to a different sample, and the neuron transfer function is:

$$p_i = \exp\left[-\frac{(X - X_i)^T (X - X_i)}{2\sigma^2}\right], \quad i = 1, 2, \cdots, n \tag{6}$$

where X represents the network input variable, X_i is the corresponding learning sample of neuron i, and σ belongs to the width coefficient of the Gaussian function, which is called the smoothing factor.

(3) summation layer. Two types of neurons are used for summation.

One kind of calculation formula is $\sum_{i-1}^{n} \exp\left[-\frac{(X-X_i)^T(X-X_i)}{2\sigma^2}\right]$, which sums up the output of all neurons in pattern layer, and the connection weight between the pattern layer and each neuron equals 1. The transfer function is

$$S_D = \sum_{i=1}^{n} P_i \tag{7}$$

Another calculation formula is $\sum_{i-1}^{n} Y_i \exp\left[-\frac{(X-X_i)^T(X-X_i)}{2\sigma^2}\right]$, which performs weighted summation on all the neurons in the pattern layer. The connection weight between the ith neuron in the pattern layer and the jth molecule in the summation layer is the jth element of ith output sample Y_i. The transfer function is

$$S_{Nj} = \sum_{i=1}^{n} y_{ij} P_i, \quad j = 1, 2, \cdots, k \tag{8}$$

(4) output layer. The number of neurons is equal to the dimension k of the output vector in the sample. Each neuron will divide the output of the summation layer, and the output of neuron j corresponds to the jth element of the estimated result $\hat{Y}(X)$, namely

$$y_j = \frac{S_{Nj}}{S_D}, \quad j = 1, 2, \cdots, k \tag{9}$$

When the smooth factor σ is very large, $\hat{Y}(X)$ is approximately the mean of all the sample-dependent variables. On the contrary, when the smooth factor tends to 0, $\hat{Y}(X)$ is very close to the training sample. When the point to be predicted is included in the training sample set, the forecasting value of the dependent variable will be very close to the corresponding dependent variable in the sample. Once encountered, the sample cannot be included in the point, and it is possible to predict a very poor performance, which indicates that the network has a poor generalization ability. When the value of σ is moderate, the dependent variable of all of the training samples is considered in the estimation $\hat{Y}(X)$, and the dependent variable corresponding to the forecasting point distance is added to the larger weight. Therefore, the value of σ has a great influence on the forecasting results of the GRNN, and the FOA is used to find the optimal processing of σ.

2.3. A GRNN Based on the FOA with Parameter Selection

The FOA is a new global optimization method based on foraging behaviors. The fruit flies themselves are superior in smell and vision to other species; specifically, they can collect all kinds of smells in the air and fly in the direction of the food or gather with companions. Thus, there are two steps for searching for food of a fruit fly swarm [30]: (1) use an olfactory organ to collect odors floating in the air and fly towards the food location; and (2) use vision to find food and other fruit flies' gathering position and fly to that direction. The iterative food searching process of a fruit fly swarm is shown in Figure 2. Compared with PSO, the FOA has strong robustness as a result of the algorithm's operation not involving multiple loops and complicated functions. An optimization problem with only one parameter can achieve very good results. In this paper, the FOA is utilized to select the best value of the smoothing factor σ in the GRNN.

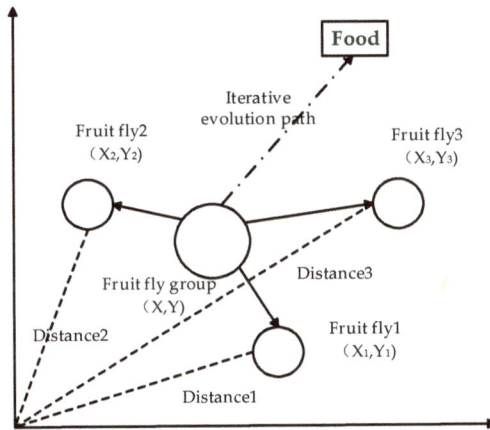

Figure 2. Iterative food searching process of a fruit fly swarm.

The wind speed forecasting model combining EMD, the FOA, and the GRNN are constructed as illustrated in Figure 3.

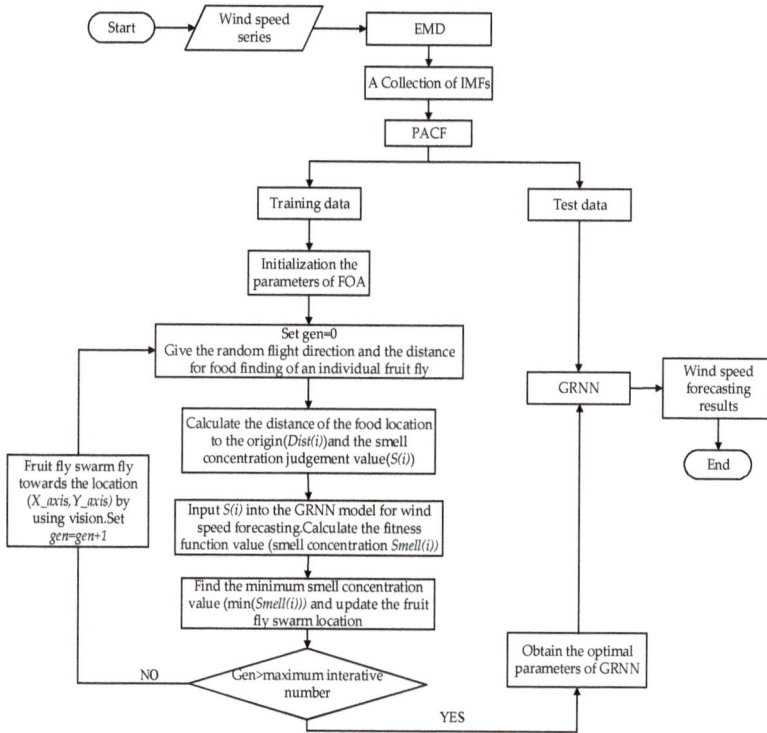

Figure 3. The flow chart of EMD-FOA-GRNN. EMD: empirical mode decomposition; IMF: intrinsic mode function; PACF: partial autocorrelation coefficient function; FOA: fruit fly optimization algorithm.

The specific steps for wind speed prediction are listed as follows:

Step 1: Decompose the original wind series into a collection of IMFs by EMD. In order to choose a proper input, the partial autocorrelation coefficient function (PACF) is applied to each of the IMFs to select the elements for the training set. For the stationary time series $\{Y_t\}$, the so-called k-order lag PACF refers to the correlation between y_{t-k} and y_t under the condition of a given middle random variable $y_{t-1}, y_{t-2}, \cdots, y_{t-k+1}$, or after eliminating the interference of the middle random variable $y_{t-1}, y_{t-2}, \cdots, y_{t-k+1}$.

Step 2: Initialize the parameters: after many attempts, the optimal population size and the maximum iteration number are respectively proposed as 20 and 50. The initial position of the fruit fly swarm is set as $X\_axis = rand()$, $Y\_axis = rand()$, where $rand()$ represents the random number generation function. Here, according to general value range of the smoothing factor σ, the range of random flight distance is set as $(-10, 10)$. To avoid overtraining, the samples are divided into two groups to carry out the cross-training.

Step 3: Start searching for the optimum: according to the operation mechanism of the FOA [29], set the number of iterations $gen = 0$, and let $[X(i), Y(i)]$ be the random direction and distance that an individual fruit fly follows to look for food.

$$X(i) = X\_axis + 20 \times rand() - 10 \tag{10}$$

$$Y(i) = Y\_axis + 20 \times rand() - 10 \tag{11}$$

Step 4: Evaluate the population: firstly, calculate the distance $Dist(i)$ from the location of the fruit fly to the origin and take the reciprocal of $Dist(i)$ as the smell concentration judgment value $S(i)$.

$$Dist(i) = \left(X(i)^2 + Y(i)^2\right)^{\frac{1}{2}} \tag{12}$$

$$S(i) = \frac{1}{Dist(i)} \tag{13}$$

Secondly, set $S(i)$ as the value of the smoothing factor in the GRNN to predict the wind speed. Thirdly, select the root mean square error as the fitness function ($Function()$) to evaluate the disparity between the actual value and forecasting result, and record the corresponding value as the smell concentration $Smell(i)$. Finally, the fruit fly with the minimal smell concentration can be found out.

$$Smell(i) = Function(S(i)) \tag{14}$$

$$\left[\begin{array}{cc} bestSmell & bestIndex \end{array}\right] = \min(Smell(i)) \tag{15}$$

Step 5: Record the optimal value. The best smell concentration value $Smellbest$, smell concentration judgment $bestS(i)$, and the x and y coordinates need to be kept as follows. Then, the fruit flies utilize vision to fly towards that location.

$$Smellbest = bestSmell \tag{16}$$

$$bestS(i) = S(bestIndex) \tag{17}$$

$$X\_axis = X(bestIndex) \tag{18}$$

$$Y\_axis = Y(bestIndex) \tag{19}$$

Step 6: Implement iteration optimization. Repeat Step 3 and Step 4 to determine whether the smell concentration is better than the previous one. If it is, go to Step 5 and set $gen = gen + 1$.

Step 7: Stop optimization and start prediction. Circulation ends at the maximum number of iterations. Here, the best value of the smoothing factor can be substituted into the GRNN model for wind speed forecasting.

3. Evaluation Criteria of Forecasting Performance

It is the primary issue to determine which forecasting model outperforms the other models, and the performance of the prediction models is usually assessed by statistical criteria: the mean absolute error (MAE), the mean absolute percentage error (MAPE), the root mean square error (RMSE), and the index of agreement (IoA). For the first three indexes, the smaller the values are, the better the forecasting performance is. For the IoA, the closer the value is to 1, the better the forecasting performance is. In addition, an MAPE <10% indicates high prediction accuracy, $10\% \leq \text{MAPE} \leq 20\%$ indicates good prediction, $20\% \leq \text{MAPE} \leq 50\%$ implies acceptable prediction, and an $\text{MAPE} \geq 50\%$ implies inaccurate prediction [9]. These four error indexes are defined as follows:

$$\text{MAE} = \frac{1}{N}\sum_{t=1}^{N}|y_t - y_t^*| \tag{20}$$

$$\text{MAPE} = \frac{1}{N}\sum_{t=1}^{N}\left|\frac{y_t - y_t^*}{y_t}\right| \times 100\% \tag{21}$$

$$\text{RMSE} = \sqrt{\frac{1}{N}\sum_{t=1}^{N}(y_t - y_t^*)^2} \tag{22}$$

$$\text{IoA} = 1 - \frac{\sum_{t=1}^{N}(y_t^* - y_t)^2}{\sum_{t=1}^{N}(|y_t^* - \bar{y}_t| - |y_t - \bar{y}_t|)^2} \tag{23}$$

where y_t and y_t^* are the actual and forecast wind speeds at time period t, respectively; N is the forecasting period; and $\bar{y}_t$ represents the mean value of the actual wind speed at time period t.

Additionally, in order to show the improvement degree of forecasting errors for different models, ζ is defined as follows:

$$\zeta = \begin{cases} \frac{\zeta^{compared} - \zeta^{proposed}}{\zeta^{compared}} \times 100\% & \text{suitable for MAE, MAPE and RMSE} \\ \frac{\zeta^{proposed} - \zeta^{compared}}{\zeta^{compared}} \times 100\% & \text{suitable for IoA} \end{cases} \tag{24}$$

where $\zeta^{proposed}$ and $\zeta^{compared}$ represent the MAE, MAPE, RMSE, and IoA generated by the proposed model and other compared models, respectively.

4. Case Study

4.1. Wind Speed Data

Gansu province, with a wealth of wind energy resources, is one of the top seven 10-million-kilowatt wind power bases heavily invested for construction in China. Located in Jiuquan City, Guazhou is known as "the World Storehouse of Wind Energy", whose geographical position is shown in Figure 4. In recent years, a series of favorable policies have been issued to promote the further development of wind power generation in this area. The installed wind capacity is expected to reach 6.45 million kW and the annual generating capacity will come to 14 billion kWh in 2015 for the Guazhou region. According to preliminary planning, by 2020, the total installed wind capacity will exceed 10 million kW. Therefore, accurate wind speed forecasting is not only the basis for wind power prediction, but also has profound significance for planning and designing a wind farm, making a schedule for operating the generator, ensuring the safe operation of the electric power system, and improving economic benefits, etc.

The wind speed data every 20 min from 28 October 2011 to 1 December 2011 were collected from a wind farm in the northwest of Guazhou, totaling 2520 records. Here, the data from 28 October 2011 to 28 November 2011 are selected as the training set and the remaining 216 data are utilized as the test set. Figure 5 shows the original wind speed time series (including 2304 samplings) with its nonlinear and nonstationary characteristics.

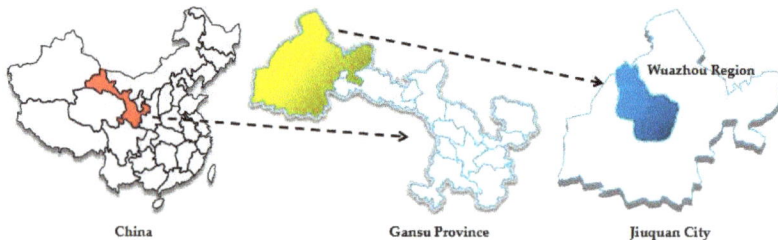

China   Gansu Province   Jiuquan City

Figure 4. The geographical location of Guazhou Region.

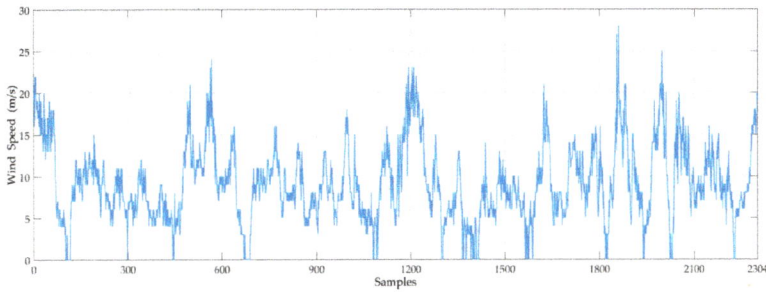

Figure 5. Original wind speed series from 28 October 2011 to 28 November 2011.

4.2. Forecasting Steps

Step 1: Wind speed decomposition. EMD is applied to decompose the original wind speed series into several IMFs to eliminate the nonstationarity of the data, which may have an impact on prediction accuracy. From Figure 6, it can be observed that until eight independent IMFs and one residue R0 are decomposed, the wind speed time series in this case satisfies the condition of EMD.

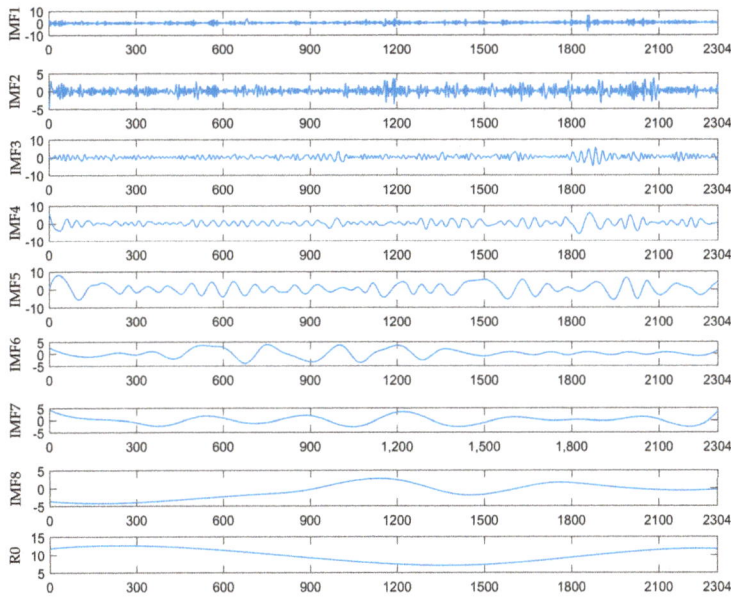

Figure 6. The EMD results of original wind speed series.

Step 2: Input variables selection based on PACF. Since the weather variables that may affect wind speed cannot be easily collected, and the wind speed shows certain time series characteristics, the wind speed data values before the forecast wind speed point are considered as the input variables of the GRNN, and the PACF is applied to determine the specific input variable number of each of the IMFs and R0. Figure 7 is the plot of the partial correlation analysis of the wind speed, where PACF1~PACF8 stand for IMFs (PACF1~PACF8), respectively, and PACF9 represents the residue (R0). Setting each decomposition wind speed time series x_i as the output variable, if the PACF at lag k is out of the 95% confidence interval, x_{i-k} is applied as one of the input variables.

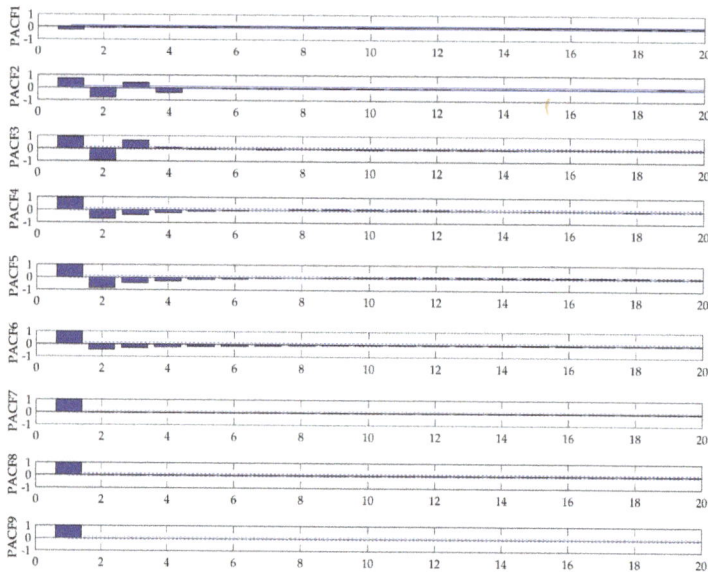

Figure 7. The PACFs of the IMFs and R0.

So, it is obvious that the input variables of these nine series for the FOA-GRNN are the ones shown as follows.

- IMF1: (x_{t-1})
- IMF2: $(x_{t-1}, x_{t-2}, x_{t-3}, x_{t-4})$
- IMF3: $(x_{t-1}, x_{t-2}, x_{t-3})$
- IMF4: $(x_{t-1}, x_{t-2}, x_{t-3}, x_{t-4}, x_{t-5})$
- IMF5: $(x_{t-1}, x_{t-2}, x_{t-3}, x_{t-4}, x_{t-5}, x_{t-6})$
- IMF6: $(x_{t-1}, x_{t-2}, x_{t-3}, x_{t-4}, x_{t-5}, x_{t-6}, x_{t-7}, x_{t-8}, x_{t-9})$
- IMF7: (x_{t-1})
- IMF8: (x_{t-1})
- R0: (x_{t-1})

Step 3: Wind speed forecasting. The FOA-GRNN is utilized to predict the corresponding sub-series with the selected input variables in Step 2. The final forecasting results for the wind speed from 29 November 2011 to 1 December 2011 are obtained by aggregating the prediction results of each sub-series. The values of the smoothing factor in the GRNN optimized by the FOA are recorded in Table 1 for these IMFs and R0.

Table 1. The values of the smoothing factor for each FOA-GRNN trained by the IMFs and R0.

IMFs	IMF1	IMF2	IMF3	IMF4	IMF5	IMF6	IMF7	IMF8	R0
Smoothing factor	0.0652	0.0330	0.0076	0.0128	0.0109	0.0273	0.0031	0.0023	0.0023

Step 4: Comparative analysis of different models. From Figure 8, eight models are presented to predict the wind speed. The ARIMA, BPNN, and GRNN models are three different basic forecasting models. Since PSO and the FOA both belong to the class of swarm intelligent optimization algorithms, there are similarities in the operating mechanism. Thus, the PSO-GRNN and the FOA-GRNN are

utilized to test whether the optimization part donates to the prediction accuracy. The EMD-GRNN, the EMD-PSO-GRNN, and the EMD-FOA-GRNN can be applied to explore the effectiveness of EMD. After many attempts, the proper parameters settings in each algorithm are displayed in Table 2. The wind speed actual and forecasting values for different models are shown in Figures 9 and 10.

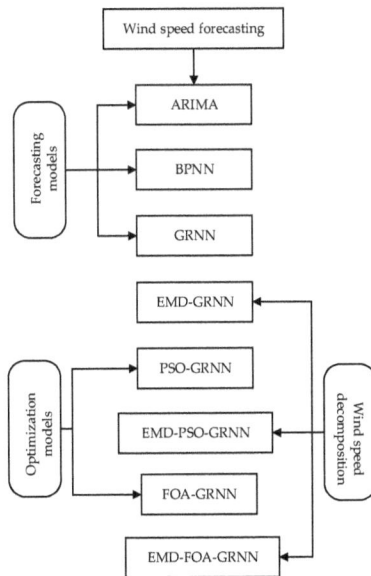

Figure 8. Comparison framework for wind speed forecasting models. ARIMA: autoregressive integrated moving average.

Table 2. The parameters settings in each of the comparison models.

Algorithm	Affiliated Comparison Model	Parameter Name	Value Setting
BPNN	BPNN	maximum iteration number	50
		learning rate	0.1
		minimum error	0.001
GRNN	GRNN PSO-GRNN EMD-GRNN EMD-PSO-GRNN	smoothing factor	0.05
PSO	PSO-GRNN EMD-PSO-GRNN	population size	20
		maximum iteration number	50
		learning factor c1, c2	0.8, 0.8
		maximum velocity	1
		minimum error	0.001

BPNN: back propagation neural network; PSO: particle swarm optimization.

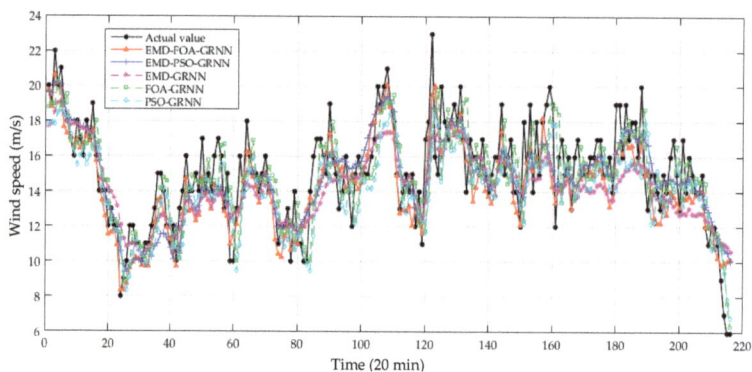

Figure 9. Prediction results of wind speed from 29 November 2011 to 1 December 2011 (I).

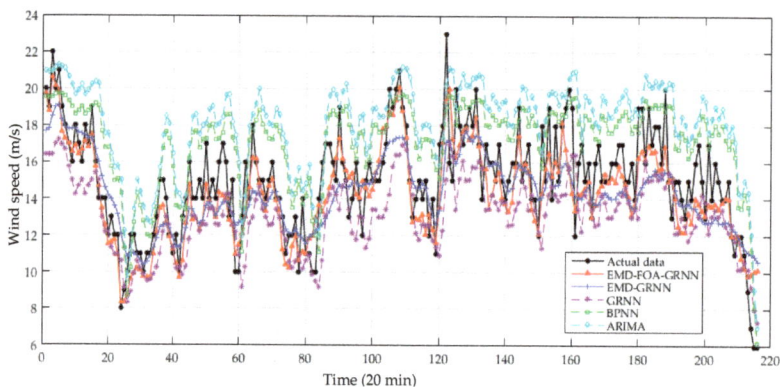

Figure 10. Prediction results of wind speed from 29 November 2011 to 1 December 2011 (II).

4.3. Results Analysis

Figure 11 displays the degree of correlation between the actual value and predicted values of the wind speed. It can be seen that the forecasted wind speed obtained by the EMD-FOA-GRNN model most correlates to the actual one compared with the FOA-GRNN and EMD-GRNN models.

As is presented in Figure 12, the absolute value of error by the EMD-FOA-GRNN model is relatively stable and there are only five error points out of 4 m/s, which means that the forecasting results can be accepted. The accuracy estimation of the predicted wind speed by different models is shown in Table 3. It can be observed that the relative errors in the hybrid models mainly concentrate on the level of less than 10%. Moreover, the number of errors less than 10% generated by the EMD-FOA-GRNN, EMD-PSO-GRNN, and EMD-GRNN models is more than 120, which shows a good performance in wind speed forecasting.

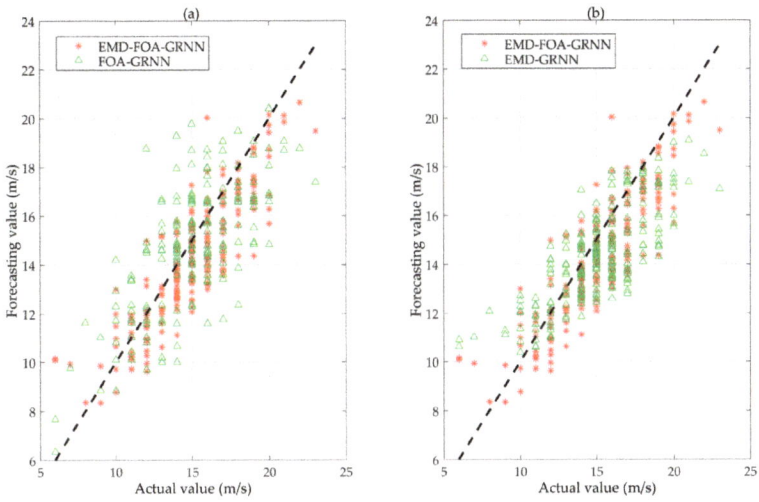

Figure 11. The correlation between forecasting and actual wind speed. (**a**) EMD-FOA-GRNN and FOA-GRNN; (**b**) EMD-FOA-GRNN and EMD-GRNN.

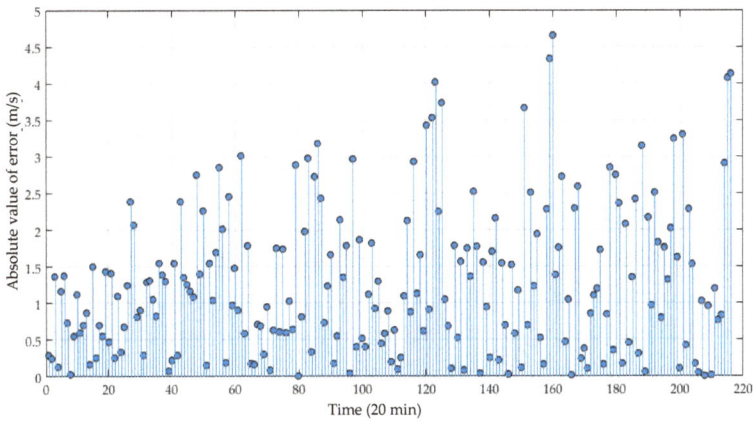

Figure 12. The absolute value of error by the EMD-FOA-GRNN model.

Table 3. Accuracy estimation of forecasting models for the test samples.

Forecasting Models	<10%		10–20%		>20%	
	Number	Percentage	Number	Percentage	Number	Percentage
EMD-FOA-GRNN	140	64.81%	65	30.09%	11	5.09%
EMD-PSO-GRNN	139	64.35%	60	27.78%	17	7.87%
EMD-GRNN	126	58.33%	61	28.24%	29	13.43%
FOA-GRNN	118	54.63%	59	27.31%	39	18.06%
PSO-GRNN	105	48.61%	80	37.04%	31	14.35%
GRNN	80	37.04%	73	33.80%	63	29.17%
BPNN	73	33.80%	57	26.39%	86	39.81%
ARIMA	49	22.69%	55	25.46%	112	51.85%

From Table 4, it can be analyzed that: (a) based on the four evaluation criteria MAE, MAPE, RMSE, and IoA, the proposed model EMD-FOA-GRNN shows the best forecasting performance among the eight models. The MAE, MAPE, RMSE, and IoA of the proposed model is 1.286 m/s, 8.95%, 0.124 m/s, and 0.9070, respectively. (b) by comparing the three single models, the GRNN and the BPNN have higher accuracy than the ARIMA model. Therefore, it can be concluded that intelligent models can obtain better forecasting results than statistical models. Additionally, the GRNN presents more satisfactory performance than the BPNN. The prediction accuracy of the BPNN is closely related to typical training samples and network structure, and it is easy to fall into a local extreme. However, the GRNN, with only one parameter to be optimized, is more suitable for forecasting nonlinear and non-stationary wind speed series. (c) when comparing the EMD-FOA-GRNN with the FOA-GRNN, the EMD-PSO-GRNN with the PSO-GRNN, and the EMD-GRNN with the GRNN, EMD improves the forecasting performance in terms of lower MAE, MAPE, RMSE, and a higher IoA, which proves it can effectively decompose the volatile signals to promote the forecasting capacity. (d) the two optimized models PSO-GRNN and FOA-GRNN produce better results than the single GRNN model. Here, the FOA and the PSO algorithm are utilized to select the appropriate value of the smoothing factor for the GRNN. These two optimization algorithms can effectively enhance the training and learning process so as to avoid falling into a local optimum and improve the global searching ability of the GRNN. Moreover, as seen from these four indexes' values, the FOA made a better optimal performance than that of PSO, which verified the optimization mechanism of the FOA.

Table 4. Statistical error measures of prediction methods in Case One.

Forecasting Models	Indexes			
	MAE (m/s)	MAPE (%)	RMSE (m/s)	IoA
EMD-FOA-GRNN	1.286	8.95	0.124	0.9070
EMD-PSO-GRNN	1.320	9.45	0.135	0.8921
EMD-GRNN	1.593	10.99	0.151	0.8195
FOA-GRNN	1.657	11.38	0.146	0.8354
PSO-GRNN	1.739	11.57	0.145	0.8124
GRNN	2.265	14.50	0.171	0.7310
BPNN	2.461	18.26	0.231	0.7257
ARIMA	3.197	23.52	0.285	0.6618

MAE: mean absolute error; MAPE: mean absolute percentage error; RMSE: root mean square error; IoA: Index of Agreement.

From Table 5, it can be found that: (a) the forecasting performance of the EMD-GRNN combined with the PSO algorithm and the FOA have been effectively improved. It can be seen that the FOA performs better than the PSO algorithm in improving prediction accuracy, mainly because it is easier to use the FOA to fulfill a global optimization goal with less parameters to be optimized. (b) For the basic EMD-GRNN model, it can be analyzed that the FOA enhances the forecasting accuracy and the MAPE promoted percentage is 18.56%. Similarly, owing to EMD, the promoted percentage of MAPE is 21.35%. (c) The EMD part makes more of a contribution than the FOA part in the EMD-FOA-GRNN model.

Table 5. Promoted percentage of errors in Case One.

Forecasting Models	Promoted Percentage of Errors (%)			
	ξ_{MAE}	ξ_{MAPE}	ξ_{RMSE}	ξ_{IoA}
EMD-FOA-GRNN versus EMD-GRNN	19.27	18.56	17.88	10.68
EMD-FOA-GRNN versus FOA-GRNN	22.39	21.35	15.07	8.57
EMD-PSO-GRNN versus EMD-GRNN	17.14	14.01	10.60	8.86
EMD-PSO-GRNN versus PSO-GRNN	24.09	18.32	6.90	9.81

5. Case Two

In order to verify that the proposed model has good adaptability in different times and places, another case which selects the wind speed data in a wind farm located in the middle of Inner Mongolia (shown in Figure 13) is provided in this paper. The study is carried out with the data from 18 February 2012 to 22 March 2012 as the training set and data from 23 March 2012 to 25 March 2012 as the test set. The forecasting results are displayed in Figures 14 and 15. The error analyses are shown in Tables 5 and 6.

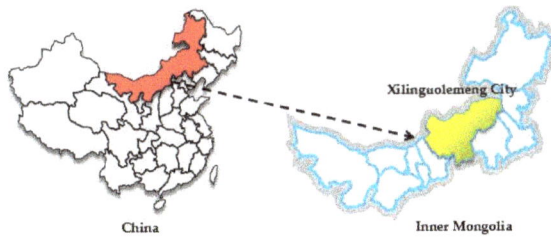

Figure 13. The geographical location of Xilinguolemeng City.

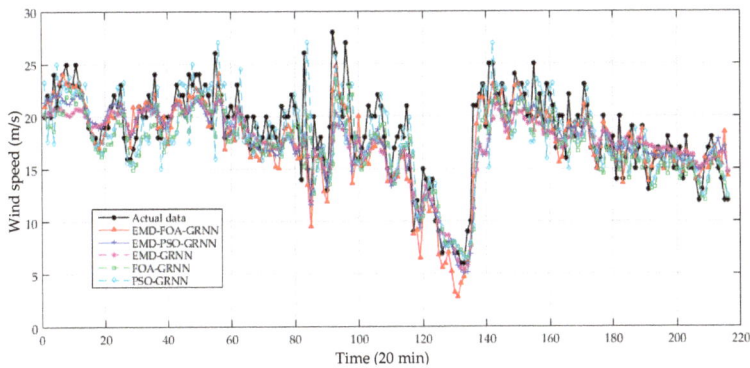

Figure 14. Prediction results of wind speed from 23 March 2012 to 25 March 2012 (III).

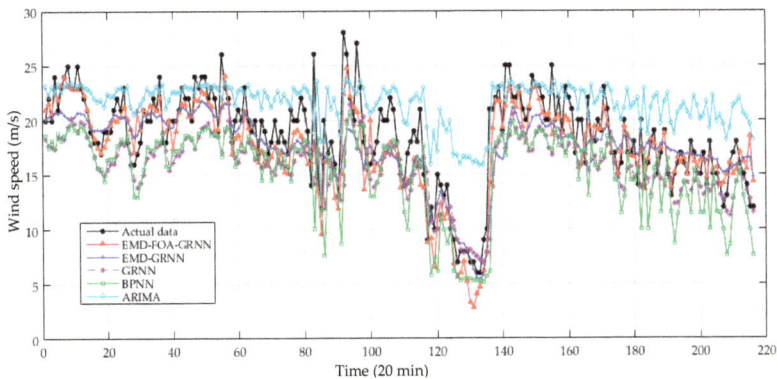

Figure 15. Prediction results of wind speed from 23 March 2012 to 25 March 2012 (IV).

Table 6. Statistical error measures of prediction methods in Case Two.

Forecasting Models	Indexes			
	MAE (m/s)	MAPE (%)	RMSE (m/s)	IoA
EMD-FOA-GRNN	1.677	9.87	0.137	0.9288
EMD-PSO-GRNN	1.880	10.16	0.142	0.8941
EMD-GRNN	2.037	10.91	0.147	0.8671
FOA-GRNN	2.300	12.47	0.164	0.8334
PSO-GRNN	2.459	13.53	0.183	0.8264
GRNN	3.164	16.38	0.194	0.7453
BPNN	3.505	18.85	0.226	0.7518
ARIMA	3.594	25.19	0.382	0.6258

As demonstrated in Tables 6 and 7: (a) intelligent algorithms have higher accuracy in wind speed forecasting than statistical models. (b) the hybrid models show better performance than the single one. (c) EMD improves the performance compared with the corresponding forecasting models that directly utilize the original wind speed series to make predictions. (d) The EMD-FOA-GRNN presents the best forecasting results among the models, and the EMD part donates much more than the FOA part in improving prediction precision. In all, the results in Case Two once again verify the feasibility and effectiveness of the proposed model.

Table 7. Promoted percentage of errors in Case Two.

Forecasting Models	Promoted Percentage of Errors (%)			
	ζ_{MAE}	ζ_{MAPE}	ζ_{RMSE}	ζ_{IoA}
EMD-FOA-GRNN vs. EMD-GRNN	19.27	18.56	17.88	10.68
EMD-FOA-GRNN vs. FOA-GRNN	22.39	21.35	15.07	8.57
EMD-PSO-GRNN vs. EMD-GRNN	17.14	14.01	10.60	8.86
EMD-PSO-GRNN vs. PSO-GRNN	24.09	18.32	6.90	9.81

6. Conclusions

This paper presents a hybrid intelligent algorithm for wind speed forecasting. Firstly, EMD is proposed to preprocess the original wind speed signals to eliminate the random fluctuations of the wind speed data. Then, the GRNN model, which is improved by the FOA, is used to forecast the set of IMFs obtained by EMD. The PACF is used to select the arguments of the GRNN and choose the lags of the historical speeds. Major conclusions are summarized as follows: (a) the EMD effectively improves the forecasting performance; (b) the optimization algorithms FOA and PSO increase the strong global searching capability of the model, and the FOA shows better performance; (c) the EMD part contributes more than the FOA in increasing the accuracy of the EMD-FOA-GRNN model; and (d) the error valuation criteria shows that the EMD-FOA-GRNN is a very promising methodology, which can provide a new idea for short-term wind speed forecasting. In addition, with the development of signal processes and intelligent algorithms, there will be more advanced models applied to predict wind speed, which is our study direction in the future.

Acknowledgments: This work is supported by the Natural Science Foundation of China (Project No. 71471059) and the Fundamental Research Funds for the Central Universities (Project No. 2017XS103). Wei-Chiang Hong thanks the research grant sponsored by the Ministry of Science & Technology, Taiwan (MOST 106-2221-E-161-005-MY2).

Author Contributions: Yi Liang designed this research and wrote this paper; Dongxiao Niu and Wei-Chiang Hong provided professional guidance.

Conflicts of Interest: The authors declare no conflict of interest.

Energies **2017**, *10*, 2001

References

1. Hu, Q.; Zhang, R.; Zhou, Y. Transfer learning for short-term wind speed prediction with deep neural networks. *Renew. Energy* **2016**, *85*, 83–95. [CrossRef]
2. Wang, J.; Hu, J. A robust combination approach for short-term wind speed forecasting and analysis—Combination of the ARIMA (Autoregressive Integrated Moving Average), ELM (Extreme Learning Machine), SVM (Support Vector Machine) and LSSVM (Least Square SVM) forecasts using a GPR (Gaussian Process Regression) model. *Energy* **2015**, *93*, 41–56.
3. Sawyer, S.; Fried, L.; Shukla, S.; Qiao, L. *Global Wind Report 2016—Annual Market Update*; Global Wind Energy Council: Brussels, Belgium, 2017.
4. National Development and Reform Commission. *National Climate Change Program (2014–2020)*; National Development and Reform Commission: Beijing, China, 2014. Available online: http://www.ndrc.gov.cn/zcfb/zcfbtz/201411/W020141104584717807138.pdf (accessed on 30 November 2017).
5. Erdem, E.; Shi, J. ARMA based approaches for forecasting the tuple of wind speed and direction. *Appl. Energy* **2011**, *88*, 1405–1414. [CrossRef]
6. Liu, H.; Tian, H.Q.; Li, Y.F. Comparison of two new ARIMA-ANN and ARIMA-kalman hybrid methods for wind speed prediction. *Appl. Energy* **2012**, *98*, 415–424. [CrossRef]
7. Hodge, B.M.; Zeiler, A.; Brooks, D.; Blau, G.; Pekny, J.; Reklatis, G. Improved Wind Power Forecasting with ARIMA Models. *Comput. Aided Chem. Eng.* **2011**, *29*, 1789–1793.
8. Wang, J.; Hu, J.; Ma, K.; Zhang, Y. A self-adaptive hybrid approach for wind speed forecasting. *Renew. Energy* **2015**, *78*, 374–385. [CrossRef]
9. Ramasamy, P.; Chandel, S.S.; Yadav, A.K. Wind speed prediction in the mountainous region of India using an artificial neural network model. *Renew. Energy* **2015**, *80*, 338–347. [CrossRef]
10. Hui, L.; Tian, H.Q.; Li, Y.F.; Zhang, L. Comparison of four Adaboost algorithm based artificial neural networks in wind speed predictions. *Energy Convers. Manag.* **2015**, *92*, 67–81.
11. Zjavka, L. Wind speed forecast correction models using polynomial neural networks. *Renew. Energy* **2015**, *83*, 998–1006. [CrossRef]
12. Babu, C.N.; Reddy, B.E. A moving-average filter based hybrid ARIMA-ANN model for forecasting time series data. *Appl. Soft Comput.* **2014**, *23*, 27–38. [CrossRef]
13. Chen, K.; Yu, J. Short-term wind speed prediction using an unscented Kalman filter based state-space support vector regression approach. *Appl. Energy* **2014**, *113*, 690–705. [CrossRef]
14. Zhou, J.; Jing, S.; Gong, L. Fine tuning support vector machines for short-term wind speed forecasting. *Energy Convers. Manag.* **2011**, *52*, 1990–1998. [CrossRef]
15. Weron, R. Electricity price forecasting: A review of the state-of-the-art with a look into the future. *Int. J. Forecast.* **2014**, *30*, 1030–1081. [CrossRef]
16. Cincotti, S.; Gallo, G.; Ponta, L.; Marco, R. Modeling and forecasting of electricity spot-prices: Computational intelligence vs classical econometrics. *AI Commun.* **2014**, *27*, 301–314.
17. Amjady, N.; Keynia, F. Day ahead price forecasting of electricity markets by a mixed data model and hybrid forecast method. *Int. J. Electr. Power Energy Syst.* **2008**, *30*, 533–546. [CrossRef]
18. Guo, Z.H.; Wu, J.; Lu, H.Y.; Wang, J.Z. A case study on a hybrid wind speed forecasting method using BP neural network. *Knowl.-Based Syst.* **2011**, *24*, 1048–1056. [CrossRef]
19. Liu, H.; Chen, C.; Tian, H.Q.; Li, Y.F. A hybrid model for wind speed prediction using empirical mode decomposition and artificial neural networks. *Renew. Energy* **2012**, *48*, 545–556. [CrossRef]
20. Zhang, W.; Wang, J.; Wang, J.; Zhao, Z.; Tian, M. Short-term wind speed forecasting based on a hybrid model. *Appl. Soft Comput.* **2013**, *13*, 3225–3233. [CrossRef]
21. Liu, D.; Wang, J.; Wang, H. Short-term wind speed forecasting based on spectral clustering and optimized echo state networks. *Renew. Energy* **2015**, *78*, 599–608. [CrossRef]
22. Liu, H.; Tian, H.; Liang, X.; Li, Y. New wind speed forecasting approaches using fast ensemble empirical model decomposition, genetic algorithm, Mind Evolutionary Algorithm and Artificial Neural Networks. *Renew. Energy* **2015**, *83*, 1066–1075. [CrossRef]
23. Liu, H.; Tian, H.Q.; Chen, C.; Li, Y.F. An experimental investigation of two Wavelet-MLP hybrid frameworks for wind speed prediction using GA and PSO optimization. *Int. J. Electr. Power Energy Syst.* **2013**, *52*, 161–173. [CrossRef]

24. Liu, D.; Niu, D.; Wang, H.; Fan, L. Short-term wind speed forecasting using wavelet transform and support vector machines optimized by genetic algorithm. *Renew. Energy* **2014**, *62*, 592–597. [CrossRef]
25. Yu, S.; Ke, W.; Wei, Y.M. A hybrid self-adaptive Particle Swarm Optimization-Genetic Algorithm-Radial Basis Function model for annual electricity demand prediction. *Energy Convers. Manag.* **2015**, *91*, 176–185. [CrossRef]
26. Rahmani, R.; Yusof, R.; Seyedmahmoudian, M.; Mekhilef, S. Hybrid technique of ant colony and particle swarm optimization for short term wind energy forecasting. *J. Wind Eng. Ind. Aerodyn.* **2013**, *123*, 163–170. [CrossRef]
27. Bahrami, S.; Hooshmand, R.A.; Parastegari, M. Short term electric load forecasting by wavelet transform and grey model improved by PSO (particle swarm optimization) algorithm. *Energy* **2014**, *72*, 434–442. [CrossRef]
28. Yeh, W.C.; Yeh, Y.M.; Chang, P.C.; Ke, Y.C.; Chung, V. Forecasting wind power in the Mai Liao Wind Farm based on the multi-layer perceptron artificial neural network model with improved simplified swarm optimization. *Int. J. Electr. Power Energy Syst.* **2014**, *55*, 741–748. [CrossRef]
29. Ren, C.; An, N.; Wang, J.; Li, L.; Hu, B.; Shang, D. Optimal parameters selection for BP neural network based on particle swarm optimization: A case study of wind speed forecasting. *Knowl.-Based Syst.* **2014**, *56*, 226–239. [CrossRef]
30. Pan, W.T. A new fruit fly optimization algorithm: Taking the financial distress model as an example. *Knowl.-Based Syst.* **2012**, *26*, 69–74. [CrossRef]
31. Li, H.Z.; Guo, S.; Li, C.J.; Sun, J.Q. A hybrid annual power load forecasting model based on generalized regression neural network with fruit fly optimization algorithm. *Knowl.-Based Syst.* **2013**, *37*, 378–387. [CrossRef]
32. Yuan, X.; Liu, Y.; Xiang, Y.; Ye, X. Parameter identification of BIPT system using chaotic-enhanced fruit fly optimization algorithm. *Appl. Math. Comput.* **2015**, *268*, 1267–1281. [CrossRef]
33. Dai, H.; Liu, A.; Lu, J.; Dai, S.; Wu, X.; Sun, Y. Optimization about the layout of IMUs in large ship based on fruit fly optimization algorithm. *Opt. Int. J. Light Electron Opt.* **2015**, *126*, 490–493. [CrossRef]
34. Wang, Y.H.; Yeh, C.H.; Young, H.W.V.; Hu, K.; Lo, M.T. On the computational complexity of the empirical mode decomposition algorithm. *Phys. Stat. Mech. Appl.* **2014**, *400*, 159–167. [CrossRef]
35. Samet, H.; Marzbani, F. Quantizing the deterministic nonlinearity in wind speed time series. *Renew. Sustain. Energy Rev.* **2014**, *39*, 1143–1154. [CrossRef]
36. Hong, Y.Y.; Yu, T.H.; Liu, C.Y. Hour-Ahead Wind Speed and Power Forecasting Using Empirical Mode Decomposition. *Energies* **2013**, *6*, 6137–6152. [CrossRef]
37. Wang, J.; Zhang, W.; Li, Y.; Wang, J.; Dang, Z. Forecasting wind speed using empirical mode decomposition and Elman neural network. *Appl. Soft Comput.* **2014**, *23*, 452–459. [CrossRef]
38. Huang, N.E. The empirical mode decomposition and the Hilbert spectrum for nonlinear and non-stationary time series analysis. *R. Soc. Lond. Proc.* **1998**, *454*, 903–995. [CrossRef]
39. Specht, D.F. A general regression neural network. *IEEE Trans. Neural Netw.* **1991**, *2*, 568–576. [CrossRef] [PubMed]

energies

MDPI

Article

A New Hybrid Wind Power Forecaster Using the Beveridge-Nelson Decomposition Method and a Relevance Vector Machine Optimized by the Ant Lion Optimizer

Sen Guo [1,2,*], Haoran Zhao [1,2,*] and Huiru Zhao [1,2]

[1] School of Economics and Management, North China Electric Power University, Beijing 102206, China; zhaohuiru@ncepu.edu.cn

[2] Beijing Key Laboratory of New Energy and Low-Carbon Development, North China Electric Power University, Beijing 102206, China

* Correspondence: guosen@ncepu.edu.cn (S.G.); haoranzhao0118@163.com (H.Z.);
Tel.: +86-10-6177-3134 (S.G); Fax: +86-10-6177-3084 (S.G)

Academic Editor: Wei-Chiang Hong
Received: 15 May 2017; Accepted: 28 June 2017; Published: 4 July 2017

Abstract: As one of the most promising kinds of the renewable energy power, wind power has developed rapidly in recent years. However, wind power has the characteristics of intermittency and volatility, so its penetration into electric power systems brings challenges for their safe and stable operation, therefore making accurate wind power forecasting increasingly important, which is also a challenging task. In this paper, a new hybrid wind power forecasting method, named the BND-ALO-RVM forecaster, is proposed. It combines the Beveridge-Nelson decomposition method (BND), relevance vector machine (RVM) and ant lion optimizer (ALO). Considering the nonlinear and non-stationary characteristics of wind power data, the wind power time series were firstly decomposed into deterministic, cyclical and stochastic components using BND. Then, these three decomposed components were respectively forecasted using RVM. Meanwhile, to improve the forecasting performance, the kernel width parameter of RVM was optimally determined by ALO, a new Nature-inspired meta-heuristic algorithm. Finally, the wind power forecasting result was obtained by multiplying the forecasting results of those three components. The proposed BND-ALO-RVM wind power forecaster was tested with real-world hourly wind power data from the Xinjiang Uygur autonomous region in China. To verify the effectiveness and feasibility of the proposed forecaster, it was compared with single RVM without time series decomposition and parameter optimization, RVM with time series decomposition based on BND (BND-RVM), RVM with parameter optimization (ALO-RVM), and Generalized Regression Neural Network with data decomposition based on Wavelet Transform (WT-GRNN) using three forecasting performance criteria, namely MAE (Mean Absolute Error), MAPE (Mean Absolute Percentage Error) and RMSE (Root Mean Square Error). The results indicate the proposed BND-ALO-RVM wind power forecaster has the best forecasting performance of all the tested options, which confirms its validity.

Keywords: wind power forecasting; Beveridge-Nelson decomposition method; relevance vector machine; ant lion optimizer; parameter intelligent optimization

---

1. Introduction

Facing the unfavorable situation of fossil energy resource depletion and environmental deterioration, people are increasingly focusing on the exploitation and utilization of renewable energy resources, such as wind power and solar photovoltaic power [1]. Nowadays, wind power has become one of the fastest growing and most promising renewable energy power sources, and the share of wind power generation in the total electricity output has been increasing yearly [2,3]. According to the data released by the Global Wind Energy Council, by the end of 2016, the installed wind power capacity around the world reached 486.7 GW. The cumulative installed wind power capacity in China amounts to 168.7 GW, which accounts for 34.7% of the world total.

Wind power is environmentally friendly. However, wind power output has the characteristics of stochastic fluctuation, intermittency and uncertainty [4]. When a wind generator is connected to the power grid, it will impose new requirements and challenges on electric power systems, such as efficient scheduling of power resources and continuing guarantee of smooth power system operation [5]. With the increase of wind power penetration, it is necessary to increase operation costs to deal with the electric energy unbalance issues due to the stochastic fluctuation of wind power [6]. Meanwhile, to coordinate with wind generators, thermal power generating units have no choice but to frequently adjust their power output, which will reduce the operational efficiency and increase running costs [7]. Accurate wind power forecasting is an effective and important way to alleviate the abovementioned adverse effects.

In past years, many researchers have developed models and methods for wind power/speed forecasting, which have made great achievements [8,9]. Currently, there are mainly two kinds of forecasting techniques related to wind power/speed, which are physical-based forecasting techniques and statistical-based forecasting techniques. Physical forecasting techniques represent a traditional forecasting approach, which needs detailed physical descriptions related to the on-site conditions of wind farms, such as the wind farm layout, wind turbines, and atmospheric conditions [10]. The main representatives are *Prediktor* developed by the Risoe National Laboratory in Denmark [11], *Previento* developed by University of Oldenburg in Germany [12], and *eWind* developed by AWS True Wind Inc. (New York, NY, USA) [13]. In the past few years, statistical forecasting techniques, which conduct wind power/speed forecasting based on historical power/speed data and other meteorological data have been developed greatly. This kind of forecasting technique includes two kinds of approaches, namely conventional statistical approaches and emerging artificial intelligent approaches. Among conventional statistical approaches, the auto-regressive moving average (ARMA) model [14,15] and auto-regressive integrated moving average (ARIMA) model [16] have been widely used to forecast wind speed and wind power. The conventional statistical approaches hold the assumption that wind power and wind speed have linear relationships with their influencing factors. However, in fact, the relationships between wind power/speed and their influencing factors are non-linear. Therefore, the conventional statistical approaches fall short of obtaining high forecasting accuracy and satisfactory forecasting results. The other statistical forecasting techniques, namely the emerging artificial intelligence approaches, do not assume a linear relationship between wind power/speed and their influencing factors can effectively cover the abovementioned shortcomings of conventional statistical approaches. Currently, there are several emerging artificial intelligence approaches which have been employed to forecast wind power and wind speed, such as artificial neural networks [17,18], support vector machine [19], and extreme learning machine [20,21].

When emerging artificial intelligence approaches are employed to forecast wind power/speed, there are several parameters that must be set first, such as neuron number of artificial neural networks and kernel parameters of support vector machines. This is very difficult for practitioners. To tackle this issue, an intelligent optimization algorithm is usually introduced to determine the optimal parameters of emerging artificial intelligence approaches. Ren, et al. [22] applied particle swam optimization (PSO) to automatically determine the parameters of a back propagation neural network (BPNN) for short-term wind speed forecasting. Amjady, et al. [23] used enhanced particle swarm optimization to optimize a modified neural network for wind power prediction. Jursa and Rohrig [24] employed PSO and differential evolution (DE) to optimize artificial neural networks for short-term wind power forecasting. Liu, et al. [25] used a genetic algorithm to select the optimal parameters of support vector machines for short-term wind speed forecasting. Salcedo-Sanz, et al. [26] used a coral reefs optimization algorithm to optimize an extreme learning machine for wind speed prediction.

Generally speaking, wind power time series have both nonlinear and nonstationary characteristics, so the decomposition of wind power time series is often needed to improve wind power forecasting accuracy [27]. Currently, there are mainly two kinds of wind power time series decomposition methods, which are the wavelet transform (WT) [28,29] and empirical mode decomposition (EMD) [30,31]. The Beveridge-Nelson decomposition (BND) method, proposed by researchers Stephen Beveridge and Charles Nelson in 1981, is a kind of non-stationary time series decomposition technique [32], which has been widely applied in economic fields, such as business cycle analysis [33], GNP and stock prices [34], and regional income fluctuations [35]. As an effective time series decomposition method, it is very regretful to find that the BND has not been used for wind power time series. To fill this gap, in this paper the BND technique is employed to decompose wind power time series, which is a new application.

In this paper, relevance vector machine (RVM), a kind of sparse and supervised learning probabilistic method, is also used for wind power forecasting. RVM has some merits compared with other machine learning methods, such as better adaptability, a need for fewer sample data, sparsity, and simplified parameter setting [36]. Nowadays, RVM is employed in many practical issues, such as silent speech classification [37], battery health monitoring [38], canal flow prediction [39], daily potential evapotranspiration forecast [40], and system fault diagnosis [41]. However, the RVM technique has rarely been used for wind power forecasting. To improve RVM-based forecasting performance of wind power, a new Nature-inspired meta-heuristic algorithm, named the ant lion optimizer (ALO) [42] is also employed in this paper to automatically determine the optimal parameters of RVM. Therefore, a new hybrid BND-ALO-RVM method for wind power forecasting is proposed in this paper. To verify the effectiveness and applicability of this proposed method, real-world hourly wind power data from the Xinjiang Uygur autonomous region in China is selected as our empirical analysis example, and the forecasting results are compared with other forecasting methods, including single RVM, BND-RVM, ALO-RVM, and WT-GRNN. The main contributions of this paper are as follows:

(1) A new hybrid BND-ALO-RVM method for wind power forecasting is proposed, which combines Beveridge-Nelson decomposition (BND), relevance vector machine (RVM) and ant lion optimizer (ALO). Empirical results indicate that the proposed method can improve wind power forecasting accuracy and shows superiority over other compared methods. The proposed method in this paper can be a promising alternative forecasting technique for wind power, which enriches the current wind power forecasting method toolbox.

(2) The Beveridge-Nelson decomposition (BND) method, which has been frequently and widely used for economic issues, is employed in energy issues for the first time. In this paper, the wind power time series are decomposed into three components, namely the deterministic, cyclical and stochastic component. Empirical results show the wind power forecasting accuracy can be improved after decomposing wind power time series by using BND, which indicates BND is an effective method for wind power time series decomposition. It can be said that this paper expands the application domains of the BND method, and enriches the data decomposition library for wind power time series.

(3) Relevance vector machine (RVM) technique is employed to forecast the different decomposed components of wind power time series. To improve the forecasting performance of RVM, a new Nature-inspired meta-heuristic algorithm, namely the ant lion optimizer (ALO), is used to optimally determine the kernel width parameter of RVM model. Forecasting results reveal the ALO is effective, which can determine the optimal kernel width parameter of RVM and improve the RVM-based wind power forecasting accuracy. In our previous study [43], has was verified that the ALO can improve GM (1,1)-based power load forecasting accuracy. Therefore, ALO, as a new intelligent optimization algorithm, can be promising with a good development foreground.

This paper makes a new attempt to use ALO for parameter optimization of RVM, which also enlarges the application scope of the ALO algorithm.

The reminder of this paper is organized as follow: Section 2 gives a brief introduction of the basic methods and algorithms used, including the Beveridge-Nelson decomposition (BND) method, relevance vector machine (RVM), and ant lion optimizer (ALO); the proposed hybrid BND-ALO-RVM forecaster for wind power is described in Section 3; Section 4 conducts an empirical analysis, and the forecasting performance of the proposed method is compared with other methods. The main conclusions are drawn in Section 5.

2. Brief Introduction of the Beveridge-Nelson Decomposition Method, Relevance Vector Machine and Ant Lion Optimizer

2.1. Beveridge-Nelson Decomposition Method (BND)

In 1981, two researchers Stephen Beveridge and Charles Nelson proposed a new general procedure for non-stationary time series decomposition, named the Beveridge-Nelson decomposition (BND) method [32]. The BND method decomposes the stationary first-order difference of original time series with first-order co-integration characteristics into permanent components and transitory (cyclical) components [44]. The permanent component is a random walk process with drift, which includes deterministic component and stochastic component. The deterministic component can be estimated using ARIMA technique, and the transitory (cyclical) component is a stationary process with a zero average value.

The first step of the BND method is to determine whether the first-order difference of a non-stationary wind power time series is stationary or not [45]. If yes, the detailed steps of wind power time series decomposition by using BND method are as follows:

The wind power time series are represented as WP. According to the Wold theorem, under the the condition of first-order stationarity, the natural logarithm of the wind power time series at time t (denoted as $\ln WP_t$) satisfies:

$$\Delta \ln WP_t = \mu + \varepsilon_t + \sum_{i=1}^{\infty} \lambda_i \varepsilon_{t-i} \tag{1}$$

where WP_t is the wind power at time t; μ is the long-run mean value of $\Delta \ln WP_t$; $\varepsilon_t \sim i.i.d.N(0, \sigma^2)$ (i.i.d. represents independently and identically distribute); λ_i is the coefficient; and $\Delta \ln WP_t = \ln WP_t - \ln WP_{t-1}$.

Taking the expectation on both sides of Equation (1), we can obtain:

$$E(\Delta \ln WP_t) = E(\mu) + E(\varepsilon_t) + E\left(\sum_{i=1}^{\infty} \lambda_i \varepsilon_{t-i}\right) = E(\mu) \tag{2}$$

where $E(\bullet)$ represents the expected computation on variables.

According to the Beveridge-Nelson decomposition theorem, the deterministic component (represented as D_t) of the wind power time series can be decomposed as:

$$D_t = \ln WP_0 + \mu t \tag{3}$$

where D_t represents the deterministic component of the wind power time series at time t; and $\ln WP_0$ is the natural logarithm value of the initial wind power data.

According to Morley [44], the time series can be forecasted by using first-order difference AR(1) model, namely:

$$(\Delta \ln WP_t - \mu) = \phi(\Delta \ln WP_{t-1} - \mu) + \varepsilon_t \tag{4}$$

where $|\phi| < 1$, and $\varepsilon_t \sim i.i.d.N(0, \sigma^2)$.

According to the Wold theorem, the expected value of minimum mean squared error (MMSE) of first-order difference $\Delta \ln WP_t$ at next j period under the assumption of normality is:

$$E_t\left[\left(\Delta \ln WP_{t+j} - \mu\right)\right] = \phi^j(\Delta \ln WP_t - \mu) \tag{5}$$

The BN trend of wind power time series, denoted as T_t, is defined as the MMSE forecast of time series long-term level, namely:

$$T_t = \lim_{j \to \infty} E_t\left[\left(\ln WP_{t+j} - j\mu\right)\right] = \ln WP_t + \lim_{j \to \infty} E_t\left[\left(\Delta \ln WP_{t+j} - \mu\right)\right] \tag{6}$$

Thus, substituting Equation (5) into Equation (6), the BN trend of $\ln WP_t$ for the case AR (1) can be obtained as:

$$T_t = \ln WP + \frac{\phi}{1-\phi}(\Delta \ln WP_t - \mu) \tag{7}$$

Meanwhile, the cyclical component C_t of wind power time series can be calculated by:

$$C_t = -\frac{\phi}{1-\phi}(\Delta \ln WP_t - \mu) \tag{8}$$

Finally, the stochastic component T_t of wind power time series can be computed as:

$$T_t = \ln WP + \frac{\phi}{1-\phi}(\Delta \ln WP_t - \mu) - (\ln WP_0 + \mu t) \tag{9}$$

2.2. Relevance Vector Machine (RVM)

Relevance vector machine (RVM), proposed by Tipping, is a kind of sparse and supervised learning probabilistic method [46]. Compared with traditional supervised learning algorithms, RVM under a Bayesian framework has a better non-linear mapping capability, which can be used in the case of a small number of samples and can also obtain a good generalization performance.

Given a training sample set $\{p_i, v_i\}_{i=1}^N$, p_i is a two-dimensional input vector, and v_i is a one-dimensional target value. The RVM model can be formulated as:

$$y(p, w) = \sum_{i=1}^N w_i K(p, p_i) + w_0 \tag{10}$$

where N is the number of training sample; w_i is the weight; and $K(p, p_i)$ is a kernel function.

There are several kinds of kernel functions, and the Gaussian kernel function is usually employed, namely:

$$K(p, p_i) = \exp\left(\frac{-\|p - p_i\|}{2\sigma^2}\right) \tag{11}$$

where σ is the width of kernel function.

Suppose that the noise ε_i obeys normal distribution with mean of 0 and variance σ^2. Then:

$$v_i = y(p_i, w) + \varepsilon_i \tag{12}$$

Since v_i is independent and identically distributed, the likelihood function of the training sample is:

$$P\left(v|w, \sigma^2\right) = \frac{1}{\sqrt{2\pi}\sigma} \exp\left(\frac{-\|v - \Phi w\|^2}{2\sigma^2}\right) \tag{13}$$

where $v = (v_1, v_2, \cdots, v_N)^T$; $w = (w_0, w_1, \cdots, w_n)^T$; $\Phi_{N \times (N+1)} = [\phi(p_1), \phi(p_2), \cdots, \phi(p_N)]^T$; and $\phi(p_i) = [1, K(p_i, p_1), \cdots, K(p_i, p_N)]$.

To avoid the issues of too many relevance vectors, over-fitting and poor generalization capability, the weight w should obey normal distribution, namely:

$$P(w|\alpha) = \prod_{i=0}^N \frac{\alpha_i}{\sqrt{2\pi}} \exp\left(-\frac{\alpha_i w_i^2}{2}\right) \tag{14}$$

According to the Bayesian criterion, it can be inferred that the a posteriori distributions of w and t are both normal distributions, namely:

$$P\left(v|\alpha,\sigma^2\right) = \frac{1}{\sqrt{(2\pi)^N \left(\sigma^2 I + \Phi A^{-1} \Phi^T\right)}} \exp\left(-\frac{v^T v}{2\left(\sigma^2 I + \Phi A^{-1} \Phi^T\right)}\right) \tag{15}$$

$$P\left(w|v,\alpha,\sigma^2\right) = \frac{P\left(v|w,\sigma^2\right)P(w|\alpha)}{P\left(v|\alpha,\sigma^2\right)} = \frac{i}{\sqrt{(2\pi)^{N+1}\sum^1}} \exp\left\{-\frac{(w-\mu)^T(w-\mu)}{2\sum^1}\right\} \tag{16}$$

where $\mu = \sigma^{-2}\left(\sigma^2 I + \Phi A^{-1}\Phi^T\right)\Phi^T v$; $\sum^1 = \left(\sigma^{-2}\Phi^T\Phi + A\right)^{-1}$; $A = diag(\alpha_0,\alpha_1,\cdots,\alpha_N)$.

To maximize the hyper-parameter likelihood distribution $P\left(v|\alpha,\sigma^2\right)$, the optimal values of α^{best} and σ^2_{best} can be obtain according to the following iterations, namely:

$$\alpha_i^{best} = \frac{1 - \alpha_i N_{ii}}{\mu_i^2} \tag{17}$$

$$\sigma^2_{best} = \frac{\|v - \Phi\mu\|^2}{N - \sum\limits_{i=0}^{N}(1 - \alpha_i N_{ii})} \tag{18}$$

where α_i^{best} is the ith value of optimal parameter α^{best}; μ_i is the ith posterior mean value of μ; and N_{ii} is the ith diagonal element of posterior variance matrix.

When the new input data p^* is given, the corresponding probability distribution of forecasting output can be obtained as:

$$P\left(v^*|v,\alpha^{best},\sigma^2_{best}\right) = \int P\left(v^*|w,\sigma^2_{best}\right)P\left(w|v,\alpha^{best},\sigma^2_{best}\right)dw \tag{19}$$

The forecasting variance is:

$$\sigma^2_* = \sigma^2_{best} + \phi(p^*)^T \sum^1 \phi(p^*) \tag{20}$$

The mean value of the forecasting output is:

$$y^* = \mu^T\phi(p^*) \tag{21}$$

During the parameter estimation process, most α_i will tend to ∞, and then the corresponding w_i will equal 0. This means that many terms of the kernel matrix will not participate in the forecasting process, and this is why RVM can achieve sparsity [36]. Compared with the support vector machine (SVM) technique, there is only one parameter which needs to be set for RVM, namely the kernel width parameter σ [36].

2.3. Ant Lion Optimizer (ALO)

In 2015, Mirjalili proposed a new Nature-inspired meta-heuristic algorithm, namely the ant lion optimizer (ALO) [42]. The ALO was put forward by the inspiration of intelligence behavior of ant lions hunting for ants. The detailed steps of the ALO algorithm are as follows:

Step 1: Set initial parameters.

When ALO is used, the initial values of five parameters need to be set, which are the number of ants and ant lions *Agents_no*; maximum iteration number *Max_iteration*; variables number *dim*; lower bound $lb = [lb_1, lb_2, \cdots]$ and upper bound $ub = [ub_1, ub_2, \cdots]$ of variables.

Step 2: Initialize the positions of ants and antlions.

The positions of ants and antlions need to be initialized, which can be represented by Equations (22) and (23).

$$M_{Ant} = \begin{bmatrix} A_{11} & A_{12} & \cdots & A_{1d} \\ A_{21} & A_{22} & \cdots & A_{2d} \\ \vdots & \vdots & \vdots & \vdots \\ A_{n1} & A_{n2} & \cdots & A_{nd} \end{bmatrix} \tag{22}$$

$$M_{Antlion} = \begin{bmatrix} AL_{11} & AL_{12} & \cdots & AL_{1d} \\ AL_{21} & AL_{22} & \cdots & AL_{2d} \\ \vdots & \vdots & \vdots & \vdots \\ AL_{n1} & AL_{n2} & \cdots & AL_{nd} \end{bmatrix} \tag{23}$$

where M_{Ant} represents each ant position; A_{ij} represents the j-th parameter's value of the i-th ant; $M_{Antlion}$ represents each antlion position; AL_{ij} represents the j-th parameter's value of the i-th antlion; $i = 1, 2, \cdots, n, j = 1, 2, \cdots, d$.

The positions of ants and antlions are generated at random. Therefore, the entry of position matrix of ant and antlion can be obtained according to Equation (24):

$$A_{*j} \text{ or } AL_{*j} = rand \times \left(ub_j - lb_j \right) + lb_j \tag{24}$$

where A_{*j} and AL_{*j} are the j-th column values of position matrix; $rand$ represents the generated random number with uniform distribution in the interval [0, 1]; lb_j and ub_j respectively, represent the lower and upper boundary of the j-th variable.

Step 3: Select initial elite.

The elite refers to the best antlion obtained in the iteration process. The best antlion can be determined according to fitness function, and the antlion with the maximal fitness value is elite, which is also called as the fittest antlion. In ALO, the fitness function of antlion is represented by $f[*]$, and matrix M_{OAL} is used to store fitness values of antlions, namely:

$$M_{OAL} = \left\{ \begin{array}{l} f\left[\begin{pmatrix} AL_{11} & AL_{12} & \cdots & AL_{1d} \end{pmatrix} \right] \\ f\left[\begin{pmatrix} AL_{21} & AL_{22} & \cdots & AL_{2d} \end{pmatrix} \right] \\ \vdots \\ f\left[\begin{pmatrix} AL_{n1} & AL_{n2} & \cdots & AL_{nd} \end{pmatrix} \right] \end{array} \right\} \tag{25}$$

where M_{OAL} is the fitness matrix of antlions.

According to Equation (25), the fitness values of antlions can be calculated, and then the initial elite can be selected.

Step 4: Start iteration.

In ALO, the ant position is influenced by both the antlion and the selected elite. The ants randomly walk around the selected elite and antlions are selected by a roulette wheel algorithm. According to this rule, the i-th ant position at the t-th iteration can be obtained by:

$$Ant_i^t = \frac{R_E^t + R_A^t}{2} \tag{26}$$

where Ant_i^t is the i-th ant position at the t-th iteration, R_E^t represents the random walk around the selected elite at the t-th iteration, and R_A^t represents the random walk around the antlion selected by the roulette wheel algorithm.

Random walk is an important strategy for modelling the positions and movements of ants and antlions in the ALO algorithm. For a detailed interpretation related to random walk in ALO readers can refer to [42,43].

In the iteration process, the ants update their positions according to Equation (26). In order to keep the random walk inside the search space, the ant position needs to be normalized at each iteration as follows:

$$\widetilde{X}_j^t = \frac{\left(X_j^t - a_j\right) \times \left(d_j^t - c_j^t\right)}{\left(b_j - a_j\right)} + c_j^t \tag{27}$$

where $\widetilde{X}_j^t$ represents the normalized value of the j-th variable at the t-th iteration; a_j and b_j respectively represent the minimum and maximum of random walk of the j-th variable; c_j^t and d_j^t are the minimum and maximum of random walk of the j-th variable at t-th iteration, respectively.

In the ALO algorithm, the random walk of ants is influenced by the traps of antlions, which can be modeled as follows:

$$\begin{cases} c_i^t = Antlion_i^t + c^t \\ d_i^t = Antlion_i^t + d^t \end{cases} \tag{28}$$

where c_i^t and d_i^t respectively represent the minimum and maximum of variables related to the i-th ant at t-th iteration; c^t and d^t respectively represent the minimum and maximum of variables at t-th iteration; and $Antlion_i^t$ represents the i-th antlion position at the t-th iteration.

During the iteration and optimization process, antlions build their pits proportional to their fitness values. The antlions with better fitness values have larger pits, which indicate these antlions have better chances of catching ants. When the antlions find that ants are trapped, they will throw sand outwards the centers of pits. Therefore, the random walk range is set to decrease adaptively to simulate the movement behavior of ants sliding towards antlions, which can be modelled as follows:

$$c^t = \frac{c^t}{I} \tag{29}$$

$$d^t = \frac{d^t}{I} \tag{30}$$

where I is a decreased ratio as follows:

$$I = \begin{cases} 10^2 \times \frac{t}{T}, & \frac{t}{T} > 0.1 \\ 10^3 \times \frac{t}{T}, & \frac{t}{T} > 0.5 \\ 10^4 \times \frac{t}{T}, & \frac{t}{T} > 0.75 \\ 10^5 \times \frac{t}{T}, & \frac{t}{T} > 0.9 \\ 10^6 \times \frac{t}{T}, & \frac{t}{T} > 0.95 \end{cases} \tag{31}$$

Step 5: Select the optimal antlion (final elite).

The antlion will catch an ant when ant reaches the bottom of the cone-shaped pit, and then it pulls the ant into the sand and consumes it. To improve the probability of catching a new ant, the antlion will update its position according to the position of the latest caught ant and then build a new pit for catching prey. In the ALO algorithm, when the ant is fitter than the antlion, it will be caught. The position of the antlion can be updated by:

$$Antlion_i^t = Ant_i^t \quad if \quad f\left(Ant_i^t\right) > f\left(Antlion_i^t\right) \tag{32}$$

For each iteration, the fitness and position of antlions can be updated according to Equation (32), and then the new elite can be redetermined. When the stopping criteria for iteration (such as maximum iteration number) is satisfied, the ALO algorithm will end. At this time, the final elite, namely the optimal antlion can be obtained.

3. The Proposed Hybrid BND-ALO-RVM Forecaster for Wind Power

In this paper, a new hybrid BND-ALO-RVM forecaster is proposed for wind power. Considering the non-stationary characteristics of wind power time series, the BND method is firstly employed to decompose the initial wind power time series into three components: the deterministic, cyclical and stochastic component. Then, the RVM technique is used to forecast these three components, respectively. To improve the forecasting performance, the kernel width parameter σ of the RVM model is optimally determined using a new swarm intelligent algorithm ALO. Finally, the forecasted wind power can be obtained by multiplying the forecasting results of three decomposed components.

The detailed procedures of the proposed hybrid BND-ALO-RVM method for wind power forecasting are elaborated as below:

Step 1: Perform unit root test.

When the Beveridge-Nelson decomposition method is used, the first thing is to examine whether the logarithmic sequence of initial wind power time series is first-order stationary or not. If the first-order difference of logarithmic sequence of initial wind power time series is stationary, then we can proceed to the next step. In this paper, the Augmented Dickey-Fuller (ADF) method is used to perform unit root tests.

Step 2: Decompose wind power time series.

Once the condition that the first-order difference of logarithmic sequence of initial wind power data is stationary is confirmed, the wind power time series can be decomposed into the deterministic, cyclical and stochastic components by using the BND method.

Step 3: Set initial parameters.

For the ALO algorithm, five parameters, namely ants and antlion numbers *Agents_no*, variables number *dim*, maximum iteration number *Max_iteration*, lower boundary $lb = [lb_1, lb_2, \cdots]$ and upper boundary $ub = [ub_1, ub_2, \cdots]$ need to be initially set. In this paper, these five parameters are set as follows: *Agents_no* = 10, *dim* = 1, *Max_iteration* = 100, *lb* = 0.001, and *ub* = 100.

Step 4: Start optimization search.

The fitness function $f[*]$ needs to be determined firstly when the ALO is used to optimize the kernel width parameter of RVM. In this study, the root mean square error (RMSE) between actual wind power data and forecasted wind power data is employed to build the fitness function, namely:

$$RMSE = \sqrt{\frac{1}{n}\sum_{k=1}^{n}\left(x(k) - \hat{x}(k)\right)^2} \tag{33}$$

where $x(k)$ is actual wind power data at time k; and $\hat{x}(k)$ is the where forecasted wind power value at time k.

The kernel width parameter of RVM is represented by the antlion's position $M_{Antlion}$, namely each column of $M_{Antlion}$. The optimal position of antlions will be updated at each iteration, and then the optimal kernel width parameter of RVM can be also updated so far. Suppose that the actual wind power decomposition data $\{x^{(0)}(1), x^{(0)}(2), ..., x^{(0)}(n)\}$ is used in the first iteration, the forecasted wind power decomposition value $\{\hat{x}^{(0)}(1), \hat{x}^{(0)}(2), ..., \hat{x}^{(0)}(n)\}$ can be calculated using the optimized RVM model. At this time, the fitness function and optimization object can be built at this iteration by minimizing RMSE as follows:

$$f = \min\sqrt{\frac{1}{n}\sum_{k=1}^{n}\left(x(k) - \hat{x}(k)\right)^2} \tag{34}$$

Step 5: Determine the optimal parameter of RVM.

During the iteration and optimization process, different RMSEs will be obtained with different kernel width parameters of RVM. When the iteration reaches a maximum, the minimum RMSE can be found, and then the optimal kernel width parameter σ of RVM can be determined, which will be used for three components forecasting of wind power time series by using the RVM technique.

Step 6: Integrate three components forecasting results and forecast wind power.

After the initial wind power time series are decomposed into three components and the optimal kernel width parameter of RVM is determined, the RVM optimized by ALO will be employed to respectively forecast the three decomposed components, namely the deterministic, cyclical and stochastic component. Then, the forecasting results of these three components are integrated. Finally, the forecasting result of wind power can be obtained by multiplying the forecasting results of the corresponding deterministic, cyclical and stochastic components.

The procedure of the proposed BND-ALO-RVM method used for wind power forecasting in this paper is shown in Figure 1.

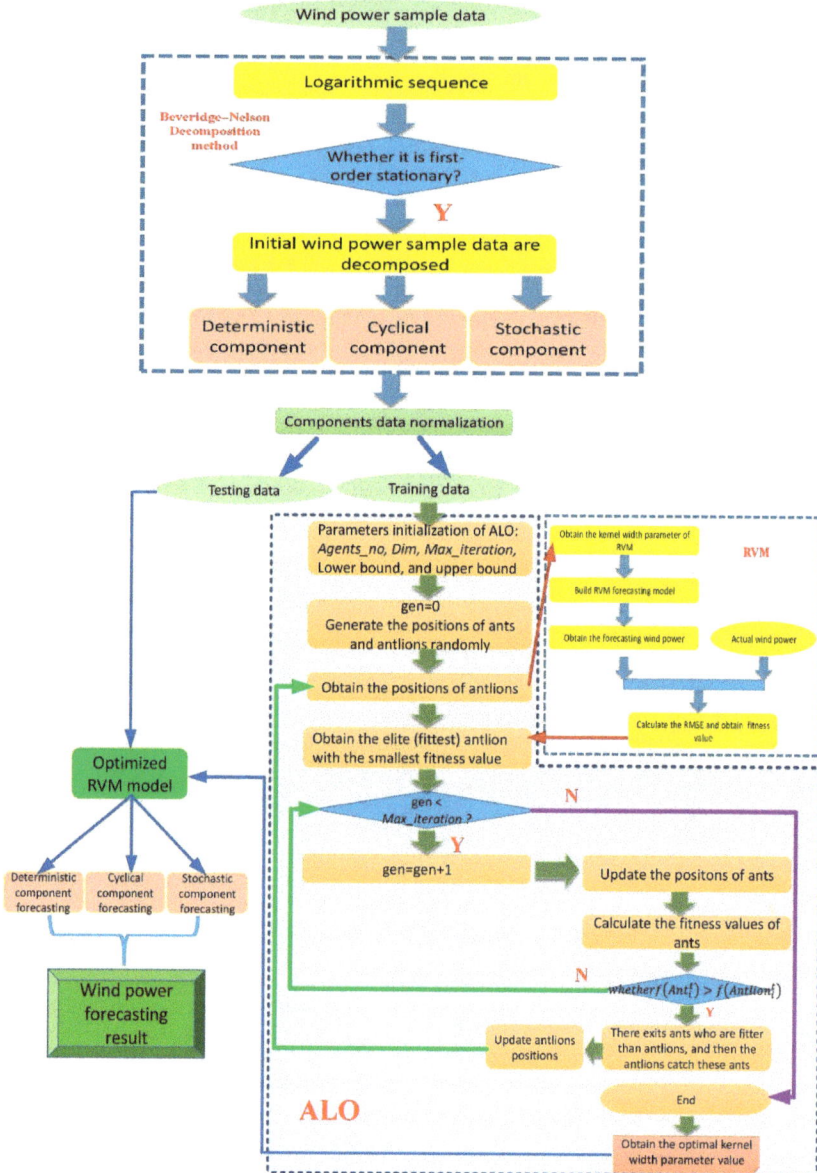

Figure 1. Flowchart of the proposed BND-ALO-RVM wind power forecaster.

4. Empirical Analysis

In this paper, hourly wind power from the Xinjiang Uygur autonomous region in China collected during June 2015 was employed for empirical analysis to validate the proposed BND-ALO-RVM forecaster. The Xinjiang Uygur autonomous region, which has plentiful wind resources, is located in the northwest of China, as shown in Figure 2. The sample set in this paper includes 408 hourly wind power points from 14 June to 30 June, which are shown in Figure 3. It can be seen that the wind power, which ranges from about 5 MW to 164 MW, fluctuates greatly. No apparent variation pattern of the wind power time series can be seen.

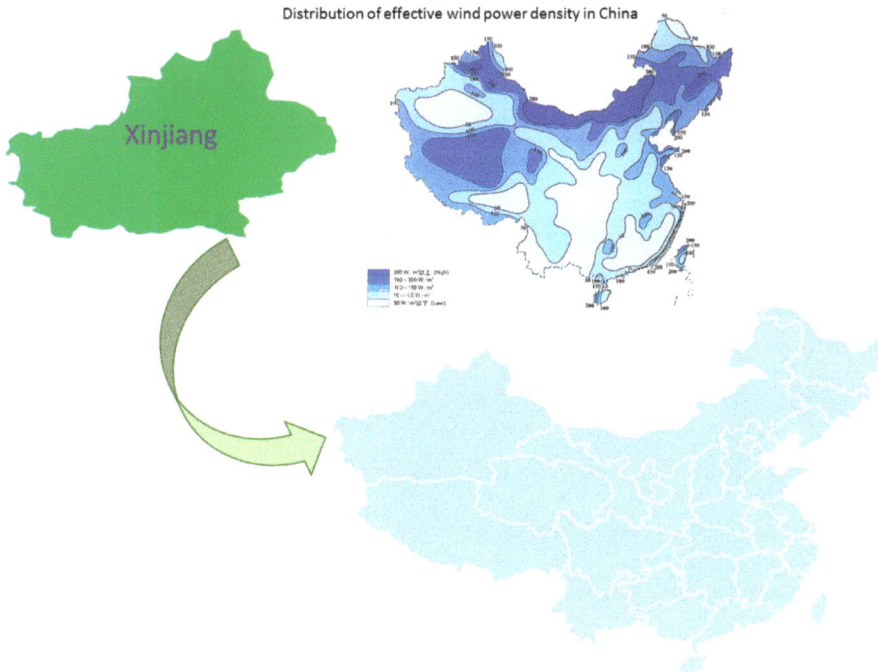

Figure 2. Geographical location of Xinjiang and its wind resources.

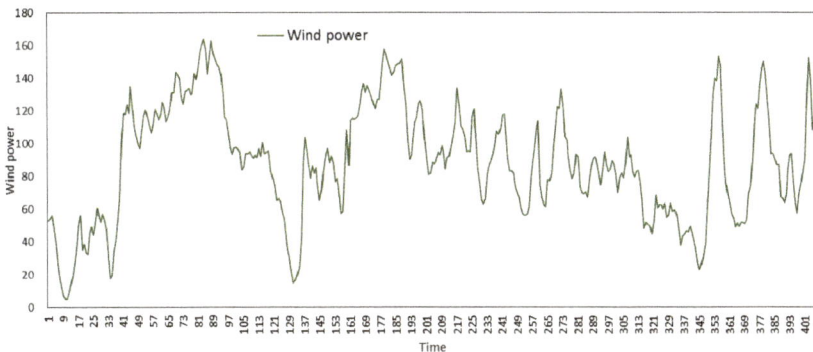

Figure 3. Wind power time series from 14 June to 30 June (408 sample points).

4.1. Wind Power Time Series Decomposition Using BND Method

The BND method is applied to decompose the original wind power time series. Before data decomposition, the unit root test based on the ADF test is necessary to judge whether the logarithmic sequence of wind power time series is first-order stationary. The ADF test results are listed in Table 1, where it can be seen that the wind power time series are stable after first-order difference. Then, the BND method can be used to decompose the wind power time series. The decomposition result of the original wind power time series is shown in Figure 4, which includes the deterministic, cyclical and stochastic components.

Table 1. ADF test result of wind power time series.

Sequence	Test Form (C,T,K)	ADF Test Value	P Value	Conclusion
ln P	(N,N,1)	−2.5705	0.1098	Unstable
Δ ln P	(N,N,0)	−11.5359	0.0000	stable

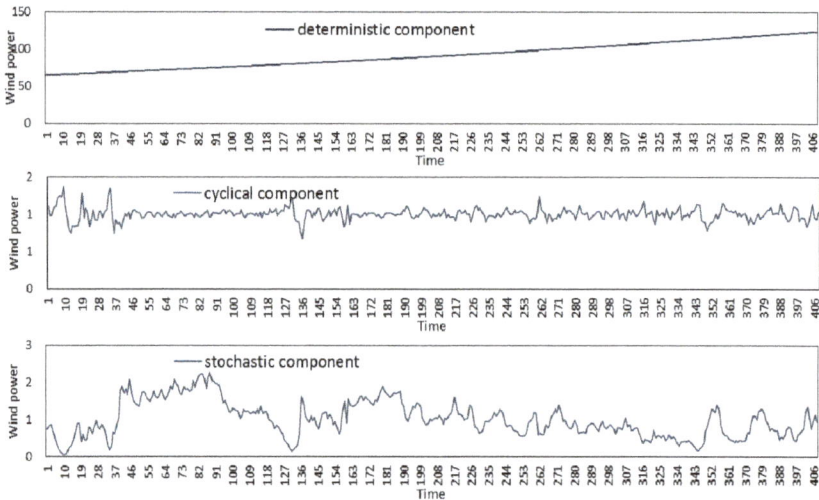

Figure 4. Decomposed deterministic component, cyclical component and stochastic component of the original wind power time series.

4.2. Forecasting Results

After the original wind power time series are decomposed, the deterministic, cyclical and stochastic components will be respectively forecasted using ALO-RVM, which means the kernel width parameter of RVM will be respectively optimized and determined by ALO. The sample set of wind power from 14 June to 30 June is divided into a training sample set and a testing sample set. Two hundred and sixteen (216) sample data points of hourly wind power from 14 June to 23 June are used as training sample, and the remaining 168 sample data points from 24 June to 30 June are treated as testing sample.

The inputs of ALO-RVM model are historical hourly wind power 1 h ahead and hourly wind power at the same time yesterday. For example, for forecasting the deterministic component of wind power time series at the 1st hour of 24 June, the deterministic component of wind power time series at 1 h ahead (i.e., deterministic component of the 24th hour wind power data at 23 June) and deterministic component of wind power time series at the same time yesterday (i.e., deterministic component of the 1st hour wind power data at 23 June) will be employed as the input variables. In this way, the deterministic components of wind power time series from the 2nd hour of 24 June to the 24th hour of 30 June can be forecasted. Meanwhile, the cyclical components and stochastic components of the wind power time series from the 1st hour of 24 June to the 24th hour of 30 June can also be forecasted. Finally, the wind power from the 1st hour of 24 June to the 24th hour of 30 June can be respectively obtained by conducting the multiplication of forecasted deterministic components, cyclical components and stochastic components from the 1st hour of 24 June to the 24th hour of 30 June. For the training period, the same way is taken in this paper. The training sample data and testing sample data will be respectively normalized before training and testing by using ALO-RVM method, and the normalization method is as follows:

$$\tilde{x} = \frac{x - x_{min}}{x_{max} - x_{min}} \tag{35}$$

where x and $\tilde{x}$ are the original and normalized wind power data, respectively; x_{max} and x_{min} respectively represent the maximum and minimum value of each input wind power time series.

During the training period, the kernel width parameter of RVM will be optimally determined using the ALO algorithm for the deterministic, cyclical and stochastic components, respectively. The optimal values of RVM kernel width parameter σ for the deterministic, cyclical and stochastic components are respectively 29.2504, 0.0137 and 11.8443, which are also the RVM kernel widths for the deterministic, cyclical and stochastic component during the testing period. The forecasting results of the deterministic, cyclical and stochastic component of the hourly wind power from 24 June to 30 June are shown in Figure 5. Then, the hourly wind power from 24 June to 30 June can be forecast by multiplying the forecasted deterministic, cyclical and stochastic components from 24 June to 30 June, which are shown in Figure 6.

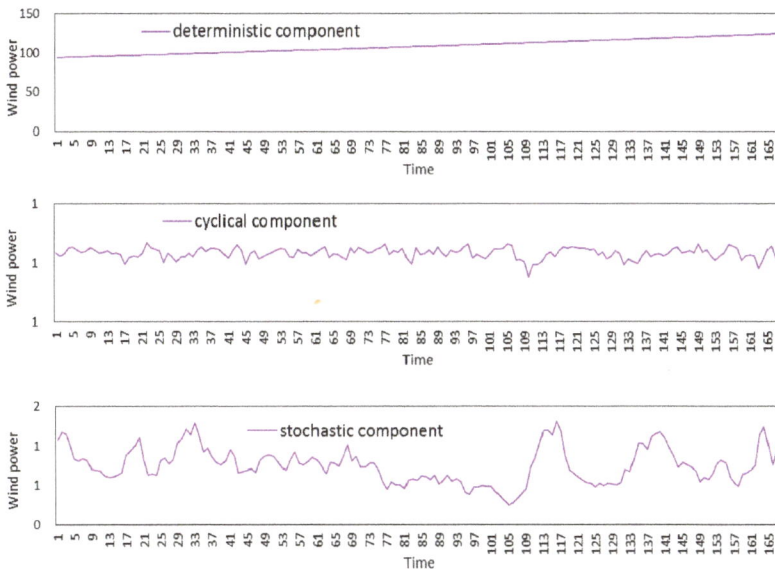

Figure 5. Forecasted deterministic component, cyclical component and stochastic component of the wind power time series from 24 June to 30 June.

4.3. Forecasting Performance Evaluation

To evaluate the forecasting performance of the proposed BND-ALO-RVM forecaster for wind power, four comparison methods are selected, which are single RVM without data decomposition and parameter optimization (referred to as RVM), RVM only with data decomposition using BND (referred to as BND-RVM), RVM only with parameter optimization (referred to as ALO-RVM), and generalized regression neural network with data decomposition using wavelet transform (referred to as WT-GRNN). The input variables and output variable of these four comparing forecasting methods are the same as those of the BDN-ALO-RVM method.

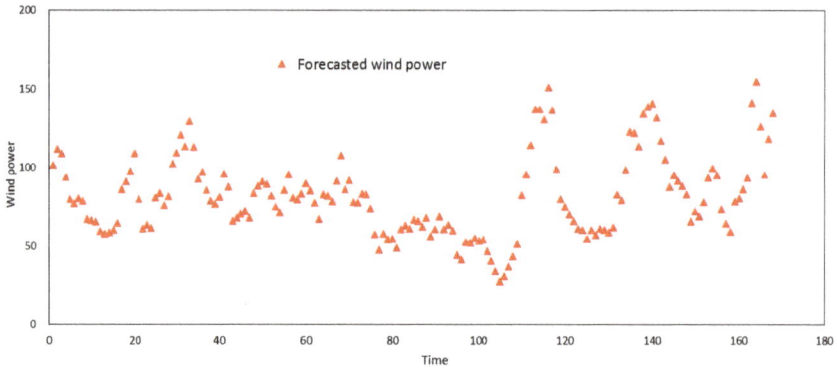

Figure 6. Forecasting result of wind power from 24 June to 30 June.

For RVM and BND-RVM, the kernel width parameter is set as 3. For ALO-RVM, through training, the optimal kernel width parameter of RVM is determined as 7.4645. For WT-GRNN, a fast discrete wavelet transform based on four filters developed by Mallat [47] is employed, and the spread parameter value of GRNN is selected as 3. The forecasted hourly wind powers from 24 June to 30 June using RVM, BND-RVM, ALO-RVM, and WT-GRNN are shown in Figure 7. The relative errors of forecasted hourly wind power using BND-ALO-RVM, RVM, BND-RVM, ALO-RVM, and WT-GRNN methods are shown in Figure 8. From Figure 8, it can be roughly seen that the proposed BND-ALO-RVM method has the best forecasting performance due to its much smaller relative errors, and RVM without data decomposition and parameter optimization has the poorest forecasting capacity due to its larger relative errors.

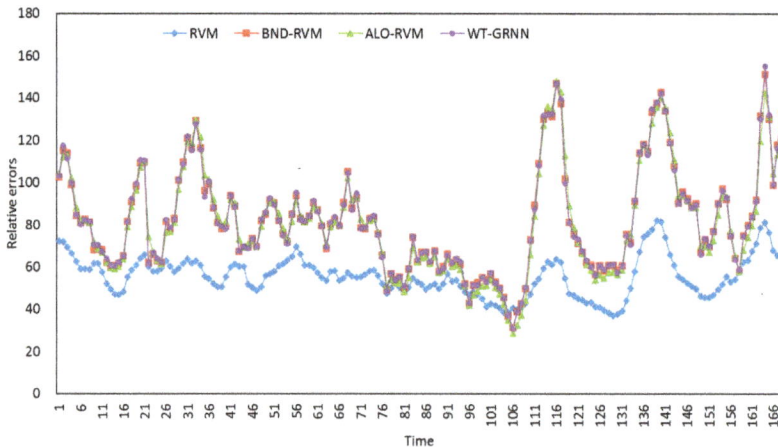

Figure 7. Forecasting results of hourly wind power by using different comparison methods.

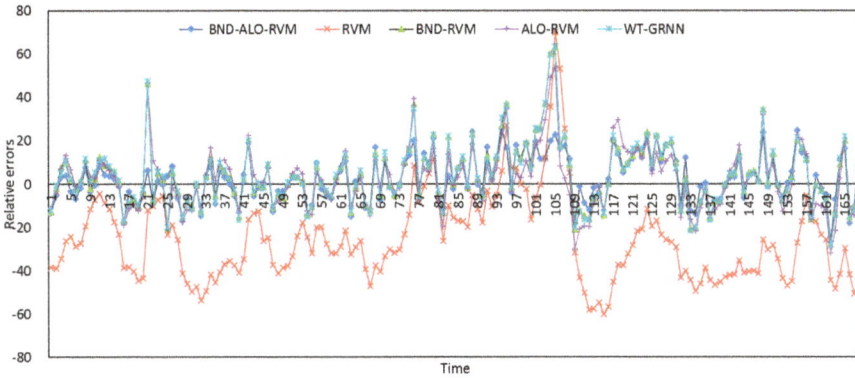

Figure 8. Relative errors of the forecasted hourly wind power of different methods.

To further compare the forecasting performances of the different methods, three forecasting error criteria are selected to evaluate the forecasting performances of the different models, which are Mean Absolute Percentage Error (MAPE, see Equation (36)), Root Mean Square Error (RMSE) and Mean Absolute Error (MAE, see Equation (37)):

$$MAPE = \frac{1}{n}\sum_{k=1}^{n}\left|\frac{x(k) - \hat{x}(k)}{x(k)}\right| \times 100\% \tag{36}$$

$$MAE = \frac{1}{n}\sum_{k=1}^{n}|x(k) - \hat{x}(k)| \times 100\% \tag{37}$$

The MAPEs, RMSEs and MAEs of different forecasting methods, namely BND-ALO-RVM, RVM, BND-RVM, ALO-RVM, and WT-GRNN at each forecasted day are listed in Table 2. From the perspective of average forecasting performance from 24 June to 30 June, the proposed BND-ALO-RVM has the best forecasting performance because it has the minimum MAPE, RMSE and MAE, which are 8.95%, 8.79 MW and 6.86 MW, respectively. The MAPE of BND-ALO-RVM model is $(29.20 - 8.95)/29.20 = 69.35\%$, $(11.01 - 8.95)/11.01 = 18.75\%$, $(10.74 - 8.95)/10.74 = 16.66\%$, and $(11.07 - 8.95)/11.07 = 19.15\%$ lower than that of RVM, BND-RVM, ALO-RVM, and WT-GRNN, respectively. The RMSE of BND-ALO-RVM model is $(33.64 - 8.79)/33.64 = 73.89\%$, $(10.29 - 8.79)/10.29 = 14.58\%$, $(11.10 - 8.79)/11.10 = 20.81\%$, and $(10.46 - 8.79)/10.46 = 15.97\%$ lower than that of RVM, BND-RVM, ALO-RVM, and WT-GRNN, respectively. The MAE of BND-ALO-RVM model is $(26.79 - 6.86)/26.79 = 74.39\%$, $(8.08 - 6.86)/8.08 = 15.07\%$, $(8.37 - 6.86)/8.37 = 18.05\%$, and $(8.13 - 6.86)/8.13 = 15.62\%$ lower than that of RVM, BND-RVM, ALO-RVM, and WT-GRNN, respectively. Therefore, it can be seen that the proposed BDN-ALO-RVM method has the best forecasting performance in terms of short-term wind power, and the single RVM has the worst forecasting performance. The MAPE of BND-RVM is larger than that of ALO-RVM, but smaller than that of WT-GRNN. The RMSE and MAE of BND-RVM are both smaller than that of ALO-RVM and WT-GRNN. Therefore, it can be said that the forecasting performance of BND-RVM is better than that of WT-GRNN, but it is hard to say which is better between BND-RVM and ALO-RVM. The MAPE of ALO-RVM is smaller than that of WT-GRNN, but the RMSE and MAE of ALO-RVM are larger than that of ALO-RVM, so for ALO-RVM and WT-GRNN, we cannot decide in this paper which one is better in term of wind power forecasting.

Table 2. Forecasting performances of different methods.

Methods/Criteria	BND-ALO-RVM			RVM			BND-RVM			ALO-RVM			WT-GRNN		
	MAPE (%)	RMSE (MW)	MAE (MW)	MAPE (%)	RMSE (MW)	MAE (MW)	MAPE (%)	RMSE (MW)	MAE (MW)	MAPE (%)	RMSE (MW)	MAE (MW)	MAPE (%)	RMSE (MW)	MAE (MW)
24 June	5.47	6.25	4.71	23.66	26.67	21.41	8.29	9.73	6.71	9.09	10.34	7.46	8.44	9.73	6.65
25 June	7.23	8.56	6.78	35.21	38.18	34.02	7.83	8.97	7.22	8.44	9.86	7.81	7.97	9.26	7.36
26 June	6.87	7.32	5.94	30.84	27.87	26.67	6.76	7.08	5.82	6.94	7.18	5.94	6.92	7.36	5.97
27 June	10.12	6.60	5.46	12.98	9.85	7.89	12.37	8.23	6.60	10.76	7.52	5.77	12.43	8.24	6.64
28 June	12.32	9.49	7.60	33.34	42.69	30.09	19.11	12.26	10.66	17.01	13.99	11.00	18.85	12.30	10.55
29 June	10.69	9.44	8.13	34.50	39.83	33.87	12.01	10.98	9.42	10.73	11.26	9.25	12.07	11.05	9.43
30 June	9.94	12.33	9.40	33.87	38.38	33.54	10.71	13.29	10.12	12.18	15.07	11.37	10.84	13.79	10.35
26–30 June	8.95	8.79	6.86	29.20	33.64	26.79	11.01	10.29	8.08	10.74	11.10	8.37	11.07	10.46	8.13

From the perspective of forecasting performance during each day, the proposed BDN-ALO-RVM method obtains the smallest MAPEs, RMSEs and MAEs among all five forecasting models on 24 June, 25 June, 27 June, 28 June, 29 June and 30 June. On 26 June, the BND-RVM obtains the smallest MAPE, RMSE, and MAE. The RVM model gets the largest MAPE, RMSE, and MAE on each day between 24 June and 30 June, which indicates it has the worst forecasting performance again. This is because the RVM without data decomposition and parameter optimization cannot grasp the characteristics of wind power time series. The forecasting performance of BND-RVM method is better than single RVM because BND-RVM method obtains smaller MAPEs, RMSEs and MAEs, which indicates BND is an effective wind power time series decomposition method to improve the wind power forecasting accuracy. Meanwhile, the forecasting performance of the ALO-RVM method is better than single RVM, which indicates the ALO is an effective algorithm for kernel width parameter optimization of RVM to improve the wind power forecasting accuracy. For BND-RVM and ALO-RVM methods, the BND-RVM method has better forecasting performance than ALO-RVM on 24 June, 25 June, 26 June and 30 June, and the ALO-RVM has better forecasting performance than BND-RVM on 27 June. It is hard to judge which one has better forecasting performance between BND-RVM and ALO-RVM because the different rankings related to MAPEs, RMSEs, and MAEs on 28 June and 29 June. For BND-RVM and WT-GRNN methods, except the MAE criterion on 24 June and 28 as well as MAPE criterion on 28 June, the BND-RVM shows better forecasting performance than WT-GRNN. On the whole, the BND-RVM has better forecasting performance than WT-GRNN, which indicates the BND employed in this paper is an effective technique for wind power time series decomposition.

To sum up the above analysis, we can conclude that the proposed BDN-ALO-RVM forecaster is an effective and applicable technique, which can improve the forecasting accuracy of hourly wind power. Meanwhile, the BND method is a valid wind power time series decomposition technique, and ALO is an efficient meta-heuristic algorithm for kernel width parameter determination of RVM in the field of wind power forecasting.

5. Conclusions

Wind power is a kind of environmentally friendly renewable energy power, which has developed rapidly in recent years. However, the large-scale penetration of wind power with its stochastic and intermittent characteristics into electric power systems will pose some threats to the stable and safe operation of these systems. Improving wind power forecasting accuracy can alleviate these threats. Therefore, a new hybrid BND-ALO-RVM forecaster for wind power is proposed in this paper, which combines the Beveridge-Nelson decomposition method, relevance vector machine and ant lion optimizer. The wind power time series are firstly decomposed into deterministic, cyclical and stochastic components. Then, these three decomposed components are respectively forecasted by using ALO-RVM method, which mean that the kernel width parameter of RVM is optimally determined by the ALO algorithm. Finally, the wind power forecasting results can be obtained by multiplying the forecasted deterministic, cyclical and stochastic components. Taking hourly wind power from the Xinjiang Uygur autonomous region in China as an example, the empirical results indicate the proposed BND-ALO-RVM forecaster obtains the best forecasting performance compared with the single RVM, BND-RVM, ALO-RVM, and WT-GRNN methods. The proposed BND-ALO-RVM method has the minimum MAPE, RMSE and MAE, which are 8.95%, 8.79 MW and 6.86 MW, respectively. However, MAPE, RMSE and MAE of single RVM method are respectively 29.20%, 33.64 MW and 26.79 MW; MAPE, RMSE and MAE of BND-RVM method are respectively 11.01%, 10.29 MW and 8.08 MW; MAPE, RMSE and MAE of ALO-RVM method are respectively 10.74%, 11.10 MW and 8.37 MW; and MAPE, RMSE and MAE of BND-RVM method are respectively 11.07%, 10.46 MW and 8.13 MW. The proposed BND-ALO-RVM method is effective and practical for short-term wind power forecasting. The BND method is a valid wind power time series decomposition technique, and ALO is an attractive meta-heuristic algorithm for RVM parameter determination. This paper enriches the methodology library related to wind power forecasting, and also extends the application domains of BDN and ALO.

In future research, the proposed BND-ALO-RVM method may also be employed for other issues, such as photovoltaic power forecasting and power load forecasting.

Acknowledgments: This study is supported by the National Key R&D Program of China under Grant No. 2016YFB0900501, the National Natural Science Foundation of China under Grant No. 71373076, and the Fundamental Research Funds for the Central Universities under Grant No. 2017MS060.

Author Contributions: Sen Guo proposed the concept of this research and completed the manuscript. Haoran Zhao analyzed the empirical data. Huiru Zhao gave some suggestions.

Conflicts of Interest: The authors declare no conflict of interest.

References

1. Zhao, H.-R.; Guo, S.; Fu, L.-W. Review on the costs and benefits of renewable energy power subsidy in China. *Renew. Sustain. Energy Rev.* **2014**, *37*, 538–549. [CrossRef]
2. Kang, J.; Yuan, J.; Hu, Z.; Xu, Y. Review on wind power development and relevant policies in China during the 11th Five-Year-Plan period. *Renew. Sustain. Energy Rev.* **2012**, *16*, 1907–1915. [CrossRef]
3. Hitaj, C. Wind power development in the United States. *J. Environ. Econ. Manag.* **2013**, *65*, 394–410. [CrossRef]
4. Wang, J.; Botterud, A.; Bessa, R.; Keko, H.; Carvalho, L.; Issicaba, D.; Sumaili, J.; Miranda, V. Wind power forecasting uncertainty and unit commitment. *Appl. Energy* **2011**, *88*, 4014–4023. [CrossRef]
5. Aigner, T.; Jaehnert, S.; Doorman, G.L.; Gjengedal, T. The effect of large-scale wind power on system balancing in Northern Europe. *IEEE Trans. Sustain. Energy* **2012**, *3*, 751–759. [CrossRef]
6. Ortega-Vazquez, M.A.; Kirschen, D.S. Assessing the impact of wind power generation on operating costs. *IEEE Trans. Smart Grid* **2010**, *1*, 295–301. [CrossRef]
7. Ummels, B.C.; Gibescu, M.; Pelgrum, E.; Kling, W.L.; Brand, A.J. Impacts of wind power on thermal generation unit commitment and dispatch. *IEEE Trans. Energy Convers.* **2007**, *22*, 44–51. [CrossRef]
8. Soman, S.S.; Zareipour, H.; Malik, O.; Mandal, P. A Review of Wind Power and Wind Speed Forecasting Methods with Different Time Horizons. In Proceedings of the North American Power Symposium (NAPS 2010), Arlington, TX, USA, 26–28 September 2010; Institute of Electrical and Electronics Engineers: New York, NY, USA, 2010; pp. 1–8.
9. Costa, A.; Crespo, A.; Navarro, J.; Lizcano, G.; Madsen, H.; Feitosa, E. A review on the young history of the wind power short-term prediction. *Renew. Sustain. Energy Rev.* **2008**, *12*, 1725–1744. [CrossRef]
10. Lange, M.; Focken, U. *Physical Approach to Short-Term Wind Power Prediction*; Springer: New York, NY, USA, 2006.
11. Landberg, L.; Watson, S.J. Short-term prediction of local wind conditions. *Bound. Layer Meteorol.* **1994**, *70*, 171–195. [CrossRef]
12. Focken, U.; Lange, M.; Waldl, H.-P. Previento-A wind power prediction system with an innovative upscaling algorithm. In Proceedings of the European Wind Energy Conference, Copenhagen, Denmark, 2–6 July 2001.
13. McGowin, C. *California Wind Energy Forecasting System Development and Testing. Phase 1: Initial Testing*; EPRI Final Report; Electric Power Reserach Institute: Palo Alto, CA, USA, 2003.
14. Erdem, E.; Shi, J. ARMA based approaches for forecasting the tuple of wind speed and direction. *Appl. Energy* **2011**, *88*, 1405–1414. [CrossRef]
15. Torres, J.L.; Garcia, A.; de Blas, M.; de Francisco, A. Forecast of hourly average wind speed with ARMA models in Navarre (Spain). *Sol. Energy* **2005**, *79*, 65–77. [CrossRef]
16. Kavasseri, R.G.; Seetharaman, K. Day-ahead wind speed forecasting using f-ARIMA models. *Renew. Energy* **2009**, *34*, 1388–1393. [CrossRef]
17. Li, G.; Shi, J. On comparing three artificial neural networks for wind speed forecasting. *Appl. Energy* **2010**, *87*, 2313–2320. [CrossRef]
18. Barbounis, T.G.; Theocharis, J.B.; Alexiadis, M.C.; Dokopoulos, P.S. Long-term wind speed and power forecasting using local recurrent neural network models. *IEEE Trans. Energy Convers.* **2006**, *21*, 273–284. [CrossRef]
19. Liu, Y.; Shi, J.; Yang, Y.; Lee, W.-J. Short-term wind-power prediction based on wavelet transform–support vector machine and statistic-characteristics analysis. *IEEE Trans. Ind. Appl.* **2012**, *48*, 1136–1141. [CrossRef]
20. Wan, C.; Xu, Z.; Pinson, P.; Dong, Z.Y.; Wong, K.P. Probabilistic forecasting of wind power generation using extreme learning machine. *IEEE Trans. Power Syst.* **2014**, *29*, 1033–1044. [CrossRef]

21. Mohammadi, K.; Shamshirband, S.; Yee, L.; Petković, D.; Zamani, M.; Ch, S. Predicting the wind power density based upon extreme learning machine. *Energy* **2015**, *86*, 232–239. [CrossRef]

22. Ren, C.; An, N.; Wang, J.; Li, L.; Hu, B.; Shang, D. Optimal parameters selection for BP neural network based on particle swarm optimization: A case study of wind speed forecasting. *Knowl. Based Syst.* **2014**, *56*, 226–239. [CrossRef]

23. Amjady, N.; Keynia, F.; Zareipour, H. Wind power prediction by a new forecast engine composed of modified hybrid neural network and enhanced particle swarm optimization. *IEEE Trans. Sustain. Energy* **2011**, *2*, 265–276. [CrossRef]

24. Jursa, R.; Rohrig, K. Short-term wind power forecasting using evolutionary algorithms for the automated specification of artificial intelligence models. *Int. J. Forecast.* **2008**, *24*, 694–709. [CrossRef]

25. Liu, D.; Niu, D.; Wang, H.; Fan, L. Short-term wind speed forecasting using wavelet transform and support vector machines optimized by genetic algorithm. *Renew. Energy* **2014**, *62*, 592–597. [CrossRef]

26. Salcedo-Sanz, S.; Pastor-Sánchez, A.; Prieto, L.; Blanco-Aguilera, A.; García-Herrera, R. Feature selection in wind speed prediction systems based on a hybrid coral reefs optimization–Extreme learning machine approach. *Energy Convers. Manag.* **2014**, *87*, 10–18. [CrossRef]

27. Foley, A.M.; Leahy, P.G.; Marvuglia, A.; McKeogh, E.J. Current methods and advances in forecasting of wind power generation. *Renew. Energy* **2012**, *37*, 1–8. [CrossRef]

28. Osório, G.; Matias, J.; Catalão, J. Short-term wind power forecasting using adaptive neuro-fuzzy inference system combined with evolutionary particle swarm optimization, wavelet transform and mutual information. *Renew. Energy* **2015**, *75*, 301–307. [CrossRef]

29. Haque, A.U.; Mandal, P.; Meng, J.; Srivastava, A.K.; Tseng, T.-L.; Senjyu, T. A novel hybrid approach based on wavelet transform and fuzzy ARTMAP networks for predicting wind farm power production. *IEEE Trans. Ind. Appl.* **2013**, *49*, 2253–2261. [CrossRef]

30. Ren, Y.; Suganthan, P.; Srikanth, N. A comparative study of empirical mode decomposition-based short-term wind speed forecasting methods. *IEEE Trans. Sustain. Energy* **2015**, *6*, 236–244. [CrossRef]

31. Liu, H.; Tian, H.; Liang, X.; Li, Y. New wind speed forecasting approaches using fast ensemble empirical model decomposition, genetic algorithm, Mind Evolutionary Algorithm and Artificial Neural Networks. *Renew. Energy* **2015**, *83*, 1066–1075. [CrossRef]

32. Beveridge, S.; Nelson, C.R. A new approach to decomposition of economic time series into permanent and transitory components with particular attention to measurement of the 'business cycle'. *J. Monet. Econ.* **1981**, *7*, 151–174. [CrossRef]

33. Zarnowitz, V.; Ozyildirim, A. Time series decomposition and measurement of business cycles, trends and growth cycles. *J. Monet. Econ.* **2006**, *53*, 1717–1739. [CrossRef]

34. Cochrane, J.H. Permanent and transitory components of GNP and stock prices. *Q. J. Econ.* **1994**, *109*, 241–265. [CrossRef]

35. Carlino, G.; Sill, K. Regional income fluctuations: Common trends and common cycles. *Rev. Econ. Stat.* **2001**, *83*, 446–456. [CrossRef]

36. Yan, J.; Liu, Y.; Han, S.; Qiu, M. Wind power grouping forecasts and its uncertainty analysis using optimized relevance vector machine. *Renew. Sustain Energy Rev.* **2013**, *27*, 613–621. [CrossRef]

37. Matsumoto, M.; Hori, J. Classification of silent speech using support vector machine and relevance vector machine. *Appl. Soft Comput.* **2014**, *20*, 95–102. [CrossRef]

38. Li, H.; Pan, D.; Chen, C.P. Intelligent prognostics for battery health monitoring using the mean entropy and relevance vector machine. *IEEE Trans. Syst. Man Cybern. Syst.* **2014**, *44*, 851–862. [CrossRef]

39. Flake, J.; Moon, T.K.; McKee, M.; Gunther, J.H. Application of the relevance vector machine to canal flow prediction in the Sevier River Basin. *Agric. Water Manag.* **2010**, *97*, 208–214. [CrossRef]

40. Torres, A.F.; Walker, W.R.; McKee, M. Forecasting daily potential evapotranspiration using machine learning and limited climatic data. *Agric. Water Manag.* **2011**, *98*, 553–562. [CrossRef]

41. Wang, T.; Xu, H.; Han, J.; Elbouchikhi, E.; Benbouzid, M.E.H. Cascaded H-bridge multilevel inverter system fault diagnosis using a PCA and multiclass relevance vector machine approach. *IEEE Trans. Power Electron.* **2015**, *30*, 7006–7018. [CrossRef]

42. Mirjalili, S. The ant lion optimizer. *Adv. Eng. Softw.* **2015**, *83*, 80–98. [CrossRef]

43. Zhao, H.; Guo, S. An optimized grey model for annual power load forecasting. *Energy* **2016**, *107*, 272–286. [CrossRef]

44. Morley, J.C. A state-space approach to calculating the Beveridge-Nelson decomposition. *Econ. Lett.* **2002**, *75*, 123–127. [CrossRef]

45. Newbold, P. Precise and efficient computation of the Beveridge-Nelson decomposition of economic time series. *J. Monét. Econ.* **1990**, *26*, 453–457. [CrossRef]

46. Tipping, M.E. Sparse Bayesian learning and the relevance vector machine. *J. Mach. Learn. Res.* **2001**, *1*, 211–244.

47. Mallat, S.G. A theory for multiresolution signal decomposition: The wavelet representation. *IEEE Trans. Pattern Anal. Mach. Inell.* **1989**, *11*, 674–693. [CrossRef]

energies

MDPI

Article

The General Regression Neural Network Based on the Fruit Fly Optimization Algorithm and the Data Inconsistency Rate for Transmission Line Icing Prediction

Dongxiao Niu, Haichao Wang *, Hanyu Chen and Yi Liang

School of Economics and Management, North China Electric Power University, Beijing 102206, China; niudx@126.com (D.N.); hbdxlhr@163.com (H.C.); louisliang@ncepu.edu.cn (Y.L.)
* Correspondence: ncepuwhc@ncepu.edu.cn; Tel.: +86-010-6177-3079

Received: 26 October 2017; Accepted: 29 November 2017; Published: 5 December 2017

Abstract: Accurate and stable prediction of icing thickness on transmission lines is of great significance for ensuring the safe operation of the power grid. In order to improve the accuracy and stability of icing prediction, an innovative prediction model based on the generalized regression neural network (GRNN) and the fruit fly optimization algorithm (FOA) is proposed. Firstly, a feature selection method based on the data inconsistency rate (IR) is adopted to select the optimal feature, which aims to reduce redundant input vectors. Then, the fruit FOA is utilized for optimization of smoothing factor for the GRNN. Lastly, the icing forecasting method FOA-IR-GRNN is established. Two cases in different locations and different months are selected to validate the proposed model. The results indicate that the new hybrid FOA-IR-GRNN model presents better accuracy, robustness, and generality in icing forecasting.

Keywords: icing prediction; general regression neural network (GRNN); fruit fly optimization algorithm (FOA); data inconsistency rate (IR)

1. Introduction

The transmission line ice coating can cause many types of accidents, including those related to flashover performance of the ice-covered insulator, breakage of the ground line, and the collapse of the tower [1]. They seriously affect the stability and security of power system operation. Since the recording of icing accidents began, cases of transmission line ice coating causing the fall of high voltage (HV) transmission line towers as well as wire breakages have been reported at home and abroad. Some accidents are serious. In January 1998, a week-long ice disaster occurred in Canada, which caused a blackout for one million users [2]. January 2008 witnessed four successive large-scale rainy and snowy storms in the south of China. The electricity grid was seriously iced and the power line was repeatedly broken, resulting in a direct economic loss of 10.45 billion CNY [3]. Therefore, establishing a prediction model of icing thickness and predicting the icing thickness of transmission lines accurately are of great significance for ensuring the security and stability of the power grid.

Currently, some scholars at home and abroad are researching icing thickness prediction of the transmission line. They have put forward a variety of forecasting models, which mainly include mathematical physics prediction models, statistical prediction models, and intelligent prediction models. The mathematical physics prediction model mostly predicts the icing thickness of the transmission line, based on the fluid motion law and the heat transfer mechanism of the wire icing [4]. The authors of [5], from the view of aerodynamics and thermodynamics, establish an icing forecasting model including the super-cooled water drop and the heat transfer process on ice. The authors of [6] point out that the icing of transmission lines is the result of coupling effect of thermodynamics, hydromechanics, and the electric current and field. On this basis, the physics prediction model of icing thickness is built. In addition, typical mathematical physics prediction methods for icing thickness include the Imai model [7], the Goodwin model [8] and the Lenhard model [9]. However, due to the fact that some of the parameters in the mathematical physics prediction model are difficult to obtain through the measurement in the actual line, such models are more difficult to apply directly to the ice prediction of the actual transmission lines. The statistical prediction model is based on the statistical laws of icing thickness of transmission lines [10], mainly including the extrema prediction model [11], Markov chain prediction model [12], and so on. However, the icing thickness prediction model based on the data statistics method cannot be extended to other transmission lines with different geographical environments, so the desired effect of this model is not satisfactory.

Therefore, under the background of rapid development of artificial intelligence technology, it is more significant to predict the icing thickness of transmission lines by using intelligent prediction methods. Intelligent prediction methods mainly include artificial neural networks (ANNs) [13] and the support vector machine (SVM) [14]. Here, the back-propagation neural network (BPNN) is typical of ANNs. Luo et al. [15] presented an icing forecasting model of BPNN based on Levenberg–Marquardt and obtained a higher prediction accuracy than the statistical forecasting model. However, the BPNN has the problem of many parameters to set, and can easily to fall into over-fitting or local optimum. For avoiding the local optimum problem, some scholars began to adopt the SVM model in the field of icing prediction. Li et al. [16] proposed a model based on the SVM for icing forecasting and its generalization ability is better than the model based on the BPNN. Ma et al. [17] introduced a short-term prediction model of icing thickness based on grey SVM, and it was pointed out that the model can achieve better prediction effect in ice-prone areas. However, it is difficult for the SVM model to deal with large-scale training samples, so it cannot obtain ideal prediction accuracy. The generalized regression neural network (GRNN) is a kind of radial basis function neural network proposed by Specht, which has a strong ability for nonlinear mapping [18]. Compared with the BPNN and the SVM, the GRNN has fewer adjustment parameters, does not easily fall into local minima, and is good at processing large-scale training samples. In addition, the GRNN has an advantage in forecasting volatile data. Therefore, the GRNN has been widely employed in the field of prediction, such as electricity price forecasting [19], energy consumption forecasting [20], and traffic flow forecasting [21]. Zhang et al. [19] introduced a novel hybrid forecasting model using the GRNN combined with wavelet transform for electricity price forecasting, and this model obtained better forecasting performance compared with the BPNN and SVM. Zhao et al. [20] utilized the GRNN model to forecast the annual energy consumption due to its good ability for dealing with the nonlinear problems. Leng et al. [21] established a short-term forecasting model of traffic flow based on the GRNN and it has stronger approximation capability and higher forecasting accuracy than the forecasting models of the radial basis function (RBF) and back-propagation (BP) neural network.

However, it is difficult to determine the smoothing factor in the GRNN model exactly and the selection of this parameter has a significant influence on its forecasting performance. Intelligent optimization algorithms such as the genetic algorithm (GA) [22] and particle swarm optimization (PSO) [23] are usually taken to select parameters for forecasting models. Gao et al. [22] proposed the GA to optimize the initial weights and thresholds of BPNN for housing price prediction, which accelerated the convergence rate of BPNN and improved the prediction accuracy of house prices. Ye [23] presented a kernel extreme learning machine model based on particle swarm optimization (PSO-KELM) to predict the power interval of wind power. PSO algorithm is utilized to optimize the output weights of KELM, and satisfactory prediction results are obtained. The above algorithms effectively improved the forecasting accuracy but also presented the malpractice of easily falling into local optimum. In order to overcome the drawbacks, the fruit fly optimization algorithm (FOA) [24], based on the behaviors of food finding, was proposed by Pan in 2011. This method only needs to set a few parameters and performs at a relatively high speed for optimum searching with wide applications [25]. Sun et al. [26]

introduced a new model based on wavelet transform and the least-squares support vector machine (LSSVM) optimized by the FOA for short-term load forecasting and compared the forecasting results between the proposed model and least-squares SVM optimized by PSO, which demonstrated that the FOA performed better than PSO. In addition, Li et al. [27] presented an LSSVM-based annual electric load forecasting model optimized by the FOA, and the proposed model obtained better forecasting effectiveness than the LSSVM optimized by the coupled simulated annealing algorithm (CSA). Hence, the FOA is utilized to adjust the appropriate smoothing factor in the GRNN model.

In addition, many factors can influence the formation of icing on the transmission line. If all the influencing factors are used as input indicators of the forecasting model, there will be a lot of redundant data [28]. Hence, the feature selection is also of great significance. Feature selection is about identifying and selecting the appropriate input vector in the prediction model to reduce redundant data and improve computational efficiency. The inconsistency rate (IR) model refers to dividing the feature set into many feature subsets and calculating the minimum inconsistency under this partition mode, so as to determine the optimal feature subsets and complete the feature selection [29]. Ma et al. [30] employed the IR model to select the input features of the short-term load forecasting model, whose simulation result demonstrated that the IR model gave the input vector of the strong pertinence of the prediction model, and reduced the redundancy of the input information, thus improving the accuracy of load forecasting. Liu et al. [31] also selected the optimal features for forecasting power load by adopting the IR model so as to reduce the redundancy of input vectors, and the IR model obtained an ideal feature selection effect. Using the IR model for feature selection can not only eliminate redundancy features by utilizing the inconsistency of the data set, but also take the correlative characteristics among the features into consideration, which does not ignore the relationship among features so that all the statistical information can be perfectly expressed by the selected optimal feature. Hence, this paper adopts the IR model for feature selection.

According to the above research, a GRNN model-integrated IR with the FOA is proposed. It is the first time these three models are combined for icing thickness forecasting and several comparing methods are utilized to validate the effectiveness of the proposed hybrid model. This paper is organized as follows: Section 2 introduces the implementation process of the IR and GRNN optimized by the FOA. Section 3 presents the evaluation criteria of the results. Section 4 provides a case to validate the proposed model. Section 5 analyzes another case in a different place at another time to prove the generalization of the forecasting method. Section 6 presents the conclusions in this paper.

2. Methodology

2.1. Fruit Fly Optimization Algorithm

The FOA is a new global optimization method based on foraging behaviors. There are two steps for searching food of fruit fly swarm: (1) use the olfactory organ to collect odors floating in the air and fly towards the food location; and (2) adopt a view to find food and other fruit flies' gathering positions and fly towards that direction. The iterative food searching process of the fruit fly swarm is presented in Figure 1.

The steps of looking for the optimal features are as follows:

(1)   Initialize the population size Sizepop, the iterations Maxgen, and the position coordinates (X_0, Y_0) of the random fruit fly population.

(2)   Give the individual fruit flies' random flight direction and step size so that they can find food by using the smell:

Where $i = 1, 2, \cdots$

$$X_i = X_0 + Random\ Value \tag{1}$$

$$Y_i = Y_0 + Random\ Value \tag{2}$$

(3) Since the fruit flies cannot obtain the food position, the distance $Dist_i$ between the individual and the origin of the flies is estimated first, and the taste concentration determination value S_i is calculated:

$$Dist_i = \sqrt{X_i^2 + Y_i^2} \tag{3}$$

$$S_i = 1/Dist_i \tag{4}$$

(4) Put the taste concentration determination value S_i into the adaptation function Fitness to determine the taste concentration $Smell_i$ of the individual position.

$$Smell_i = Fitness(S_i) \tag{5}$$

(5) Identify the individual of the highest concentration among the fruit fly populations including the concentration and coordinates:

$$[bestSmell \; bestIndex] = max(Smell_i) \tag{6}$$

(6) Retain the maximum taste concentration value best $Smell$ and its individual coordinates. The fruit fly population uses vision to fly in that direction:

$$Smellbest = bestSmell \tag{7}$$

$$X_0 = X(bestIndex) \tag{8}$$

$$Y_0 = Y(bestIndex) \tag{9}$$

Then, the stage of iterative refinement is entered; repeat steps (2)~(5), and judge that whether the maximum taste concentration is superior to the previous generation, and whether the current iteration is less than the maximum number of iteration Maxgen, and if so, then execute step (6).

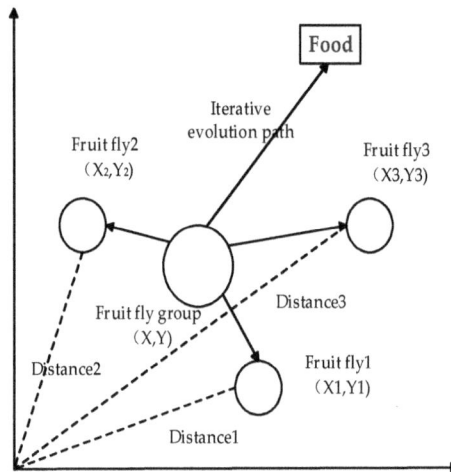

Figure 1. Iterative food searching process of the fruit fly swarm.

2.2. Data Inconsistency Rate

The aim of feature selection under large amounts of historical data of transmission line icing is to distinguish the data characteristics of the strongest correlation with respect to the icing thickness, so as to ensure that the input vector of the icing prediction model has strong pertinence, reducing the redundancy of input information and consequently improving the accuracy of icing prediction of the transmission lines. The inconsistent rate of the data can accurately describe the discrete characteristics of the input feature. Different feature patterns can be obtained by different division modes and different frequency distributions can be obtained by different partition patterns. The calculation of IR can be used to distinguish the distinguishing ability of the data category. The smaller the data IR is, the stronger the classification ability of the eigenvector is.

It is necessary to know the computing methods of the inconsistent rate if we want to perform feature selection by using the inconsistency method. Therefore, assuming that the collected icing thickness data has g characteristics (such as temperature, humidity, wind speed, etc.), which are respectively expressed as G_1, G_2, ..., G_g, Γ stands for feature set and L stands for the feature subset of Γ. Then, it is stipulated that qualification M has c categories and N data instances according to the degree of severity of the lines' icing. Z_{ji} stands for the eigenvalue which corresponds to the feature F_i, and M stands for the value λ_i, so the data instances can be expressed as $[z_j, \lambda_i]$. Thereinto, $Z_j = [Z_{j1}, Z_{j2}, Z_{j3}, \Lambda, Z_{jg}]$. Therefore, the calculation formula of the data inconsistency rate is:

$$\tau = \frac{\sum\limits_{k=1}^{p}\left(\sum\limits_{l=1}^{c} f_{kl} - \max\limits_{l}\{f_{kl}\}\right)}{N} \tag{10}$$

In the formula, f_{kl} is the number of data instances in the feature subset of the X_K mode in the data set; and X_k means that there are in total P patterns of feature partition range ($k = 1, 2, \ldots, p; p \le N$). The steps for using the inconsistent rate to perform the feature selection are as follows:

(1) Initialize the optimal feature subset as null set $\Gamma = \{\}$.
(2) Calculate the inconsistent rate of the data sets $G_1, G_2, ..., G_g$ in the feature subsets which are made up with the remaining feature of each subset.
(3) Select the feature G_i which corresponds to the minimum inconsistent rate as the optimum feature, and then update the optimum feature subset to $\Gamma = \{\Gamma, G_i\}$.
(4) Calculate the inconsistent rate statistics table of the feature subsets and arrange them from small to large.
(5) Select the feature subsets L with the smallest number of features, which can be selected as the optimal feature subsets if they satisfy the condition that $\tau_L \approx \tau_\Gamma$ or $\tau_{L'}/\tau_L$ is the minimum of the inconsistent rate of all adjacent feature subsets. L' is an adjacent feature subset of L.

Using calculating inconsistent rate can not only eliminate redundancy features by utilizing the inconsistency of the data set, but also take the correlative characteristics among the features into consideration, which does not ignore the relationship among features so that all the statistic information can be perfectly expressed by the selected optimal feature.

2.3. Generalized Regression Neural Network

The general regression neural network (GRNN) was proposed by the American scholar Donald F. Specht in 1991, with the theoretical basis of nonlinear regression analysis. As shown in Figure 2, the GRNN constitutes four components:

(1) The input layer: the original variables enter the network which correspond to the neurons one by one and are submitted to the next layer.
(2) The pattern layer: nonlinear transformation is applied to the values received from the input layer. The transfer function of the ith neuron in the pattern layer is:

$$P_i = \exp[-(X - X_i)^T(X - X_i)/2\sigma^2] \quad i = 1, 2, \cdots n, \tag{11}$$

where X represents input variable, X_i is the learning sample corresponding to the ith neuron; and σ is the smoothing parameter.

(3) The summation layer: calculate the sum and weighted sum of the pattern outputs.

The summation layer contains two types of neurons, in which one neuron S_A makes arithmetic summation of the output of all pattern layer neurons, and the connection weight of each neuron in the pattern layer to this neuron is 1. Its transfer function is:

$$S_A = \sum_{i=1}^{n} P_i \tag{12}$$

The outputs of all neurons in the pattern layer were weighted and summed to gain the other neurons S_{Nj} in the summation layer. The transfer function of the other neurons in the summation layer is:

$$S_{Nj} = \sum_{i=1}^{n} y_{ij} P_i \quad j = 1, 2, \cdots, k, \tag{13}$$

where y_{ij} is the connection weight between the ith neuron in the pattern layer and the jth neuron in the summation layer. y_{ij} is the jth element in the ith output sample y_i.

(4) The output layer: the forecasting results can be derived. The output of each neuron is:

$$y_j = \frac{S_{Nj}}{S_A} \quad j = 1, 2, \cdots, k, \tag{14}$$

where y_j is the output of the jth neuron.

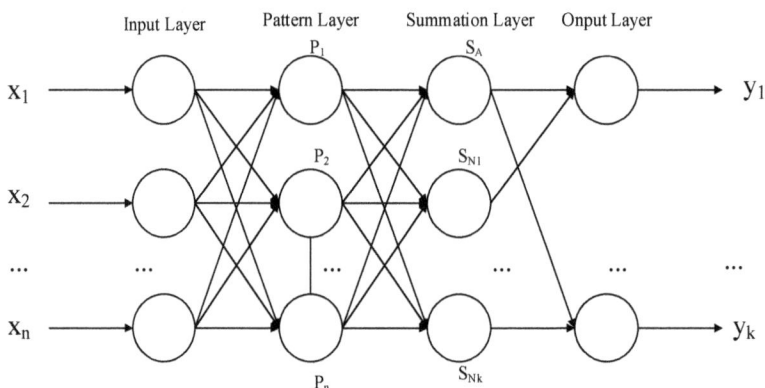

Figure 2. The structure of the generalized regression neural network (GRNN).

2.4. The Forecasting Model of FOA-IR-GRNN

The icing thickness forecasting model combining FOA, IR, and GRNN are constructed as illustrated in Figure 3. It can be seen from the figure that the model of icing prediction proposed in this paper mainly includes three parts: The first part is the feature selection based on the inconsistent rate, the second part is the sample training based on the GRNN model, and the third part is the icing prediction based on the GRNN model. When the established feature subset L cannot satisfy the algorithm stopping criteria, the program will continue to cycle until reaching the expected precision and then output the optimal feature subset. Therefore, in the model of icing prediction proposed in this paper, the purpose of the first part is to find the optimal feature subset and the best value of smoothing factor in the GRNN by iterative calculation. The purpose of the second part is to calculate the prediction accuracy of the training samples in every process of iteration, so that the fitness function can be calculated. In the third part, we will utilize the optimum feature subsets and parameters obtained from the above two parts and perform the final prediction of the icing thickness of the test samples by retraining the GRNN model.

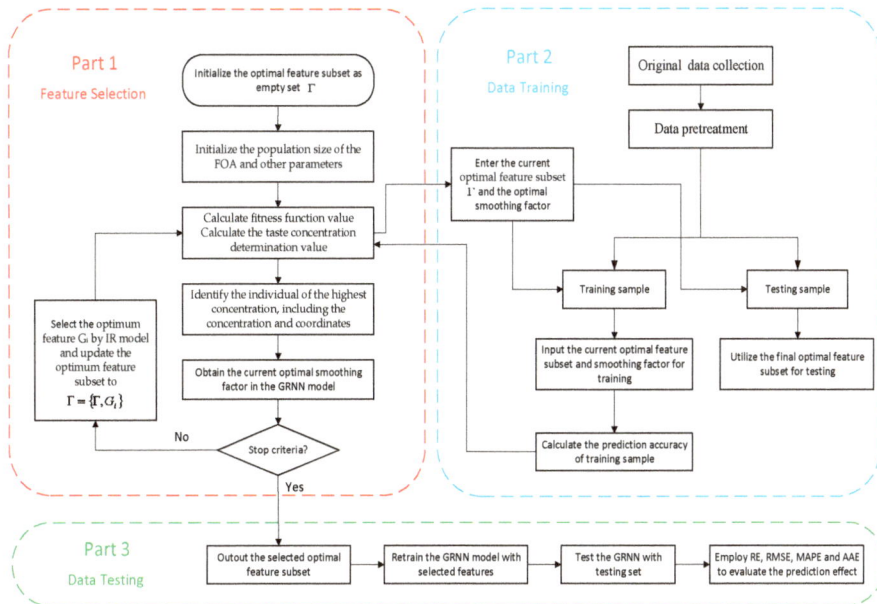

Figure 3. The flow chart of FOA-IR-GRNN. FOA: fruit fly optimization algorithm; IR: inconsistency rate; RE: relative error; RMSE: root mean square error; MAPE: mean absolute percentage error; AAE: average absolute error.

The specific steps for icing thickness prediction are listed as follows:

(1) Determine the initial candidate feature. In this paper, we choose ambient temperature, relative humidity, wind speed, wind direction, light intensity, atmospheric pressure, altitude, condensation height, conductor direction, the height of conductor suspension, load current, precipitation, and conductor surface temperature, all of which are selected as the candidate features of the factors that influence icing. In addition, when it reaches the point t-i (i = 1, 2, 3, 4), thickness value, temperature, relative humidity and wind speed are also selected as the main influencing factors of line icing. All the initial candidate features are shown in Table 1. In the IR algorithm, the optimal feature subset needs to be initialized as an empty set $\Gamma = \{\}$.

(2) Initialize the parameters of FOA. Suppose the population size is 20, the maximum iteration number is 200 and the range of random flight distance is set as $[-10, 10]$.

(3) Calculate the inconsistent rate. After completing steps (1) and (2), put the candidate features into the IR feature selection model gradually. Calculate the inconsistent rate of the data sets G_1, G_2, ..., G_g in the feature subsets which are made up of the remaining features of each subset and then select the feature G_i which corresponds to the minimum inconsistent rate as the optimum feature, and update the optimum feature as $\Gamma = \{\Gamma, G_i\}$.

(4)   Get the optimal feature subset and the best value of smoothing factor in GRNN. Put the current feature subsets into the GRNN model, and calculate the prediction accuracy during the learning process of the circular training samples. Then, the fitness function *Fitness(j)* can be worked out. We can get the optimum feature subset by comparing the fitness function among each generation and judge whether all iterations have achieved the algorithm stopping conditions. If not, re-initialize a new feature subset and put it into a new circulation until the optimum feature subset which meets all the conditions is obtained. It should be noted that the smoothing factor of the GRNN also needs to be optimized, and the initial value of smoothing factor will be assigned randomly. In this paper, a fitness function is established based on the two factors of prediction accuracy and feature selection:

$$Fitness(j) = -(a + r(j) + b \times \frac{1}{Numfeature(j)})$$
(15)

In the formula, *Numfeature(x_i)* is the number of optimum feature which is selected by each iteration and both a and b are constants between [0, 1]; *r(j)* represents the prediction accuracy of ice cover thickness at each iteration. The optimal number of features is proportional to the fitness function for all iterations, and the accuracy of the icing prediction is inversely proportional to the fitness function. Different smoothing factors will result in different forecasting results and lead to different prediction accuracy, indicating that the smoothing factor of the GRNN also influences the value of fitness function *Fitness(j)*. Hence the optimal feature subset and the best value of smoothing factor in the GRNN will be obtained at the same time in this step.

(5)   Stop optimization and start prediction. Circulation ends at the maximum number of iteration. Here, the optimum feature subset and the best value of smoothing factor can be substituted into the GRNN model for icing thickness forecasting.

Table 1. The full candidate features.

$C_1, \dots, C_4$	IT_{t-i}, $i = 1, 2, 3, 4$ represent the *t-i*th time point's icing thickness
$C_5, \dots, C_9$	T_{t-i}, $i = 0, 1, 2, 3, 4$ represent the *t-i*th time point's ambient temperature
$C_{10}, \dots, C_{14}$	H_{t-i}, $i = 0, 1, 2, 3, 4$ represent the *t-i*th time point's relative air humidity
$C_{15}, \dots, C_{19}$	WS_{t-i}, $i = 0, 1, 2, 3, 4$ represent the *t-i*th time point's wind speed
C_{20}	WD_t represents the *t*th time point's wind direction
C_{21}	SI_t represents the *t*th time point's sunlight intensity
C_{22}	AP_t represents the *t*th time point's air pressure
C_{23}	AL represents the altitude
C_{24}	CH represents the condensation height
C_{25}	LD represents the transmission line direction
C_{26}	LSH represents the transmission line suspension height
C_{27}	LC represents the load current
C_{28}	R represents the rainfall
C_{29}	ST represents the surface temperature on the transmission line

3. Performance Evaluation Index

The primary issue is to determine which forecasting model outperforms the other models, and the performance of the prediction models is usually assessed by statistical criteria: the relative error (*RE*), root mean square error (*RMSE*), mean absolute percentage error (*MAPE*) and average absolute error (*AAE*). The smaller the values of these four indicators are, the better the forecasting performance is. Furthermore, the indicators named *RMSE*, *MAPE*, and *AAE* can reflect the overall error of the prediction model and the degree of error dispersion. The smaller the values of these three indicators are, the more concentrated the distribution of errors is. These four error indexes are defined as follows:

$$RE = \frac{y_t - y_t^*}{y_t} \times 100\% \tag{16}$$

$$RMSE = \sqrt{\frac{1}{N}\sum_{t=1}^{N}\left(\frac{y_t - y_t^*}{y_t}\right)^2} \tag{17}$$

$$MAPE = \frac{1}{N}\sum_{t=1}^{N}\left|\frac{y_t - y_t^*}{y_t}\right| \times 100\% \tag{18}$$

$$AAE = \frac{1}{N}\left(\sum_{i=1}^{N}|y_t - y_t^*|\right)/\left(\frac{1}{N}\sum_{i=1}^{N}y_t\right) \tag{19}$$

where y_t and y_t^* are the actual and forecast icing thickness at the time point t, respectively. N refers to the groups of data.

4. Empirical Analysis

4.1. Data Collection and Pretreatment

In 2008, China was hit by a disaster of frozen rain and snow rarely seen in history. It brought huge losses to life, and seriously affected the national economy. Hunan Province was one of the worst hit provinces in this icing disaster. During the frozen period, the icing accident led to 182 towers with 500-kV power transmission lines falling down, 633 towers with 220-kV power transmission lines falling down, 1427 towers with 110-kV power transmission lines falling down, 1064 towers with 35-kV power transmission lines falling down, and 63,036 towers with 10-kV power transmission lines falling down. As for 10-kV and above, 50,000 wires were broken. Yueyang and Loudi (cities in Hunan Province) as well as other areas had large-area power outages. The Hunan power grid suffered the most serious threat in history, and the direct economic losses were up to more than 1 billion CNY. Therefore, this paper chooses the transmission line of the Hunan Province power grid to carry on the empirical analysis.

In this paper, the power transmission line, named "Kunxia line" in YueYang of Hunan Province is selected as the case to verify the effectiveness of the proposed model. All the data are provided by the Key Laboratory of Disaster Prevention and Mitigation of Power Transmission and Transformation Equipment (Changsha, China).

The data from the "Kunxia line" are from 10 January 2008 to 12 January 2008, and include 288 data groups. Here, taking 15 min as the data collection frequency, the first 230 groups are adopted as the training samples and the latter 58 are utilized as the testing samples in Case 1. The main micro-meteorology data, including temperature, wind speed, and humidity are shown in Figure 4.

In order to better train the proposed model and ensure the prediction accuracy, it is of significance to normalize all the original data in the range of [0, 1], and the processing equation is as follows:

$$Z = \{z_i\} = \frac{x_i - x_{min}}{x_{max} - x_{min}} \quad i = 1, 2, 3, \ldots, n \tag{20}$$

where x_i is the actual value; x_{min} and x_{max} are the minimum and maximum values of the sample data respectively; and z_i represents the value of the adjusted ith sample point.

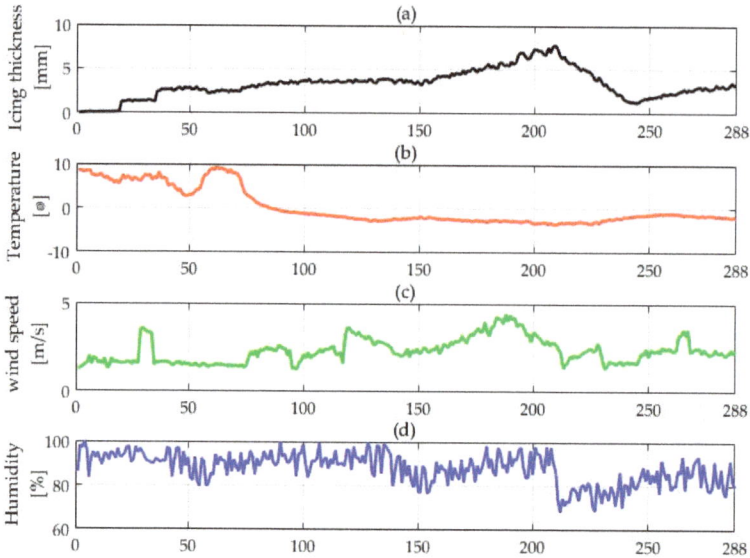

Figure 4. Original data chart of icing thickness, temperature, wind speed and humidity. Note: (**a**) represents the original data of icing thickness; (**b**) represents the original data of temperature; (**c**) represents the original data of wind speed; and (**d**) represents the original data of humidity.

4.2. Feature Selection

Based on the IR model, this section is about the selection of the optimal feature subset, and the determination of the input index of the prediction model. This paper uses Matlab R2014b for programming, and as for the test platform environment, we use the Intel Core i5-6300U, with 4G memory and the Windows 10 Professional Edition system.

Figure 5 presents the iteration process of the FOA-IR-GRNN model for training sample feature extraction. The accuracy curve shown in the figure describes the prediction accuracy of the training samples which were made by the GRNN in different iterations. The fitness curve describes the fitness function values calculated during the process of iteration. The number of selected features indicates the optimal number of features calculated by the IR model in the convergence process. The number of feature reductions is the number of features that the FOA eliminates during the convergence process.

It can be seen from Figure 5 that the FOA converges when the number of iterations is 51, and the optimal fitness function is −0.88; at this time the prediction accuracy of the training sample is up to 98.6%. This shows that through the learning and training of the algorithm, the fitting ability of the GRNN is strengthened, and the prediction accuracy of training samples is the highest. Moreover, when the FOA runs the 51st time, the number of selected features also tends to be stable. It can be concluded that the algorithm eliminates 23 redundant features from 29 candidate features, and the final input features are the tth time point's ambient temperature, relative air humidity, wind speed and $t − 1$th time point's icing thickness, ambient temperature, relative air humidity.

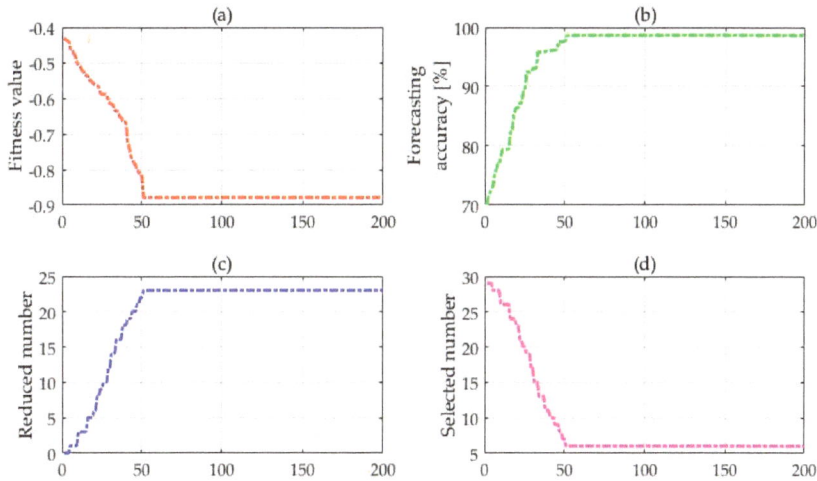

Figure 5. The curve of convergence for feature selection. Note: (**a**) represents the fitness value; (**b**) represents the forecasting accuracy; (**c**) represents the reduced number of candidate feature; and (**d**) represents the selected number of optimization feature.

4.3. The GRNN for Icing Forecasting

After the optimal feature subset is obtained, put the input vector into the model proposed in this paper to train and test. The smoothing factor of the GRNN model is 0.0031, which is calculated by the running program.

A k-fold cross validation (K-CV) test is conducted here, so as to show whether the forecasting results of the proposed model is obtained at local optimal or global optimal location and whether this proposed model can be generalized to other unseen data. The K-CV test method divides the samples into k disjoint subsets randomly, each of which is roughly equal in size. Using k − 1 subsets, a model is established for a given set of parameters, and the RMSE of the remaining last subset is employed to evaluate the performance of the parameters. Repeat the procedure K times, and each subset has the opportunity to be tested. Hence, the 288 sets of data are randomly divided into 12 datasets, each of which has 24 groups of data, and they do not intersect with each other. After 12 operations, each sub data set is tested and the RMSE of the sample is obtained, which can be seen in Table 2.

Table 2. Results of the k-fold cross-validation.

Fold Number	1	2	3	4	5	6	7	8	9	10	11	12	Average	Standard Deviation
RMSE	0.0126	0.0127	0.0128	0.0121	0.0103	0.0101	0.0123	0.0133	0.0128	0.0126	0.013	0.0115	0.0122	0.0010

From Table 2, it can be found that the average RMSE value and the RMSE standard deviation of the proposed model is 0.0122 and 0.0010, respectively. It is indicated that the validation error of the icing prediction model proposed in this paper can obtain its global minimum.

In order to verify the performance of the proposed model, this paper employs the GRNN model which is not optimized by FOA. The mature BP neural network model and SVM model do the contrast experiments, supported by the test sample data in Section 4.1. In addition, the FOA-GRNN model without considering IR model for feature selection is also utilized for icing forecasting so as to demonstrate the effects of the IR and the FOA. The smoothing factor of the single GRNN model is 0.2, while the smoothing factor of the FOA-GRNN model without considering the IR model is 0.1026. The topological structure of the BPNN model is 9-7-1, and the hidden layer transfer function is expressed by the tansig function. The output layer transfer function is expressed as purelin function. The maximum number of trainings is 100 and the minimum error of the training target is 0.0001. The training rate is 0.1. The initial weights and thresholds are obtained by their own training. In the SVM model, the penalty parameter c is 9.236 which is obtained by the training, the kernel function parameter g is 0.0026, and the ε loss function parameter p is 2.3572.

The actual values and forecasting values of the GRNN, BPNN, SVM, FOA-GRNN and the model presented in this paper are presented in Figure 6. The relative error of each model is shown in Figure 7. Figure 8 displays the RMSE, MAPE, and AAE of each prediction model. Table 3 displays part of the predicted values and errors.

Figure 6 and Table 3 describe the forecasting results of the five prediction models and the actual icing thickness. It can be seen from Figure 6 that the relative distance between the predicted and actual values of each prediction model. In general, the overall forecasting trends of the five models are close to the actual values. The forecasting curve of the proposed model is the closest to the actual curve, whereas the other prediction curves have some deviation. The forecasting curve of the FOA-GRNN is closer than that of the GRNN alone, demonstrating that the FOA makes the GRNN forecast better than the GRNN model without the FOA. However, the prediction accuracy of the FOA-GRNN model is not as good as the FOA-IR-GRNN model, indicating that feature selection method named the IR model can further improve the forecasting effectiveness of the GRNN. In addition, it can be found that the forecasting curve of the GRNN model is closer than the BPNN model and SVM model, indicating that the GRNN performs better than the BPNN and SVM for icing forecasting.

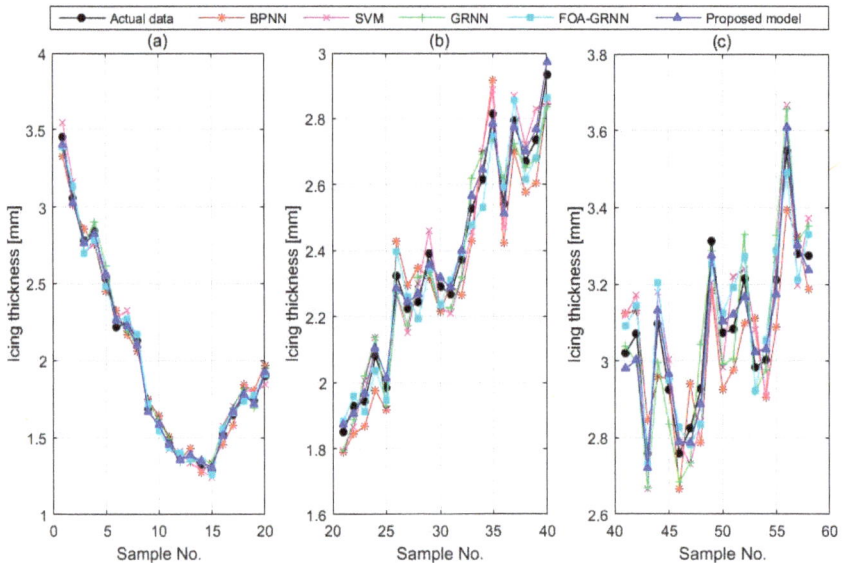

Figure 6. The forecasting values of the proposed method and the comparison methods. Note: (**a**) the forecasting value from sample points 1–20; (**b**) the forecasting value from sample points 20–40; (**c**) the forecasting values from sample points 41–58. BPNN: back-propagation neural network; SVM: support vector machine.

Figure 7 reflects the relative error distribution of the four models. From Figure 7, the difference of prediction effect among different models can be seen more clearly. The RE ranges [−3%, 3%] and [−1%, 1%] are popularly regarded as a standard to evaluate the performance of a prediction model [32].

From Figure 7, we can obtain that: (1) there are only nine relative error values of BPNN model in the range of [−3%, 3%] and only one value in the range of [−1%, 1%]; the maximum relative error is 4.99% at the 24th sample point, while the minimum is −4.98% at the sixth sample point; (2) the relative error of the SVM model has 35 forecasting points belonging to the range of [−3%, 3%], and there exist three forecasting points in the range of [−1%, 1%]; the maximum relative error value is 3.48% at the 15th sample point, and the minimum is −4.41% at the 51st point; (3) in the GRNN model, the relative errors of 43 sample points are in the range of [−3%, 3%], and the relative errors of five sample points are in the range of [−1%, 1%]; the maximum value is 3.38% at the 40th predicted point, while the minimum is −3.95% at the 23rd point; (4) there are 52 relative error values of the FOA-GRNN model in the range of [−3%, 3%] and seven values in the range of [−1%, 1%]; the maximum relative error is 3.25% at the 34th sample point, while the minimum is −3.57% at the 51st sample point; and (5) the relative errors of the FOA-IR-GRNN model are all in the range of [−3%, 3%], and there exist 12 relative errors in the range of [−1%, 1%]; the maximum relative error is 2.17% at the 42nd point, while the minimum value is −1.89% at the sixth sample point. We can also find from Figure 7 that the RE curve of the FOA-IR-GRNN model is the most stable and its values are all distributed within [−2%, 2%]. Moreover, the RE curve of the FOA-GRNN model is more stable than the GRNN's; the RE curve of the GRNN model is more stable than the SVM's; and the RE curve of the SVM model is more stable than the BPNN's. Based on the above analysis of relative error data, it can be concluded that the prediction accuracy and stability of the FOA-IR-GRNN model is the best. The input indexes obtained by IR model can help satisfactorily predict when the relative errors of the FOA-GRNN and the FOA-IR-GRNN are compared. It is also demonstrated that the FOA enhances the training and learning process effectively so as to avoid falling into a local optimum and improves the global searching ability of the GRNN by comparing the relative errors of the FOA-GRNN and GRNN. Hence, both the IR model and the FOA are significant for improving the forecasting performance of the GRNN. Additionally, the GRNN presents more satisfactory performance than the SVM and BPNN. This result indicates that the GRNN with only one parameter to be adjusted and fast calculation is more suitable for forecasting nonlinear and non-stationary icing thickness.

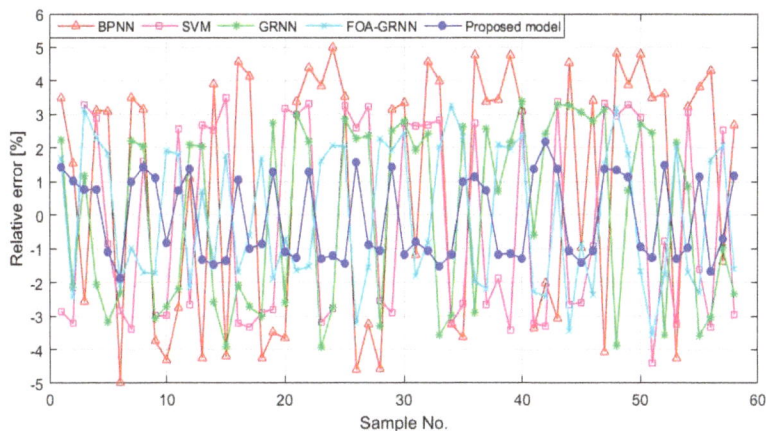

Figure 7. The relative error curve of each method.

The RMSE, MAPE, and AAE of BPNN, SVM, GRNN, FOA-GRNN, and FOA-IR-GRNN are shown in Figure 8. From Figure 8, we can conclude that the RMSE, MAPE, and AAE of the proposed model are 1.2326%, 1.2006%, and 1.2059%, respectively, which are all the smallest among the above four models. In addition, the RMSE, MAPE, and AAE of the FOA-GRNN model are 2.0485%, 1.9462%, and 1.9994% respectively; the RMSE, MAPE, and AAE of the GRNN model are 2.6514%, 2.5375%, and

2.5086% respectively; the RMSE, MAPE, and AAE of the SVM model are 2.8999%, 2.8295%, and 2.8200% respectively; and the RMSE, MAPE, and AAE of the BPNN model are 3.6889%, 3.5612% and 3.5252% respectively. These indicators can reflect the overall error of the prediction model and the degree of error dispersion. Hence it can be further proved that the overall prediction effect of the GRNN model is better than that of the SVM model and the BPNN model, while the overall prediction effect of the SVM model is better than that of the BPNN model. The prediction accuracy of the FOA-GRNN model is better than that of the GRNN model, which demonstrates that adopting the FOA to choose the smoothing parameter in the GRNN model has achieved a satisfactory optimization effect. Meanwhile, the FOA-IR-GRNN model obtains better overall forecasting accuracy than the FOA-GRNN model. This result proves that the IR model not only reduces the redundant data, but also ensures the integrity of the input information, thus obtaining the ideal prediction results.

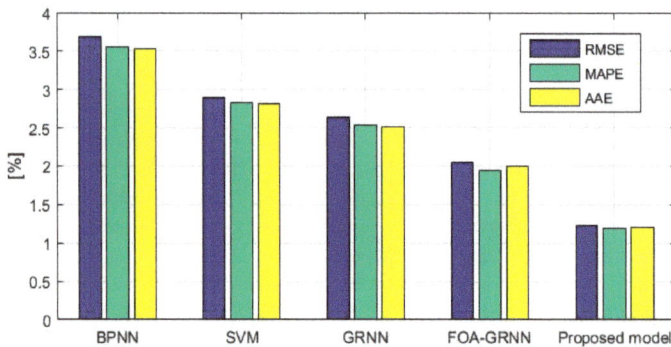

Figure 8. Values of root-mean-square error (RMSE), mean absolute percentage error (MAPE) and average absolute error (AAE).

Table 3. Part of the forecasting value and relative errors of each model.

Data Point Number	Actual Value (mm)	BPNN Forecast Value (mm)	BPNN Error (%)	SVM Forecast Value (mm)	SVM Error (%)	GRNN Forecast Value (mm)	GRNN Error (%)	FOA-GRNN Forecast Value (mm)	FOA-GRNN Error (%)	Proposed Model Forecast Value (mm)	Proposed Model Error (%)
1	3.45	3.33	3.48	3.55	−2.86	3.37	2.23	3.50	−1.3410	3.40	1.43
2	3.06	3.01	1.54	3.16	−3.22	3.12	−2.15	3.15	−3.0763	3.03	1.01
3	2.78	2.86	−2.58	2.69	3.29	2.75	1.16	2.77	0.7042	2.76	0.75
4	2.85	2.76	3.13	2.76	2.87	2.90	−2.05	2.76	3.1303	2.82	0.77
5	2.53	2.45	3.09	2.55	−0.85	2.61	−3.19	2.46	2.8785	2.56	−1.10
6	2.22	2.33	−4.98	2.28	−2.83	2.27	−2.32	2.24	−0.9358	2.26	−1.89
7	2.25	2.17	3.49	2.32	−3.40	2.20	2.21	2.30	−2.4674	2.22	0.99
8	2.13	2.06	3.15	2.10	1.61	2.09	2.05	2.19	−2.7489	2.10	1.43
9	1.68	1.75	−3.73	1.74	−2.99	1.74	−3.08	1.74	−3.0226	1.67	1.10
10	1.57	1.64	−4.32	1.62	−2.99	1.62	−2.71	1.64	−3.9586	1.59	−0.85
11	1.46	1.50	−2.75	1.42	2.57	1.49	−2.19	1.41	3.4128	1.45	0.72
12	1.37	1.36	1.10	1.41	−2.66	1.34	2.09	1.33	2.9047	1.35	1.38
13	1.37	1.43	−4.26	1.33	2.68	1.34	2.05	1.36	0.4503	1.39	−1.31
14	1.33	1.27	3.89	1.29	2.53	1.36	−2.59	1.29	2.3738	1.34	−1.47
15	1.28	1.34	−4.21	1.24	3.48	1.33	−3.91	1.32	−2.9158	1.30	−1.36

5. Case Study 2

In order to verify the proposed model has good adaptability in different time and places, another case which selects the relevant data of the "Tianshang line" located in Loudi, Hunan Province, is provided in this paper. The study is carried out with data from 17 January 2008 to 10 February 2008 as the training set and data from 11 February 2008 to 15 February 2008 as the testing set. Here, we take

2 h as data collection frequency, and there are 360 data groups in total. The icing thickness data and the main micro-meteorology data are shown in Figure 9.

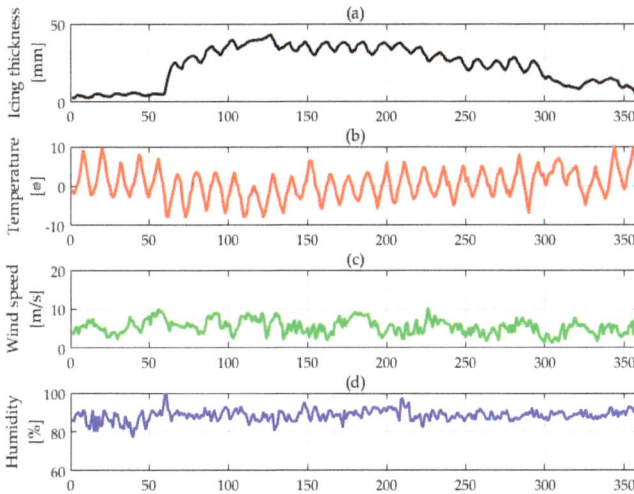

Figure 9. Original data chart of icing thickness, temperature, wind speed and humidity. Note: (**a**) represents the original data of icing thickness; (**b**) represents the original data of temperature; (**c**) represents the original data of wind speed; and (**d**) represents the original data of humidity.

The iterative process of sample data of "Tianshang line" by employing the FOA-IR-GRNN model is presented in Figure 10. From Figure 10, we can conclude that the optimal fitness function calculated by the IR model is −0.91. When the FOA achieves the optimum in the 47th iteration, the prediction accuracy of the sample reaches 98.3%. It can also be seen that 25 redundant features are eliminated from 29 candidate features, and the final input features include the tth time point's ambient temperature, relative air humidity, wind speed and the $t − 1$th time point's icing thickness. In addition, the smoothing factor of the GRNN was 0.0056, optimized by the FOA.

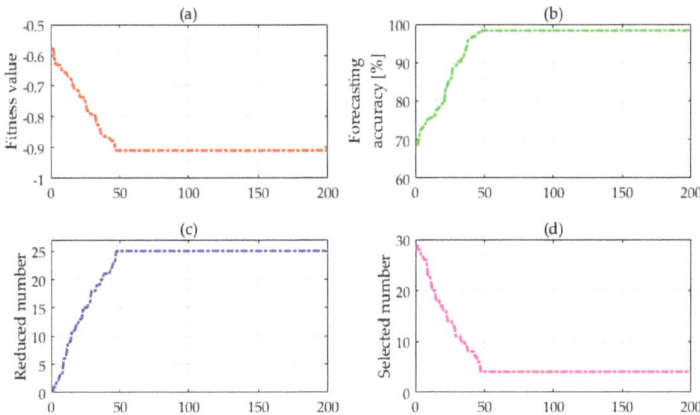

Figure 10. The curve of convergence for feature selection. Note: (**a**) represents the fitness value; (**b**) represents the forecasting accuracy; (**c**) represents the reduced number of candidate feature; and (**d**) represents the selected number of optimization feature.

The results of the k-fold cross-validation for the icing prediction model proposed in this paper are described in Table 4. The forecasting results are displayed in Figure 11 and Table 5. The error analyses are presented in Figures 12 and 13.

Table 4. Results of the k-fold cross-validation.

Fold Number	1	2	3	4	5	6	7	8	9	10	11	12	Average	Standard Deviation
RMSE	0.0115	0.0128	0.0117	0.0125	0.0133	0.011	0.0129	0.0102	0.0105	0.0132	0.0103	0.0122	0.0118	0.0011

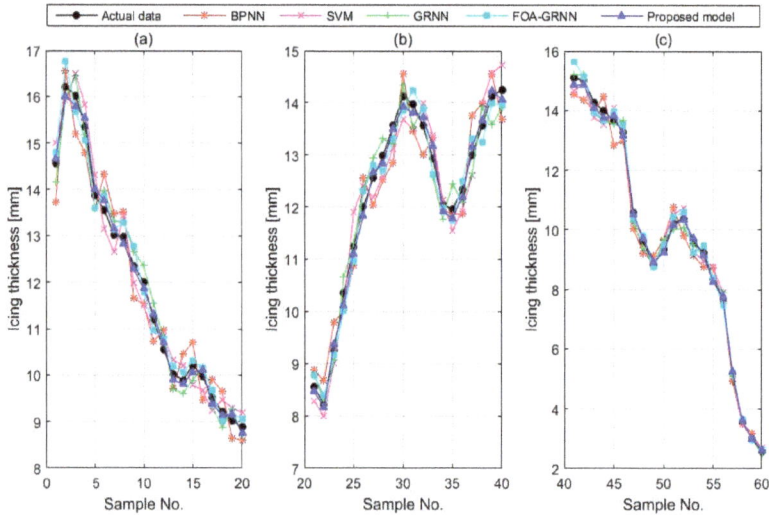

Figure 11. The forecasting values of the proposed method and the comparison methods. Note: (**a**) the forecasting value from sample points 1–20; (**b**) the forecasting value from sample points 21–40; and (**c**) the forecasting value from sample points 41–60.

Table 5. Part of the forecasting value and relative errors of each model.

Data Point Number	Actual Value (mm)	BPNN Forecast Value (mm)	BPNN Error (%)	SVM Forecast Value (mm)	SVM Error (%)	GRNN Forecast Value (mm)	GRNN Error (%)	FOA-GRNN Forecast Value (mm)	FOA-GRNN Error (%)	Proposed Model Forecast Value (mm)	Proposed Model Error (%)
1	14.56	13.71	5.81	15.01	−3.07	14.16	2.75	14.79	−1.59	14.66	−0.67
2	16.21	16.56	−2.16	15.92	1.77	15.96	1.55	16.76	−3.40	16.00	1.30
3	16.02	15.20	5.12	16.50	−2.99	16.44	−2.62	15.67	2.16	15.80	1.40
4	15.35	14.79	3.65	15.82	−3.05	15.18	1.12	15.06	1.88	15.54	−1.21
5	13.87	13.58	2.08	14.32	−3.27	13.59	2.05	13.59	2.00	14.00	−0.96
6	13.55	14.33	−5.74	13.14	3.02	13.98	−3.18	13.86	−2.26	13.77	−1.62
7	13.01	13.48	−3.60	12.65	2.75	13.31	−2.32	13.13	−0.92	13.15	−1.07
8	12.98	13.52	−4.17	13.43	−3.46	13.29	−2.38	13.28	−2.31	12.83	1.19
9	12.35	11.65	5.66	11.99	2.89	12.66	−2.51	12.77	−3.36	12.28	0.55
10	12.01	11.54	3.91	11.50	4.24	12.36	−2.93	11.79	1.84	11.86	1.22
11	11.21	10.72	4.34	10.93	2.46	11.55	−3.05	10.97	2.10	11.31	−0.92
12	10.56	10.97	−3.88	10.84	−2.63	10.81	−2.38	10.79	−2.17	10.70	−1.35
13	10.02	9.70	3.17	10.32	−3.00	9.70	3.21	10.19	−1.69	9.91	1.13
14	9.89	10.45	−5.66	10.21	−3.22	9.60	2.93	10.05	−1.63	9.80	0.93
15	10.21	10.70	−4.76	9.78	4.22	9.89	3.16	10.30	−0.92	10.06	1.47

As is shown in Table 4, the average RMSE value and RMSE standard deviation of the proposed model are 0.0118 and 0.0011, respectively. These data illustrate the fact again that the generalization performance of the icing prediction model proposed in this paper has been improved.

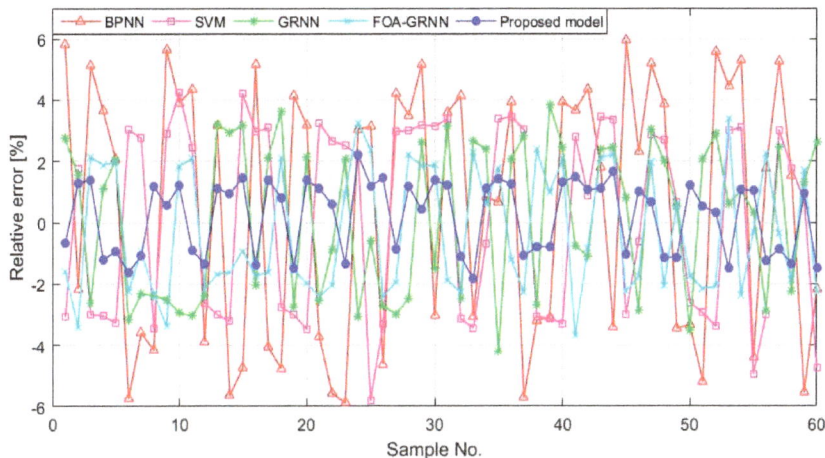

Figure 12. The relative error curves of each method.

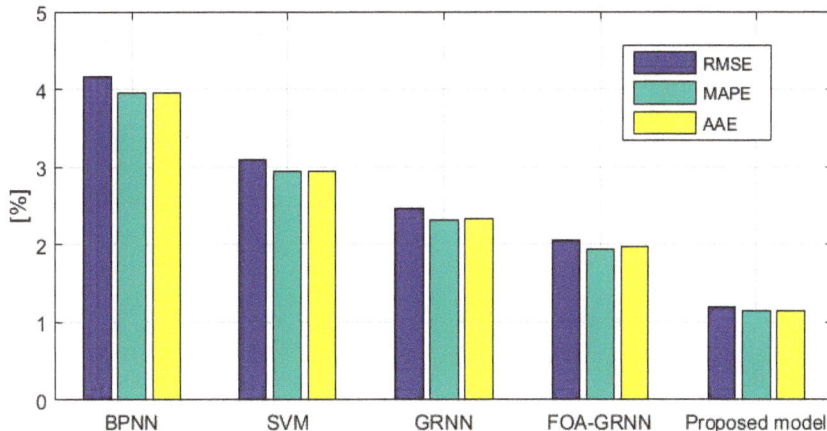

Figure 13. Values of root-mean-square error (RMSE), mean absolute percentage error (MAPE) and average absolute error (AAE).

It can be concluded from Figure 11 and Table 5 that the predicted value of the FOA-IR-GRNN model is the closest to the actual value, which demonstrates that the proposed model is not only accurate but also has robustness. When comparing the forecasting curves of the FOA-IR-GRNN model and the FOA-GRNN model, we can conclude that adopting the IR model for feature selection can significantly improve the prediction accuracy, in that this feature selection method can enhance the effectiveness of input information. Furthermore, the forecasting curve of the FOA-GRNN model is closer than that of the GRNN model, indicating that in addition to the IR model, the FOA also makes a significant contribution to the improvement of GRNN prediction accuracy. Compared with SVM and BPNN, the forecasting value of the GRNN model is closer to the actual ice thickness, which demonstrates once again that the approximation and classification ability of the GRNN model is better than that of the SVM model and the BPNN model, and the GRNN emerges with better performance in dealing with unstable data.

Figure 12 presents the relative error of the four models. As the calculation results shown, we can conclude that: (1) the fitting and learning ability of the FOA-IR-GRNN model is the strongest, in that its relative errors are all in the range of [−3%, 3%] and there exist 16 sample points belonging to the range of [−1%, 1%]; the maximum relative error is 2.21% at the 24th point, and the minimum value is −1.85% at the 33rd point; (2) there exist 55 relative error values of the FOA-GRNN model in the range of [−3%, 3%] and nine values in the range of [−1%, 1%]; the maximum relative error is 3.37% at the 33rd sample point, while the minimum is −3.70% at the 41st sample point; (3) the GRNN model emerges with 49 sample points in the range of [−3%, 3%], while seven sample points are in the range of [−1%, 1%]; the maximum value is 3.84% at the 39th point, and the minimum is −4.19% at the 35th point; (4) the SVM model emerges with 27 sample points in the range of [−3%, 3%], and there are five sample points in the scope of [−1%, 1%]; the maximum value is 4.24% at the tenth point, while the minimum is −5.82% at the 25th point; and (5) the BPNN model has nine points belonging to the range of [−3%, 3%], and there are only two points in the scope of [−1%, 1%]; the maximum value is 5.94% at the 45th point, while the minimum is −5.89% at the 23rd point. This further demonstrates that the nonlinear fitting ability of the proposed model is the strongest so that its prediction accuracy and robustness are both the most satisfactory.

The RMSE, MAPE and AAE of the four prediction models are shown in Figure 13. It can be concluded that the RMSE, MAPE, and AAE values of the FOA-IR-GRNN model are still the lowest, which are 1.2016%, 1.1534% and 1.1535%, respectively. It is proved that the proposed model can obtain the highest prediction accuracy and the best stability under different conditions. This model can eliminate the interference of redundant factors through feature selection, so as to ensure the accuracy and stability of prediction. This result is consistent with the results obtained in Section 4.3.

In summary, the proposed model optimizes the GRNN model with the FOA, and obtains the appropriate smoothing parameter in the GRNN model, which can effectively reduce the icing prediction error. The IR model can not only reduce the noise data of the input variables to improve the effectiveness of input information, but also ensure the integrity of the input information, thus improving the accuracy and robustness of icing prediction. The validity of the proposed ice prediction model is proved by the data calculation results.

6. Conclusions

This paper presents a hybrid icing forecasting model that combines IR with GRNN optimized by FOA. First, in order to predict the icing thickness, the IR combined with the FOA is employed to select the input feature. Furthermore, the FOA is adopted to optimize the smoothing factor of the GRNN. Finally, after obtaining the optimized feature subset and the best value of smoothing factor, the proposed model is utilized for icing forecasting. Several conclusions based on the studies can be obtained as follows: (1) by the utilization of IR, the influence of unrelated noises can be reduced and the forecasting performance can be effectively improved; (2) the optimization algorithm FOA adds strong global searching capability to the model, and the GRNN model optimized by FOA shows good performance; (3) based on the error valuation criteria, the FOA-IR-GRNN model is a more promising methodology in icing forecasting as compared with the three classical icing forecasting models (SVM, BPNN, and GRNN); and (4) according to the empirical analysis of two cases, it is found that the model proposed in this paper still has good prediction performance for forecasting the icing thickness of transmission lines at different times and places. Hence, the proposed icing forecasting method of the FOA-IR-GRNN model is effective and feasible, and it may be an effective alternative for icing forecasting in the electric-power industry.

Acknowledgments: This work is supported by the Natural Science Foundation of China (Project No. 71471059) and the Fundamental Research Funds for the Central Universities (Project No. 2017XS103).

Energies **2017**, *10*, 2066

Author Contributions: Haichao Wang designed this research and wrote this paper; Dongxiao Niu and Yi Liang provided professional guidance; Hanyu Chen translated and revised this paper.

Conflicts of Interest: The authors declare no conflict of interest.

References

1. Tian, H.; Liang, N.; Zhao, P.; Wang, X.; Zhu, C.; Zhang, S.; Wang, W. Model updating approach for icing forecast of transmission lines by using particle swarm optimization. *IOPscience* **2017**, *207*. [CrossRef]

2. Lamraoui, F.; Fortin, G.; Benoit, R.; Perron, J.; Masson, C. Atmospheric icing severity: Quantification and mapping. *Atmos. Res.* **2013**, *128*, 57–75. [CrossRef]

3. Zhang, W.L.; Yu, Y.Q.; Su, Z.Y.; Fan, J.B.; Li, P.; Yuan, D.L.; Wu, S.Y.; Song, G.; Deng, Z.F.; Zhao, D.L.; et al. Investigation and analysis of icing and snowing disaster happened in hunan power grid in 2008. *Power Syst. Technol.* **2008**, *32*, 1–5.

4. Zhuang, W.; Zhang, H.; Zhao, H.; Wu, S.; Wan, M. Review of the Research for Power line Icing Prediction. *Adv. Meteorol. Sci. Technol.* **2017**, *7*, 6–12.

5. Liang, X.; Li, Y.; Zhang, Y.; Liu, Y. Time-dependent simulation model of ice accretion on transmission line. *High Volt. Eng.* **2014**, *40*, 336–343.

6. Liu, C.C.; Liu, J. Ice accretion mechanism and glaze loads model on wires of power transmission lines. *High Volt. Eng.* **2011**, *37*, 241–248.

7. Imai, I. Studies on ice accretion. *Res. Snow Ice* **1953**, *3*, 35–44.

8. Goodwin, E.J.I.; Mozer, J.D.; Digioia, A.M.J.; Power, B.A. Predicting Ice and Snow Loads for Transmission Line Design. Available online: http://www.dtic.mil/docs/citations/ADP001696 (accessed on 11 August 2017).

9. Lenhand, R.W. An indirect method for estimating the weight of glaze on wires. *Bull. Am. Meteorol. Soc.* **1995**, *36*, 1–5.

10. Huang, X.B.; Li, H.B.; Zhu, Y.C.; Wang, Y.X.; Zheng, X.X.; Wang, Y.G. Transmission line icing short-term forecasting based on improved time series analysis by fireworks algorithm. In Proceedings of the International Conference on Condition Monitoring and Diagnosis, Xi'an, China, 25–28 September 2016.

11. Yang, J.L. Impact on the Extreme Value of Ice Thickness of Conductors from Probability Distribution Models. In Proceedings of the International Conference on Mechanical Engineering and Control Systems, Wuhan, China, 23–25 January 2015.

12. Liu, C.; Liu, H.W.; Wang, Y.S.; Lu, J.Z.; Xu, X.J.; Tan, Y.J. Research of icing thickness on transmission lines based on fuzzy Markov chain prediction. In Proceedings of the IEEE International Conference on Applied Superconductivity and Electromagnetic Devices (ASEMD), Beijing, China, 25–27 October 2013.

13. Chen, S.; Dong, D.; Huang, X.; Sun, M. Short-Term Prediction for Transmission Lines Icing Based on BP Neural Network. In Proceedings of the Asia-Pacific Power and Energy Engineering Conference, Shanghai, China, 27–29 March 2012.

14. Ma, T.; Niu, D.; Fu, M. Icing Forecasting for Power Transmission Lines Based on a Wavelet Support Vector Machine Optimized by a Quantum Fireworks Algorithm. *Appl. Sci.* **2016**, *6*, 54. [CrossRef]

15. Luo, Y.; Yao, Y.; Ying, L.I.; Wang, K.; Qiu, L. Study on Transmission Line Ice Accretion Mode Based on BP Neural Network. *J. Sichuan Univ. Sci. Eng.* **2012**, *25*, 63–66.

16. Li, P.; Li, Q.M.; Ren, W.P.; He, R.; Gong, Y.Y.; Li, Y. SVM-based prediction method for icing process of overhead power lines. *Int. J. Model. Identif. Control* **2015**, *23*, 362.

17. Ma, X.M.; Gao, J.; Wu, C.; He, R.; Gong, Y.Y.; Li, Y. A prediction model of ice thickness based on grey support vector machine. In Proceedings of the IEEE International Conference on High Voltage Engineering and Application, Chengdu, China, 19–22 September 2016.

18. Lee, G.E.; Zaknich, A. A mixed-integer programming approach to GRNN parameter estimation. *Inf. Sci.* **2015**, *320*, 1–11. [CrossRef]

19. Zhang, J.; Tan, Z.; Li, C. A Novel Hybrid Forecasting Method Using GRNN Combined with Wavelet Transform and a GARCH Model. *Energy Sources Part B* **2015**, *10*, 418–426. [CrossRef]

20. Zhao, H.; Guo, S. Annual Energy Consumption Forecasting Based on PSOCA-GRNN Model. *Abstr. Appl. Anal.* **2014**, *2014*, 217630. [CrossRef]

21. Leng, Z.; Gao, J.; Qin, Y.; Liu, X.; Yin, J. Short-term forecasting model of traffic flow based on GRNN. In Proceedings of the 25th Chinese Control and Decision Conference (CCDC), Guiyang, China, 25–27 May 2013.

22. Gao, Y.; Zhang, R. Analysis of House Price Prediction Based on Genetic Algorithm and BP Neural Network. *Comput. Eng.* **2014**, *40*, 187–191.

23. Yang, X.Y.; Guan, W.Y.; Liu, Y.Q.; Xiao, Y.Q. Prediction Intervals Forecasts of Wind Power based on PSO-KELM. *Proc. CSEE* **2015**, *35*, 146–153.

24. Wang, L.; Zheng, X.L.; Wang, S.Y. A novel binary fruit fly optimization algorithm for solving the multidimensional knapsack problem. *Knowl.-Based Syst.* **2013**, *48*, 17–23. [CrossRef]

25. Wang, L.; Liu, R.; Liu, S. An effective and efficient fruit fly optimization algorithm with level probability policy and its applications. *Knowl.-Based Syst.* **2016**, *97*, 158–174. [CrossRef]

26. Sun, W.; Liang, Y. Least-Squares Support Vector Machine Based on Improved Imperialist Competitive Algorithm in a Short-Term Load Forecasting Model. *J. Energy Eng.* **2014**, *141*. [CrossRef]

27. Li, H.; Guo, S.; Zhao, H.; Su, C.; Wang, B. Annual Electric Load Forecasting by a Least Squares Support Vector Machine with a Fruit Fly Optimization Algorithm. *Energies* **2012**, *5*, 4430–4445. [CrossRef]

28. Liu, D.; Wang, J.; Wang, H. Short-term wind speed forecasting based on spectral clustering and optimised echo state networks. *Renew. Energy* **2015**, *78*, 599–608. [CrossRef]

29. Chen, T.; Ma, J.; Huang, S.H.; Cai, A. Novel and efficient method on feature selection and data classification. *J. Comput. Res. Dev.* **2012**, *49*, 735–745.

30. Ma, T.; Niu, D.; Huang, Y.; Du, Z. Short-Term Load Forecasting for Distributed Energy System Based on Spark Platform and Multi-Variable L2-Boosting Regression Model. *Power Syst. Technol.* **2016**, *40*, 1642–1649.

31. Liu, J.P.; Li, C.L. The Short-Term Power Load Forecasting Based on Sperm Whale Algorithm and Wavelet Least Square Support Vector Machine with DWT-IR for Feature Selection. *Sustainability* **2017**, *9*, 1188. [CrossRef]

32. Jose, V.R.R. Percentage and Relative Error Measures in Forecast Evaluation. *Oper. Res.* **2017**, *65*, 200–211. [CrossRef]

energies

MDPI

Article

Applications of the Chaotic Quantum Genetic Algorithm with Support Vector Regression in Load Forecasting

Cheng-Wen Lee [1] and Bing-Yi Lin [2,*]

[1] Department of International Business, Chung Yuan Christian University, 200 Chung Pei Rd., Chungli District, Taoyuan City 32023, Taiwan; chengwen@cycu.edu.tw

[2] Ph.D. Program in Business, College of Business, Chung Yuan Christian University, 200 Chung Pei Rd., Chungli District, Taoyuan City 32023, Taiwan

* Correspondence: g10304612@cycu.edu.tw; Tel.: +886-2-2596-8821; Fax: +886-2-2552-8844

Received: 20 October 2017; Accepted: 8 November 2017; Published: 10 November 2017

Abstract: Accurate electricity forecasting is still the critical issue in many energy management fields. The applications of hybrid novel algorithms with support vector regression (SVR) models to overcome the premature convergence problem and improve forecasting accuracy levels also deserve to be widely explored. This paper applies chaotic function and quantum computing concepts to address the embedded drawbacks including crossover and mutation operations of genetic algorithms. Then, this paper proposes a novel electricity load forecasting model by hybridizing chaotic function and quantum computing with GA in an SVR model (named SVRCQGA) to achieve more satisfactory forecasting accuracy levels. Experimental examples demonstrate that the proposed SVRCQGA model is superior to other competitive models.

Keywords: chaotic mapping function; support vector regression (SVR); quantum genetic algorithm (QGA); electricity demand forecasting

---

1. Introduction

With rapid economic development, accurate electricity load forecasting has become essential for many energy applications, such as energy generation, power system operation security, load unit commitment, and energy marketing. For example, power system decision makers can optimize load dispatch and adjust the electricity supply/price based on the forecasted loads, i.e., improve the power system management efficiency. As indicated in Xiao et al. [1], in China, there would be a year-long operational benefit with a 1% increase in the forecasting accuracy level. In addition, accurate load forecasting could also help managers set up well electrical power scheduling and successfully reduce system management risks. On the customer side, accurate load forecasting also facilitates the power usage decisions of customers to avoid load usage during the peak times and paying higher electricity prices. This usage balance between peak and bottom periods would lead to reliable power system operation of a utility. On the contrary, inaccurate forecast results would lead to inefficient power system operations and increased operating costs. As mentioned in the literature, a 1% increase in load forecasting error can lead to a loss of millions of dollars [2]. Therefore, as electricity prices also play a critical role in electricity production decisions, there are also several scholars who have proposed electricity price forecasting models in the literature [3,4]. Readers may refer to Weron [5] for more comprehensive overviews.

The electricity load data are influenced by lots of factors, such as socio-economical activities, population, weather conditions, holidays, policy, and so on [6]. Therefore, the electric load data reveal nonlinearity, seasonality, and chaos in nature, so finding a robust load forecasting model with superior performance would be an important issue in the power load management field.

Researchers have developed and proposed lots of electricity load forecasting models. These forecasting models are often classified into two categories: traditional statistical models and artificial intelligence models. The first one are also called stochastic time series approach models, i.e., only historical data is used, which is easily to apply. These various famous time series models include the well-known Box–Jenkins' ARIMA models [7], regression models [8], exponential smoothing models [9], Kalman filtering models [10], Bayesian estimation models [11], and so on. However, the embedded drawbacks of those models are that they are defined theoretically to deal with linear relationships among electricity load and other stochastic factors such as socio-economical activities and policy effects, thus, they have difficulties to effectively capture the complicate nonlinear relationships among load data and these factors, eventually, producing high unpredictable load forecasting performance errors [12].

The artificial intelligence models such as artificial neural networks (ANNs) [12], expert system models [13], and fuzzy methodologies [14] have been well explored to improve the accuracy of load forecasting since the 1980s. In recent years, the development of artificial intelligence approaches has focused on novel hybrid or combined models, obtained by hybridizing or combining these models with each other [15], with traditional statistical tools [16], and with superior evolutionary algorithms [17]. However, similarly, these artificial intelligence models also suffer from some shortcomings during the modeling processes, such as the fact they are very dependent on the collected data, and often are unstable. Thus, it is difficult to determine the network structural parameters [18]. It is also time consuming to extract knowledge from data sets [19], and they are easily trapped in local minima [20], for more insightful discussions of AI approaches in load forecasting readers may refer to [21].

Due to the superiority in modeling nonlinear data by mapping into the high dimensional feature space, support vector regression (SVR) [22] has been applied to solve forecasting issues many research fields in the late 1990s. For load forecasting problems, Hong [23,24] proposed a valuable series exploration by integrating advanced algorithms and chaotic function with an SVR-based model to determine its three parameters, and thus achieved satisfactory forecasting performance. According to Hong's series research conclusions, good determination of parameters for the SVR model is important to achieve high forecasting accuracy levels and overcome the drawbacks of the hybrid evolutionary algorithms, such as becoming trapped in local optima, and this will ensure achieving more suitable parameter combinations. In the meanwhile, Bhunia [25] indicated that quantum computing principles can be embedded in intelligent systems to improve their performance; moreover, Dey et al. [26] also concluded that the use of both of quantum approaches and soft computing techniques in a combined form can provide a new computer science and engineering paradigm. Huang [27] proposed a novel forecasting model by hybridizing a chaotic function and a quantum PSO algorithm to receive higher forecast accuracy levels. Recently, Lee and Lin [28] also applied quantum concepts to propose the hybrid tabu search algorithm with the SVR model to adjust the three parameters and eventually obtain more accurate load forecasting performances.

The genetic algorithm (GA) is a famous algorithm which generates new offspring by finite iterative operations, including selection, crossover, mutation, and so on. It has attracted lots of attention to find satisfactory solutions and is applied in many fields. However, along with the increase of the data scale and more complicated problem, it often suffers from similar problem of becoming trapped in local optima and slow convergence to the global optimum. Dey et al. [26] claimed that an efficient quantum-based GA can be modeled to solve NP-hard problems and others. To continue exploring the feasibility of hybrid quantum-behaved approaches with advanced algorithms, and to overcome the embedded drawbacks of genetic algorithm mentioned above, this paper would like to apply quantum computing concepts to propose hybridizing chaotic function and quantum GA (namely CQGA) with the SVR model, creating the so-called SVRCQGA model to achieve more satisfactory load forecasting accuracy levels, by comparing the forecasts with other alternative models proposed in Huang [27] and

Lee and Lin [28]. The main innovative contribution of this paper is hybridizing the chaotic mapping function and quantum computing technique with GA into a SVR model, to improve the problems as mentioned above, and thus achieve improved forecasting accuracy levels.

The remainder of this paper is organized as follows: the implementation details of the proposed SVRCQGA model are demonstrated in Section 2. Brief illustrations of the SVR model and the proposed CQGA are also clearly addressed. Section 3 demonstrates an experimental example and provides a statistical comparison among other benchmarking models proposed in existing papers. Conclusions are provided in Section 4.

2. The Proposed SVRCQGA Model

2.1. Brief Description of the SVR Model

The principal modeling processes of the SVR model are briefly summarized as follows: the training data set, $\{(x_i, y_i)\}_{i=1}^{N}$, is mapped to a feature space, $\Re^{n_h}$, by the defined function, $\varphi(x) : \Re^n \to \Re^{n_h}$. The SVR function, f, is employed to linearly formulate the relationship between feature values (i.e., training data, x_i) and forecast values (y_i), and it is shown as Equation (1):

$$f(x) = w^T \varphi(x) + b \tag{1}$$

where, $f(x)$ is the forecasted values; the weight, w ($w \in \Re^{n_h}$) and coefficient, b ($b \in \Re$), could be determined during the minimization process of the empirical risk function, Equation (2):

$$R(f) = C \frac{1}{N} \sum_{i=1}^{N} \mathcal{L}_\varepsilon(y_i, w^T \varphi(x_i) + b) + \frac{1}{2} w^T w \tag{2}$$

$$\mathcal{L}_\varepsilon(y_i, f(x)) = \begin{cases} 0, & if |f(x) - y_i| \le \varepsilon \\ |f(x) - y_i| - \varepsilon, & otherwise \end{cases} \tag{3}$$

where, $\mathcal{L}_\varepsilon(y_i, f(x))$ represents the main empirical risk, it is also the so-called ε-insensitive loss function; C and ε are the essential parameters. When the forecasting error is smaller than ε, the loss would be zero (refer to Equation (3)). The second term, $\frac{1}{2} w^T w$, is the weight of the SVR function as mentioned, it determines the steepness. Therefore, C represents a trade-off role to balance the empirical risk and the steepness. For quadratic programming, two slack variables, ξ and ξ^*, are introduced to measure the length between the actual values and the edge values of ε-tube. Then, Equation (2) could be transformed to the standard programming form with constraints, as shown in Equation (4):

$$\text{Min } R(w, \xi, \xi^*) = \frac{1}{2} w^T w + C \sum_{i=1}^{N} (\xi_i + \xi_i^*)$$
$$y_i - w^T \varphi(x_i) - b \le \varepsilon + \xi_i^*,$$
$$-y_i + w^T \varphi(x_i) + b \le \varepsilon + \xi_i, \tag{4}$$
$$\xi_i^* \ge 0,$$
$$\xi_i \ge 0,$$
$$i = 1, 2, \dots, N$$

The solution weight vector, w, in the quadratic programming problem (Equation (4) is optimized by using the Lagrange multipliers method, as shown in Equation (5):

$$w^* = \sum_{i=1}^{N} (\gamma_i^* - \gamma_i) \varphi(x_i) \tag{5}$$

where γ_i^* and γ_i are the Lagrangian multipliers and satisfy the equality $\gamma_i^* \times \gamma_i = 0$. Eventually, the SVR function is formulated as Equation (6):

$$f(x) = \sum_{i=1}^{N} (\gamma_i^* - \gamma_i) K(x_i, x_j) + b \tag{6}$$

where, $K(\mathbf{x}_i, \mathbf{x}_j)$ is the so-called kernel function, its value could be calculated by the inner product of $\varphi(\mathbf{x}_i)$ and $\varphi(\mathbf{x}_j)$, i.e., $K(\mathbf{x}_i, \mathbf{x}_j) = \varphi(\mathbf{x}_i) \cdot \varphi(\mathbf{x}_j)$. There are several kinds of kernel function, the most widely used kernel function is Gaussian function, $K(\mathbf{x}_i, \mathbf{x}_j) = \exp\left(-0.5\|\mathbf{x}_i - \mathbf{x}_j\|^2 / \sigma^2\right)$, due to its excellence in complex nonlinear relationships mapping capability. Therefore, this paper employs a Gaussian function as the kernel function.

The most important job for improving the performance of an SVR model is adjusting well the parameter values, i.e., the three parameters, C, ε, and σ. However, there are no structural methods to efficiently set up the SVR parameters. This paper will continue exploring the feasibility of a chaotic quantum-behaved approach to overcome the disadvantages of genetic algorithms, namely CQGA; and, hybridizing CQGA with the SVR model, producing the SVRCQGA model, to determine the three parameters to improve the forecasting accuracy level.

2.2. Chaotic Quantum Genetic Algorithm (CQGA)

2.2.1. Introduction of QGA

GA generates new individuals by its advanced operations, including selection, crossover, and mutation operations. Particularly, the mutation operation is effective for making individuals have more satisfactory fitness values, and plays a critical role in maintaining the evolution quality for the population. Therefore, it has been applied to deal with many optimization problems. However, the population diversity would be reduced after repeated iterative computations and this leads to several major drawbacks, such as being time consuming, slow convergence, and becoming trapped in local optima.

Recently, quantum computing techniques have been hybridized with genetic algorithms, i.e., QGA [29]. By applying the main computing techniques of quantum computing, including qubit, quantum superposition, and quantum entanglement, the chromosomes in QGA have been presented by qubit coding. In addition, quantum rotation gate operation for the chromosomes is employed during the whole evolutionary process. Therefore, it has lots of superior advantages during searching, such as speedy convergence, time saving, little population scale, and robustness. The applications of QGA also receive attentions in recent years, including traveling salesmen problems, personal scheduling problems, and dynamic economic dispatch problems, as well as improvements [30]. For more application details of QGA, readers should refer to Lahoz-Beltra [31].

2.2.2. Quantum Computing Concepts

The quantum computing concepts are briefly described as follows: a quantum bit, abbreviated as qubit, is defined as the smallest information unit. In the quantum system, a qubit may be in the state "0", in the state "1", or in any superposition of these two states. The state of a qubit can be shown as Equation (7):

$$|\psi\rangle = \alpha_1 |0\rangle + \alpha_2 |1\rangle \tag{7}$$

where $|0\rangle$ and $|1\rangle$ are the values of traditional bits 0 and 1, respectively; α_1 and α_2 are the probability of their associate states and meet the normalization condition, as illustrated in Equation (8):

$$|\alpha_1|^2 + |\alpha_2|^2 = 1 \tag{8}$$

where $|\alpha_1|^2$ is the probability that the qubit is in "0" state, and $|\alpha_2|^2$ is the probability that the qubit is in "1" state. For generalization, if a system has n qubits and totally 2^n states, then, the linear superposition of all states can be presented as shown in Equation (9):

$$|\psi_i\rangle = \sum_{k=1}^{2^n} p_k |S_k\rangle \tag{9}$$

where p_k is the probability of its associate state, S_k, and meets the normalization condition, $|p_1|^2 + |p_2|^2 + \ldots + |p_{2^n}|^2 = 1$.

The probability of a qubit individual as a string with n qubits is presented as Equation (10):

$$q = \left[\begin{array}{c|c|c|c|c} \alpha_1 & \alpha_2 & \cdots & \alpha_i & \cdots & \alpha_n \\ \beta_1 & \beta_2 & \cdots & \beta_i & \cdots & \beta_n \end{array} \right] \tag{10}$$

where $|\alpha_i|^2 + |\beta_i|^2 = 1$, $i = 1, 2, \ldots, n$.

Therefore, in QGA, the chromosome, with n qubits, could be presented as, $\mathbf{P} = (q_1, q_2, \ldots, q_n)$, where q_j ($j = 1, 2, \ldots, n$) is an individual qubit of population as shown in Equation (10).

The quantum gate is an operator for qubits to implement unitary transformations, in which, the operation is represented by matrices. The basic quantum gates with a single qubit are the identity gate $\mathbf{I}$ and Pauli gates $\mathbf{X}$, $\mathbf{Y}$, and $\mathbf{Z}$, as shown in Equation (11):

$$\mathbf{I} = \left[\begin{array}{cc} 1 & 0 \\ 0 & 1 \end{array} \right] ; \mathbf{X} = \left[\begin{array}{cc} 0 & 1 \\ 1 & 0 \end{array} \right] ; \mathbf{Y} = \left[\begin{array}{cc} 0 & -i \\ i & 0 \end{array} \right] ; \mathbf{Z} = \left[\begin{array}{cc} 1 & 0 \\ 0 & -1 \end{array} \right] \tag{11}$$

The identity gate $\mathbf{I}$ keeps a qubit unchanged, i.e., $\mathbf{I} \cdot |0\rangle = |0\rangle$ and $\mathbf{I} \cdot |1\rangle = |1\rangle$ (Equation (12)); Pauli $\mathbf{X}$ gate performs a Boolean NOT operation, i.e., $\mathbf{X} \cdot |0\rangle = |1\rangle$ and $\mathbf{X} \cdot |1\rangle = |0\rangle$ (Equation (13)); Pauli $\mathbf{Y}$ gate maps $|0\rangle \rightarrow i \cdot |1\rangle$ and $|1\rangle \rightarrow -i \cdot |0\rangle$ (Equation (14)); and Pauli $\mathbf{Z}$ gate changes the phase of a qubit, i.e., $|0\rangle \rightarrow |0\rangle$ and $|1\rangle \rightarrow -1 \cdot |1\rangle$ (Equation (15)):

$$\mathbf{I} \cdot |0\rangle = \left[\begin{array}{cc} 1 & 0 \\ 0 & 1 \end{array} \right] \left[\begin{array}{c} 1 \\ 0 \end{array} \right] = \left[\begin{array}{c} 1 \\ 0 \end{array} \right] = |0\rangle ; \mathbf{I} \cdot |1\rangle = \left[\begin{array}{cc} 1 & 0 \\ 0 & 1 \end{array} \right] \left[\begin{array}{c} 0 \\ 1 \end{array} \right] = \left[\begin{array}{c} 0 \\ 1 \end{array} \right] = |1\rangle \tag{12}$$

$$\mathbf{X} \cdot |0\rangle = \left[\begin{array}{cc} 0 & 1 \\ 1 & 0 \end{array} \right] \left[\begin{array}{c} 1 \\ 0 \end{array} \right] = \left[\begin{array}{c} 0 \\ 1 \end{array} \right] = |1\rangle ; \mathbf{X} \cdot |1\rangle = \left[\begin{array}{cc} 0 & 1 \\ 1 & 0 \end{array} \right] \left[\begin{array}{c} 0 \\ 1 \end{array} \right] = \left[\begin{array}{c} 1 \\ 0 \end{array} \right] = |0\rangle \tag{13}$$

$$\mathbf{Y} \cdot |0\rangle = \left[\begin{array}{cc} 0 & -i \\ i & 0 \end{array} \right] \left[\begin{array}{c} 1 \\ 0 \end{array} \right] = \left[\begin{array}{c} 0 \\ i \end{array} \right] = i \cdot |1\rangle ; \mathbf{Y} \cdot |1\rangle = \left[\begin{array}{cc} 0 & -i \\ i & 0 \end{array} \right] \left[\begin{array}{c} 0 \\ 1 \end{array} \right] = \left[\begin{array}{c} -i \\ 0 \end{array} \right] = -i \cdot |0\rangle \tag{14}$$

$$\mathbf{Z} \cdot |0\rangle = \left[\begin{array}{cc} 1 & 0 \\ 0 & -1 \end{array} \right] \left[\begin{array}{c} 1 \\ 0 \end{array} \right] = \left[\begin{array}{c} 1 \\ 0 \end{array} \right] = |0\rangle ; \mathbf{Z} \cdot |1\rangle = \left[\begin{array}{cc} 1 & 0 \\ 0 & -1 \end{array} \right] \left[\begin{array}{c} 0 \\ 1 \end{array} \right] = \left[\begin{array}{c} 0 \\ -1 \end{array} \right] = -1 \cdot |1\rangle \tag{15}$$

To obtain more results, it is feasible to use the trigonometric function with a phase angle θ, i.e., the so-called quantum rotation gate. The quantum rotation gate (cf. Equation (16)), is employed to update as the better solution in its current state:

$$\mathbf{P}' = \left[\begin{array}{cc} \cos\theta & -\sin\theta \\ \sin\theta & \cos\theta \end{array} \right] \mathbf{P} \tag{16}$$

where $\mathbf{P}'$ is the updated chromosome; θ is the designate angle to be used in the quantum rotation gate.

2.2.3. Implementation Steps of CQGA

The outstanding property of QGA is using quantum mechanics, such as qubits and their state superposition as mentioned above to represent the chromosomes (instead of traditional binary strings). The chromosome is represented as the superposition of all possible states. In the meanwhile, to keep the diversity of the population to avoid premature convergence is also an important issue. Chaos has two advantages: (1) it is sensitive to the initial conditions, i.e., minute changes in initial conditions steer subsequent simulations towards radically different final states; and (2) any variable in the chaotic space can travel ergodically over the whole space of interest, i.e., the so-called ergodicity property. Therefore, employing chaotic sequences to keep the diversity of population in the whole optimization procedures, will lead to very different future solution-finding behaviors, due to the ergodicity property. Eventually, chaotic sequences can help to enrich the search behavior and to avoid premature convergence. Considering the above mentioned statements, this paper also applies the chaotic variable to be hybridized with QGA (namely CQGA) to prevent the premature convergence problem. Furthermore, for the better chaotic distribution characteristics of cat function, it is used to generate the chaotic sequence. The two-dimensional cat function [32] is commonly used and is employed in this paper, as shown in Equation (17):

$$
\begin{cases}
x_{n+1} = frac(x_n + y_n) \\
y_{n+1} = frac(x_n + 2y_n)
\end{cases}
\tag{17}
$$

where *frac* function is used to keep the decimal parts of a real number x by reducing an approximate integer. The complete processes of the proposed CQGA model is demonstrated in what follows and a brief flowchart is shown in Figure 1.

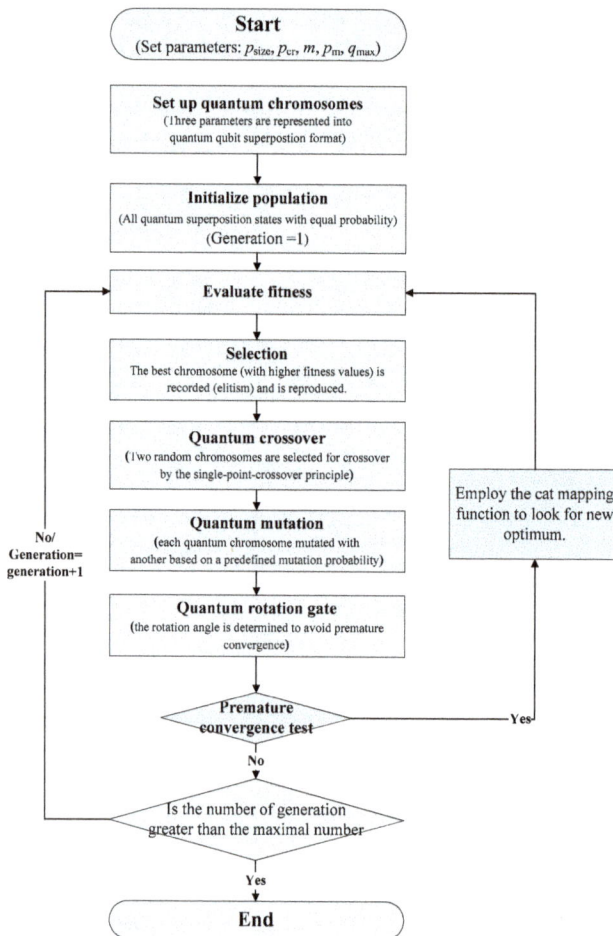

Figure 1. Quantum genetic algorithm flowchart.

Step 1. Set up quantum chromosomes. In this paper, the quantum chromosome is composed of a string of *m* qubits (superposition of all possible states), as shown in Figure 2. These SVR's three parameters, *C*, *ε*, and *σ*, are presented into the quantum qubit superposition format, i.e., each chromosome has three genes to represent it. Based on the authors' practical trials and experience, choosing a gene with 40 bits could produce more satisfactory results, thus, a chromosome in total contains 120 qubits (i.e., *m* = 120). A gene that contains more qubits would be associated with better partitioning around the space.

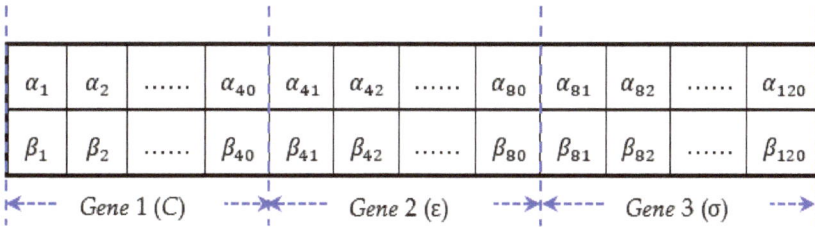

α_1	α_2	......	α_{40}	α_{41}	α_{42}	......	α_{80}	α_{81}	α_{82}	......	α_{120}
β_1	β_2	......	β_{40}	β_{41}	β_{42}	......	β_{80}	β_{81}	β_{82}	......	β_{120}

$\longleftarrow$ Gene 1 (*C*) $\longrightarrow\!\!\longleftarrow$ Gene 2 (*ε*) $\longrightarrow\!\!\longleftarrow$ Gene 3 (*σ*) $\longrightarrow$

Figure 2. Quantum chromosome structure for three parameters.

Step 2. Initialize population. The population of the quantum chromosome is initialized by setting all the amplitudes of qubits as $\frac{1}{\sqrt{2}}$ [30], i.e., all superposition states has equal probability in the initial population.

Step 3. Evaluate fitness (forecasting errors). Evaluate the objective fitness (forecasting errors) by using the values of each quantum chromosome. The mean absolute percentage error (MAPE), illustrated in Equation (18), is employed to measure the forecasting errors:

$$MAPE = \frac{1}{N} \sum_{i=1}^{N} \left| \frac{y_i - f_i}{y_i} \right| \times 100\% \tag{18}$$

where *N* is the total number of forecasting results; y_i is the actual value at each forecasting point *i*; f_i is the forecasted value at each forecasting point *i*.

Step 4. Selection. In each generation, an elitist selection mechanism is used to select the best chromosome (with smallest MAPE value), i.e., the competition strategy is applied and as mentioned the best chromosome with the smallest MAPE value is recorded as the elitist and is reproduced as the initial chromosome for the next generation.

Step 5. Quantum crossover. To keep the population diversity, a quantum crossover operation is employed. Based on predefined crossover probability, P_{cr} (set as 0.9 [26]), the single-point-crossover principle is applied to randomly select two chromosomes to conduct crossover operation at any random position. For each generation, a new chromosomes pool would be generated after the quantum crossover operation is finished. Figure 3 illustrates the processed results of the quantum crossover operation.

Step 6. Quantum mutation. This is a useful approach to ensure population diversity. In this operation, each selected position of the participated quantum chromosome would be mutated with other real numbers according to the designate mutation probability, P_m (set as 0.1 [26]). Figure 4 shows an example of the quantum mutation operation.

before crossover

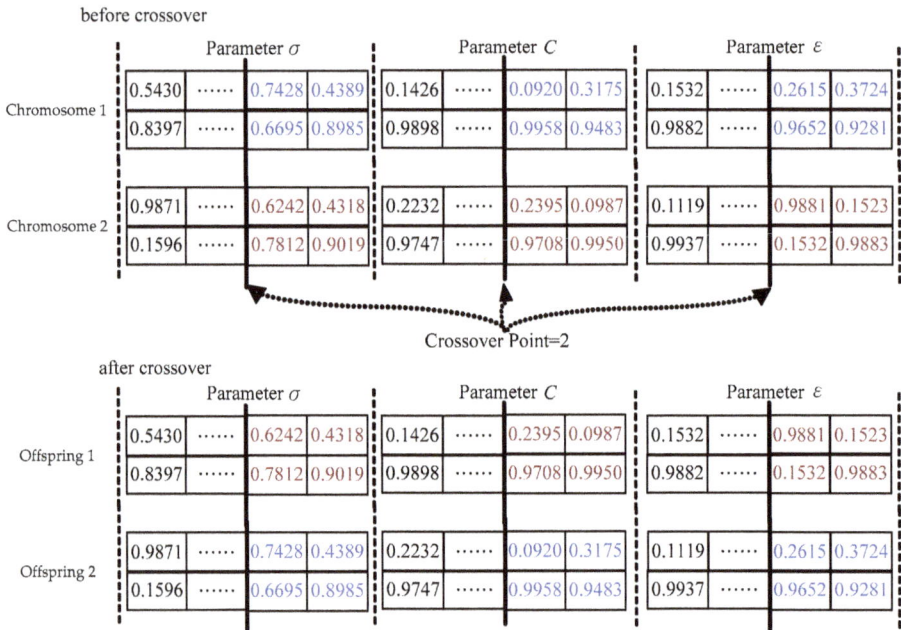

Figure 3. Example of the chromosome form of the parameters for the quantum crossover operation.

before mutation

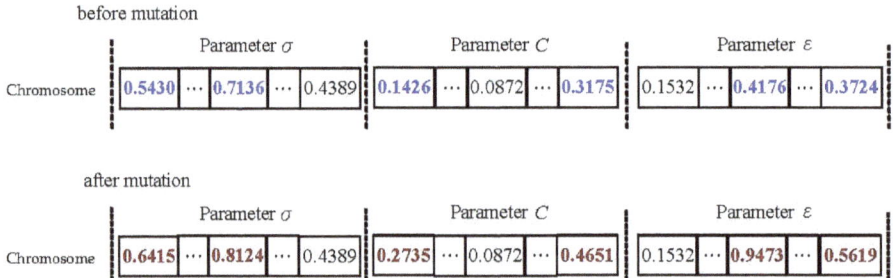

Note: blue color numbers are selected for mutation; brown color numbers are mutated.

Figure 4. Example of the chromosome form of the parameter for the quantum mutation operation.

Step 7. Quantum rotation gate. This operation modifies the oscillation ranges of individuals to improve the performance by changing the state of each qubit. It is performed by using a quantum rotation gate (as shown in Equation (16)), in which the rotation angle θ is a function of the oscillation amplitudes (α_i, β_i), and the value of the individual qubit located at the position i would also be modified accordingly [33]. The rotation angle θ is updated by Equation (19):

$$\theta = 0.005\pi + (0.1\pi - 0.005\pi)\frac{|f_i - f_{avg}|}{\max\{f_i, f_{avg}\}} \tag{19}$$

where f_i is the current forecasting error; f_{avg} is the average value of all previous forecasting errors. Based on quantum genetic algorithm performance, a general criterion to set θ values between 0.1π and 0.005π [31].

Step 8. Premature convergence test. Compute the mean square error (MSE), given by Equation (20), to test the level of premature convergence [34], and set up the criterion value, δ:

$$\text{MSE} = \frac{1}{S}\sum_{i=1}^{S}\left(\frac{f_i - f_{avg}}{f}\right)^2 \tag{20}$$

where f is given by Equation (21):

$$f = \max\left\{1, \max_{\forall i \in S}\{|f_i - f_{avg}|\}\right\} \tag{21}$$

If the value of the calculated MSE is less than δ, it implies premature convergence occurs. Hence, the chaotic cat function (Equation (17)) is employed to escape the local optimum, i.e., finding out new optimum, and set the new optimum as the best solution.

Step 9. Stop criteria. If the number of generations is greater than a given scale, then, the best solution could be the presented quantum chromosomes; otherwise, go back to *Step 3* and continue searching the next generation.

3. Experimental Examples

3.1. Data Sets of Experimental Examples

To compare the performances from the hybrid quantum-behaved evolutionary algorithms with an SVR model, this paper employs the same experimental examples used in Huang [27] and Lee and Lin [28]. These three experimental examples are: (1) the regional electricity load data in Taiwan from a published paper [23]; (2) the annual electricity load data in Taiwan from a published paper [23]; and (3) the electricity load data per hour from the 2014 Global Energy Forecasting Competition [35]. The data setting details for each examples are summarized in the following. The data characteristics of these three examples are summarized in Table 1.

Table 1. Data characteristics summary of three examples.

Examples	Data Type	Data Length	Data Size	Data Characteristics
Example 1	Regional and annual	From 1981 to 2000	4 regions and 20 years	Increment with fluctuation caused by some accidental event (921 earthquake)
Example 2	Annual	From 1945 to 2003	59 years	Increment with the continuous economic development in Taiwan
Example 3	Hourly	From 1 December 2011 to 1 January 2012	744 h	Cyclic fluctuation

3.1.1. Regional Electricity Load Data in Taiwan: Example 1

For Example 1, there are in total 20 years of regional electricity load values (from 1981 to 2000) for four regions in Taiwan. Based on the same forecasting performance comparison conditions, the modeling sub-data set division is the same as in a previous paper [23]. Thus, three subsets are obtained: a training subset (12 years of load data in total, from 1981 to 1992), a validation subset (a total of 4 years of data, from 1993 to 1996), and a testing subset (a total of 4 years of data, from 1997 to 2000). The well-known window-rolling procedure is employed during the whole process including the electricity load forecasts produced. For details of the window-rolling forecasting procedure readers should refer to Hong [23] and Lee and Lin [28]. Three parameters are determined by CQGA, while the validation error is also calculated. The most suitable parameters are finalized only when the smallest validation errors occur. Finally, the four-step (year) load forecasting for each region in Taiwan is implemented by the proposed SVRCQGA model.

3.1.2. Annual Electricity Load Data in Taiwan: Example 2

For Example 2, there are in total 59 annual electricity load values (from 1945 to 2003). Similarly, to make sure the same forecasting performance comparison conditions are used, the modeling sub-dataset division is the same as in a previous paper [23], i.e., a training subset (40 years of data, from 1945 to 1984), a validation subset (10 years of data, from 1985 to 1994), and a testing subset (9 years of data, from 1995 to 2003). The modeling processing details are almost as the same as in Example 1: the window-rolling procedure is applied, then, three parameters are also selected by CQGA. The most suitable parameters are finalized only based on the smallest validation errors. Eventually, the one-step (year) load forecasting in Taiwan is implemented using the proposed model.

3.1.3. 2014 Global Energy Forecasting Competition (GEFCOM 2014) Electricity Load Data: Example 3

For Example 3, there are a total of 744 h of electricity load data (from 00:00 1 December 2011 to 00:00 1 January 2012). Similarly, to be based on the same forecasting performance comparison conditions, the modeling sub-data set division is the same as in a previous paper [27,28]. Thus, we have a training subset (552 h of load data, from 01:00 1 December 2011 to 00:00 24 December 2011), a validation subset (96 h of load data, from 01:00 24 December 2011 to 00:00 28 December 2011), and a testing subset (96 h of load data, from 01:00 28 December 2011 to 00:00 1 January 2012). The modeling processing details are almost as the same as in the two previous examples: a window-rolling procedure is still used, and the most suitable three parameters must be finalized based only on the smallest validation errors. Finally, the one-step (hour) load forecasting results are obtained using the proposed model.

3.2. Parameters Setting & Forecasting Results and Analysis

3.2.1. Setting the CQGA Parameters

The parameters of CQGA for the three experimental examples are set practically: the population scale (P_{scale}) is set to be 200; the generations of the population (q_{max}) are no larger than 500; the qubit string length of a quantum chromosome (m) is set as 120; the probabilities of quantum crossover (P_{cr}) and quantum mutation (P_m) are set as 0.5 and 0.1 [26], respectively. Some controlled parameters during the modeling procedure are set as follows: the maximal iteration for each example is all set as 10,000 in each generation; $\sigma \in [0, 5]$, $\varepsilon \in [0, 100]$ in all examples, $C \in [0, 20,000]$ in Example 1, $C \in [0, 3 \times 10^{10}]$ in Examples 2 and 3; δ is fixed as 0.001.

3.2.2. Forecasting Accuracy Indexes

To comprehensively compare the forecasting accuracy for each models, the mean absolute percentage error (MAPE; as shown in Equation (18)), the root mean squared error (RMSE; as shown in Equation (22)), and the mean absolute error (MAE; as shown in Equation (23)) are employed:

$$RMSE = \sqrt{\frac{\sum_{i=1}^{N} (y_i - f_i)^2}{N}} \tag{22}$$

$$MAE = \frac{\sum_{i=1}^{N} |y_i - f_i|}{N} \tag{23}$$

where N is the total number of forecasting results; y_i is the actual value at each forecasting point i; f_i is the forecasted value at each forecasting point i.

3.2.3. Forecasting Performance Superiority Tests

To ensure the forecasting superiority of the proposed model is statistically significant, it is necessary to conduct some statistical tests to verify the significance of the proposed model. Based on

Diebold and Mariano's [36] and Derrac et al.'s [37] suggestions, two tests are conducted in this paper, they are Wilcoxon signed-rank test [38] and Friedman test [39].

(A) Wilcoxon Signed-rank Test

The Wilcoxon signed-rank test is used to detect the significance of a difference in the central tendency of two data series when the size of the two data series is equal. The statistic W is represented as Equation (24):

$$W = \min\{S^+, S^-\} \tag{24}$$

where:

$$S^+ = \sum_{i=1}^{N} I^+(d_i) \tag{25}$$

$$S^- = \sum_{i=1}^{N} I^-(d_i) \tag{26}$$

$$I^+(d_i) = \begin{cases} 1 & if\ d_i > 0 \\ 0 & otherwise \end{cases} \tag{27}$$

$$I^-(d_i) = \begin{cases} 1 & if\ d_i < 0 \\ 0 & otherwise \end{cases} \tag{28}$$

$$d_i = (\text{forecasting series I})_i - (\text{forecasting series II})_i \tag{29}$$

where N is the total number of forecasting results.

(B) Friedman test

The Friedman test is used to measure the ANOVA in nonparametric statistical procedures; thus, it is a multiple comparisons test that aims to detect significant differences between the behaviors of two or more algorithms. The statistic F is represented as Equation (30):

$$F = \frac{12N}{k(k+1)} \left[\sum_{j=1}^{k} R_j^2 - \frac{k(k+1)^2}{4} \right] \tag{30}$$

where N is the total number of forecasting results; k is the number of compared models; R_j is the average rank sum obtained in each forecasting value for each algorithm as shown in Equation (31):

$$R_j = \frac{1}{N} \sum_{i=1}^{N} r_i^j \tag{31}$$

where r_i^j is the rank sum from 1 (the smallest forecasting error) to k (the worst forecasting error) for ith forecasting result, for jth compared model.

The null hypothesis for Friedman's test is that equality of forecasting errors among compared models. The alternative hypothesis is defined as the negation of the null hypothesis.

3.2.4. Results and Analysis: Example 1

For Example 1, SVR's three parameter values determine the most suitable model for each region, which are computed by the QGA algorithm and CQGA algorithm, respectively, and with the smallest testing error (MAPE value). These determined parameters for each region are illustrated in Table 2.

For forecasting results comparison details, Table 3 demonstrates the forecasting accuracy indexes of the proposed SVRCQGA and other competitive models [27,28] for each region. Figure 5 illustrates the cumulative differences of MAE for each competitive models in four regions. The competitive models include SVRCQPSO (hybrid SVR with chaotic quantum PSO) [27], SVRQPSO (hybrid SVR with quantum PSO) [27], SVRCQTS (hybrid SVR with chaotic quantum tabu search) [28], and SVRQTS (hybrid SVR with quantum tabu search) [28] models.

Table 2. Determined parameters of SVRCQGA and SVRQGA models (Example 1).

Regions	SVRCQGA Parameters			MAPE of Testing (%)
	σ	C	ε	
Northern	4.0000	0.6×10^{10}	0.6500	1.0760
Central	6.0000	0.9×10^{10}	0.3500	1.2130
Southern	8.0000	1.2×10^{10}	0.4800	1.1650
Eastern	12.0000	1.0×10^{10}	0.2800	1.5180
Regions	SVRQGA Parameters			MAPE of Testing (%)
	σ	C	ε	
Northern	3.0000	1.2×10^{10}	0.3400	1.3150
Central	10.0000	1.8×10^{10}	0.4800	1.6830
Southern	6.0000	0.9×10^{10}	0.3500	1.3640
Eastern	4.0000	0.5×10^{10}	0.6800	1.9680

Table 3. Forecasting indexes of SVRCQGA, SVRQGA, and other models (Example 1).

Indexes	SVRCQGA	SVRQGA	SVRCQTS	SVRQTS	SVRCQPSO	SVRQPSO
			Northern region			
MAPE (%)	**1.0760**	1.3150	1.0870	1.3260	1.1070	1.3370
RMSE	**131.48**	159.26	132.79	159.43	142.62	160.28
MAE	**130.00**	157.50	141.00	158.50	132.25	159.00
			Central region			
MAPE (%)	**1.2130**	1.6830	1.2650	1.6870	1.2840	1.6890
RMSE	**64.46**	90.18	67.69	90.67	67.70	89.87
MAE	**64.00**	89.25	67.00	89.75	67.50	89.25
			Southern region			
MAPE (%)	**1.1650**	1.3640	1.1720	1.3670	1.1840	1.3590
RMSE	**75.44**	87.82	75.57	88.84	76.03	88.05
MAE	**74.75**	87.50	75.25	88.00	75.75	87.25
			Eastern region			
MAPE (%)	**1.5180**	1.9680	1.5430	1.9720	1.5940	1.9830
RMSE	**6.12**	7.86	6.38	7.95	6.30	7.79
MAE	**6.00**	7.75	6.00	7.75	6.25	7.75

From Table 3 and Figure 5, it is obvious from the comparison that the proposed SVRCQGA model outperforms the other quantum-SVR-based models. Thus, it once again demonstrates the superiority of an SVR model in that it could obtain a more satisfactory forecasting performance by hybridizing quantum computing mechanics with a genetic algorithm. In the same time, the super capability of the cat mapping function in looking for a closer solution to the theoretical global optimum while suffering from premature convergence is noted. The QGA almost has done its best to look for the best solutions for each region, however, these solutions are still unsatisfactory by comparison with the performances of other alternatives. These solutions could be improved by employing a chaotic mapping function (this paper uses the cat mapping function), i.e., the CQGA, to achieve satisfactory solutions.

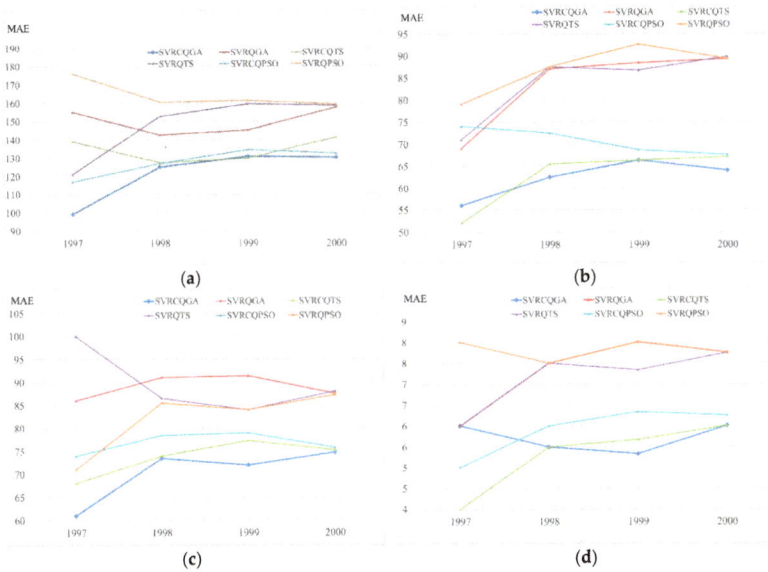

Figure 5. The cumulative differences of MAE for each competitive models in four regions (Example 1). (**a**) Northern region; (**b**) Central Region; (**c**) Southern region; (**d**) Eastern region.

Then, two forecasting performance superiority tests are conducted. Table 4 shows the test results under a one-tail-test at $\alpha = 0.05$ significance level, which point out that the proposed model achieves significantly better performance, except versus the SVRCQTS model.

Table 4. Wilcoxon signed-rank test and Friedman test (Example 1).

Compared Models	Wilcoxon Signed-Rank Test $\alpha = 0.05$; Wilcoxon W Statistic = 0	Friedman Test $\alpha = 0.05$
Northern region		
SVRCQGA vs. SVRQPSO	$W = 0$ *	$H_0: e_1 = e_2 = e_3 = e_4 = e_5 = e_6$
SVRCQGA vs. SVRCQPSO	$W = 0$ *	$F = 12.46$;
SVRCQGA vs. SVRQTS	$W = 0$ *	$p = 0.028$ (reject H_0)
SVRCQGA vs. SVRCQTS	$W = 1$	
SVRCQGA vs. SVRQGA	$W = 0$ *	
Central region		
SVRCQGA vs. SVRQPSO	$W = 0$ *	$H_0: e_1 = e_2 = e_3 = e_4 = e_5 = e_6$
SVRCQGA vs. SVRCQPSO	$W = 0$ *	$F = 13.43$;
SVRCQGA vs. SVRQTS	$W = 0$ *	$p = 0.021$ (reject H_0)
SVRCQGA vs. SVRCQTS	$W = 1$	
SVRCQGA vs. SVRQGA	$W = 0$ *	
Southern region		
SVRCQGA vs. SVRQPSO	$W = 0$ *	$H_0: e_1 = e_2 = e_3 = e_4 = e_5 = e_6$
SVRCQGA vs. SVRCQPSO	$W = 0$ *	$F = 15.57$;
SVRCQGA vs. SVRQTS	$W = 0$ *	$p = 0.013$ (reject H_0)
SVRCQGA vs. SVRCQTS	$W = 1$	
SVRCQGA vs. SVRQGA	$W = 0$ *	
Eastern region		
SVRCQGA vs. SVRQPSO	$W = 0$ *	$H_0: e_1 = e_2 = e_3 = e_4 = e_5 = e_6$
SVRCQGA vs. SVRCQPSO	$W = 0$ *	$F = 11.34$;
SVRCQGA vs. SVRQTS	$W = 0$ *	$p = 0.035$ (reject H_0)
SVRCQGA vs. SVRCQTS	$W = 1$	
SVRCQGA vs. SVRQGA	$W = 0$ *	

* Denotes that the SVRCQGA model significantly outperforms other competitive models.

3.2.5. Results and Analysis: Example 2

For Example 2, similarly, with the smallest MAPE values in the testing set, the SVR's parameters are determined by the QGA algorithm and CQGA algorithm, respectively. These most suitable parameter values for the annual electricity load data are listed in Table 5. To benchmark the results with other research approaches, Table 5 also provides the forecasting index values from other competitive models [27,28].

Table 5. Determined parameters of SVRCQGA and SVRQGA models (Example 2).

Optimization Algorithms	Parameters			MAPE of Testing (%)
	σ	C	ε	
QPSO algorithm [27]	12.0000	0.8×10^{11}	0.380	1.3460
CQPSO algorithm [27]	10.0000	1.5×10^{11}	0.560	1.1850
QTS algorithm [28]	5.0000	1.3×10^{11}	0.630	1.3210
CQTS algorithm [28]	6.0000	1.8×10^{11}	0.340	1.1540
QGA algorithm	9.0000	1.4×10^{11}	0.480	1.3180
CQGA algorithm	12.0000	1.2×10^{11}	0.650	1.1160

The forecasting accuracy indexes values, MAPE and RMSE, are shown in Table 6. Figure 6 illustrates the cumulative differences of MAE for each competitive model. The competitive models also include the SVRCQPSO [27], SVRQPSO [27], SVRCQTS [28], and SVRQTS [28] models. Similarly, the proposed model receives an outstanding performance among other competitive models. The contributions of the quantum computing concepts and the chaotic cat mapping capability are again excellent. It is obviously that the CQGA algorithm excels at finding another better solution.

Table 6. Forecasting indexes of SVRCQGA, SVRQGA, and other models (Example 2).

Years	SVRCQGA	SVRQGA	SVRCQTS	SVRQTS	SVRCQPSO	SVRQPSO
MAPE (%)	**1.1160**	1.3180	1.1540	1.3210	1.1850	1.3460
RMSE	**1502.66**	1774.62	1631.48	1778.74	1618.34	1812.51
MAE	**1466.33**	1731.78	1554.89	1735.78	1575.67	1768.78

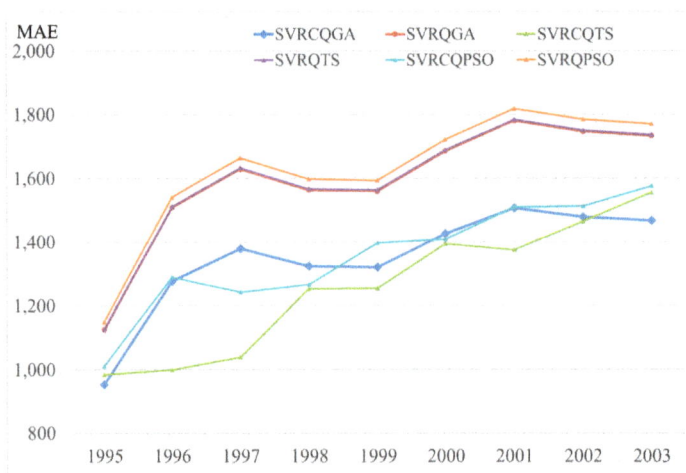

Figure 6. The cumulative differences of MAE for each competitive model (Example 2).

In Example 2, similarly, a Wilcoxon signed-rank test and Friedman test are also conducted to test the significance of the proposed model against the other competitive models. Both tests results are illustrated in Table 7, and demonstrate clearly that the proposed model significantly outperforms the other quantum-behaved algorithms with SVR-based forecasting models.

Table 7. Wilcoxon signed-rank test and Friedman test (Example 2).

Compared Models	Wilcoxon Signed-Rank Test $\alpha =$ 0.05; Wilcoxon W Statistic = 8	Friedman Test $\alpha = 0.05$
SVRCQGA vs. SVRQPSO	$W = 4$ *	
SVRCQGA vs. SVRCQPSO	$W = 2$ *	H_0: $e_1 = e_2 = e_3 = e_4 = e_5 = e_6$
SVRCQGA vs. SVRQTS	$W = 4$ *	$F = 13.35$;
SVRCQGA vs. SVRCQTS	$W = 4$ *	$p = 0.022$ (reject H_0)
SVRCQGA vs. SVRQGA	$W = 4$ *	

* Denotes that the SVRCQGA model significantly outperforms other competitive models.

3.2.6. Results and Analysis: Example 3

For Example 3, based on the similar modeling processes, the SVR's three parameters are eventually selected by the QGA algorithm and CQGA algorithm, respectively. The details of the most suitable parameters of all employed compared models for GEFCOM 2014 data set are shown in Table 8. Because references [27,28] also use GEFCOM 2014 load data set for analysis, therefore, those models shown in [27,28] are also compared with the proposed models.

Table 8. Determined parameters of SVRCQGA, SVRQGA, and other models (Example 3).

Optimization Algorithms	Parameters			MAPE of Testing (%)
	σ	C	ε	
QPSO algorithm [27]	9.000	42.000	0.1800	1.9600
CQPSO algorithm [27]	19.000	35.000	0.8200	1.2900
QTS algorithm [28]	25.000	67.000	0.0900	1.8900
CQTS algorithm [28]	12.000	26.000	0.3200	1.3200
QGA algorithm	5.000	79.000	0.3800	1.7500
CQGA algorithm	6.000	54.000	0.6200	1.1700

To achieve a meaningful comparison, the authors only selected three quantum- and SVR-based forecasting models, i.e., the SVRCQGA, SVRCQTS, and SVRCQPSO models, to compare with each other. Table 9 provides the forecasting accuracy indexes of the proposed SVRCQGA and other competitive models [27,28], and clearly illustrates that the proposed SVRCQGA model achieves results closer to the actual load values than the SVRCQTS and SVRCQPSO models.

Table 9. Forecasting indexes of SVRCQGA, SVRQGA, and other models (Example 3).

Indexes	SVRCQGA	SVRQGA	SVRCQTS	SVRQTS	SVRCQPSO	SVRQPSO
MAPE (%)	**1.1700**	1.7500	1.3200	1.8900	1.2900	1.9600
RMSE	**1.4927**	1.6584	1.9909	2.8507	1.9257	2.9358
MAE	**1.4522**	1.6174	1.8993	2.7181	1.8474	2.8090

Figure 7 also provides a good illustration of the cumulative differences of MAE for each competitive model. The competitive models also include the SVRCQPSO [27], SVRQPSO [27], SVRCQTS [28], and SVRQTS [28] models.

Finally, Table 10 illustrates both the Wilcoxon signed-rank test and Friedman test results for Example 3, which demonstrate that the proposed approach significantly, with a critical *p*-value, outperforms the other competitive models proposed in [27,28].

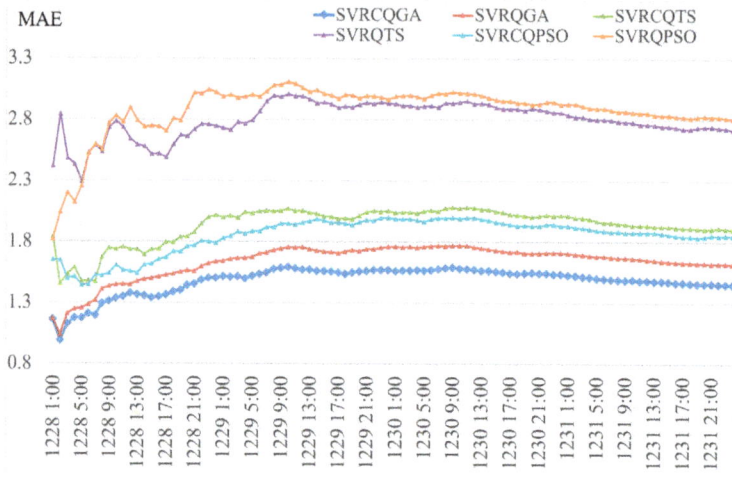

Figure 7. The cumulative differences of MAE for each competitive model (Example 3).

Table 10. Wilcoxon signed-rank test and Friedman test (Example 3).

Compared Models	Wilcoxon Signed-Rank Test		Friedman Test
	$\alpha = 0.05$; Wilcoxon W Statistic = 2328	p-Value	$\alpha = 0.05$
SVRCQGA vs. SVRQPSO	$W = 1278.0$ *	0.00012	
SVRCQGA vs. SVRCQPSO	$W = 1152.5$ *	0.00000	$H_0: e_1 = e_2 = e_3 = e_4 = e_5 = e_6$
SVRCQGA vs. SVRQTS	$W = 1256.0$ *	0.00000	$F = 71.266$;
SVRCQGA vs. SVRCQTS	$W = 1263.0$ *	0.00010	$p = 0.000$ (reject H_0)
SVRCQGA vs. SVRQGA	$W = 2134.5$ *	0.00720	

* Denotes that the SVRCQGA model significantly outperforms the other competitive models.

4. Conclusions

This paper proposes a novel electricity load forecasting model created by hybridizing a quantum-behaved algorithm with an SVR-based model. The results have completely shown that the proposed CQGA has superiority to address the embedded drawbacks of the original GA and quantum GA algorithms that suffer from getting trapped into local optima. This paper uses quantum computing mechanics to enrich the diversity of the population during the GA modeling processes, which eventually improves the forecasting accuracy level. The cat function is employed to avoid premature convergence while the QGA algorithm is processing. This paper provides support to continue the exploration of integrating quantum computing concepts and chaotic mapping techniques to enrich the search space with the limitations from conventional Newtonian dynamics, and a more effective approach to solve the trapping in local optima problem.

Author Contributions: Cheng-Wen Lee and Bing-Yi Lin conceived and designed the experiments; Bing-Yi Lin performed the experiments, analyzed the data and wrote the paper. Bing-Yi Lin is under Cheng-Wen Lee's supervision.

Conflicts of Interest: The author declares no conflict of interest.

References

1. Xiao, L.; Shao, W.; Liang, T.; Wang, C. A combined model based on multiple seasonal patterns and modified firefly algorithm for electrical load forecasting. *Appl. Energy* **2016**, *167*, 135–153. [CrossRef]
2. Fan, S.; Chen, L. Short-term load forecasting based on an adaptive hybrid method. *IEEE Trans. Power Syst.* **2006**, *21*, 392–401. [CrossRef]
3. Cincotti, S.; Gallo, G.; Ponta, L.; Raberto, M. Modelling and forecasting of electricity spot-prices: Computational intelligence vs. classical econometrics. *AI Commun.* **2014**, *27*, 301–314.
4. Amjady, N.; Keynia, F. Day ahead price forecasting of electricity markets by a mixed data model and hybrid forecast method. *Int. J. Electr. Power Energy Syst.* **2008**, *30*, 533–546. [CrossRef]
5. Weron, R. Electricity price forecasting: A review of the state-of-the-art with a look into the future. *Int. J. Forecast.* **2014**, *30*, 1030–1081. [CrossRef]
6. Fan, G.; Peng, L.-L.; Hong, W.-C.; Sun, F. Electric load forecasting by the SVR model with differential empirical mode decomposition and auto regression. *Neurocomputing* **2016**, *173*, 958–970. [CrossRef]
7. Hussain, A.; Rahman, M.; Memon, J.A. Forecasting electricity consumption in Pakistan: The way forward. *Energy Policy* **2016**, *90*, 73–80. [CrossRef]
8. Vu, D.H.; Muttaqi, K.M.; Agalgaonkar, A.P. A variance inflation factor and backward elimination based Robust regression model for forecasting monthly electricity demand using climatic variables. *Appl. Energy* **2015**, *140*, 385–394. [CrossRef]
9. Maçaira, P.M.; Souza, R.C.; Oliveira, F.L.C. Modelling and forecasting the residential electricity consumption in Brazil with pegels exponential smoothing techniques. *Procedia Comput. Sci.* **2015**, *55*, 328–335. [CrossRef]
10. Al-Hamadi, H.M.; Soliman, S.A. Short-term electric load forecasting based on Kalman filtering algorithm with moving window weather and load model. *Electr. Power Syst. Res.* **2004**, *68*, 47–59. [CrossRef]
11. Hippert, H.S.; Taylor, J.W. An evaluation of Bayesian techniques for controlling model complexity and selecting inputs in a neural network for short-term load forecasting. *Neural Netw.* **2010**, *23*, 386–395. [CrossRef] [PubMed]
12. Kelo, S.; Dudul, S. A wavelet Elman neural network for short-term electrical load prediction under the influence of temperature. *Int. J. Electr. Power Energy Syst.* **2012**, *43*, 1063–1071. [CrossRef]
13. Lahouar, A.; Slama, J.B.H. Day-ahead load forecast using random forest and expert input selection. *Energy Convers. Manag.* **2015**, *103*, 1040–1051. [CrossRef]
14. Chaturvedi, D.K.; Sinha, A.P.; Malik, O.P. Short term load forecast using fuzzy logic and wavelet transform integrated generalized neural network. *Int. J. Electr. Power Energy Syst.* **2015**, *67*, 230–237. [CrossRef]
15. Zhai, M.-Y. A new method for short-term load forecasting based on fractal interpretation and wavelet analysis. *Int. J. Electr. Power Energy Syst.* **2015**, *69*, 241–245. [CrossRef]
16. Coelho, V.N.; Coelho, I.M.; Coelho, B.N.; Reis, A.J.R.; Enayatifar, R.; Souza, M.J.F.; Guimarães, F.G. A self-adaptive evolutionary fuzzy model for load forecasting problems on smart grid environment. *Appl. Energy* **2016**, *169*, 567–584. [CrossRef]
17. Bahrami, S.; Hooshmand, R.-A.; Parastegari, M. Short term electric load forecasting by wavelet transform and grey model improved by PSO (particle swarm optimization) algorithm. *Energy* **2014**, *72*, 434–442. [CrossRef]
18. Aras, S.; Kocakoç, İ.D. A new model selection strategy in time series forecasting with artificial neural networks: IHTS. *Neurocomputing* **2016**, *174*, 974–987. [CrossRef]
19. Kendal, S.L.; Creen, M. *An Introduction to Knowledge Engineering*; Springer: London, UK, 2007.
20. Cherroun, L.; Hadroug, N.; Boumehraz, M. Hybrid approach based on ANFIS models for intelligent fault diagnosis in industrial actuator. *J. Control Electr. Eng.* **2013**, *3*, 17–22.
21. Hahn, H.; Meyer-Nieberg, S.; Pickl, S. Electric load forecasting methods: Tools for decision making. *Eur. J. Oper. Res.* **2009**, *199*, 902–907. [CrossRef]
22. Drucker, H.; Burges, C.C.J.; Kaufman, L.; Smola, A.; Vapnik, V. Support vector regression machines. *Adv. Neural Inf. Process. Syst.* **1997**, *9*, 155–161.
23. Hong, W.C. Chaotic particle swarm optimization algorithm in a support vector regression electric load forecasting model. *Energy Convers. Manag.* **2009**, *50*, 105–117. [CrossRef]
24. Hong, W.C. Electric load forecasting by seasonal recurrent SVR (support vector regression) with chaotic artificial bee colony algorithm. *Energy* **2011**, *36*, 5568–5578. [CrossRef]
25. Bhunia, C.T. Quantum computing & information technology: A tutorial. *IETE J. Educ.* **2006**, *47*, 79–90.

26. Dey, S.; Bhattacharyya, S.; Maulik, U. Quantum inspired genetic algorithm and particle swarm optimization using chaotic map model based interference for gray level image thresholding. *Swarm Evol. Comput.* **2014**, *15*, 38–57. [CrossRef]

27. Huang, M.-L. Hybridization of chaotic quantum particle swarm optimization with SVR in electric demand forecasting. *Energies* **2016**, *9*, 426. [CrossRef]

28. Lee, C.-W.; Lin, B.-Y. Application of hybrid quantum tabu search with support vector regression (SVR) for load forecasting. *Energies* **2016**, *9*, 873. [CrossRef]

29. Shor, P.W. Algorithms for quantum computation: Discrete logarithms and factoring. In Proceedings of the 35th Annual Symposium on Foundations of Computer Science, Santa Fe, NM, USA, 20–22 November 1994; pp. 124–134.

30. Han, K.-H. Quantum-inspired evolutionary algorithms with a new termination criterion, Hε gate, and two-phase scheme. *IEEE Trans. Evol. Comput.* **2004**, *8*, 156–169.

31. Lahoz-Beltra, R. Quantum genetic algorithms for computer scientists. *Computers* **2016**, *5*, 24. [CrossRef]

32. Chen, G.; Mao, Y.; Chui, C.K. Asymmetric image encryption scheme based on 3D chaotic cat maps. *Chaos Solitons Fract.* **2004**, *21*, 749–761. [CrossRef]

33. Hang, B.; Jiang, J.; Gao, Y.; Ma, Y. A Quantum Genetic Algorithm to Solve the Problem of Multivariate. In *Information Computing and Applications, Proceedings of the Second International Conference ICICA 2011, Qinhuangdao, China, 28–31 October 2011*; Liu, C., Chang, J., Yang, A., Eds.; Springer: Berlin/Heidelberg, Germany, 2011; pp. 308–314.

34. Su, H. Chaos quantum-behaved particle swarm optimization based neural networks for short-term load forecasting. *Procedia Eng.* **2011**, *15*, 199–203. [CrossRef]

35. 2014 Global Energy Forecasting Competition Site. Available online: http://www.drhongtao.com/gefcom/ (accessed on 9 November 2017).

36. Diebold, F.X.; Mariano, R.S. Comparing predictive accuracy. *J. Bus. Econ. Stat.* **1995**, *13*, 134–144.

37. Derrac, J.; García, S.; Molina, D.; Herrera, F. A practical tutorial on the use of nonparametric statistical tests as a methodology for comparing evolutionary and swarm intelligence algorithms. *Swarm Evol. Comput.* **2011**, *1*, 3–18. [CrossRef]

38. Wilcoxon, F. Individual comparisons by ranking methods. *Biom. Bull.* **1945**, *1*, 80–83. [CrossRef]

39. Friedman, M. A comparison of alternative tests of significance for the problem of m rankings. *Ann. Math. Stat.* **1940**, *11*, 86–92. [CrossRef]

MDPI

St. Alban-Anlage 66

4052 Basel

Switzerland

Tel. +41 61 683 77 34

Fax +41 61 302 89 18

www.mdpi.com

Energies Editorial Office

E-mail: energies@mdpi.com

www.mdpi.com/journal/energies

www.ingramcontent.com/pod-product-compliance
Lightning Source LLC
Chambersburg PA
CBHW051728210326
41597CB00032B/5646